Praise for *Manufacturing for Survival*

"AT&T's Manufacturing Operations have be... rac-
tical, down-to-earth approach to manufactu ...res
much of that experience."

D0862646

..LL BLVD. S.
FORT WORTH, TEXAS 76118
(817) 272-5922   Fax: (817) 272-5977

> W.D. Seelig
> Vice President
> Global Manufacturing
> AT&T

"This is the book that both practitioners and their managers have needed for the past
10 years. It is written by a real practitioner, for real practitioners, in simple straight-
forward language that tells it like it is. This is not a text book, it is a "How To" book.
If you are involved in any area of Production & Inventory Control, this book could
become your Bible. Keep a copy on your desk alongside of your dictionary."

> Richard (Rick) Titone
> President
> Why How Consulting Company

"Having read, taught from and used dozens of books on how to harness the process
of manufacturing planning and control, I am saturated with the word from 'on high.'
Blair Williams has written the only book I've seen in recent memory that speaks from
and to the practitioner and middle manager. His 'dos' and 'don'ts' focus on practical
answers to real problems . . . quickly and effectively. [Blair's book] should be required
reading for anyone who wants to be a participant in world-class manufacturing into
the 21st century."

> Donald N. Frank
> President
> D.N. Frank Associates

"Blair finally committed to print what he practices. This book is based on a solid
understanding of manufacturing and represents tried and proven methods that have
been implemented over the years. This book is a recipe for success!"

> Michael Greenstein
> Adjutant Assistant Professor of Manufacturing
> Polytechnic University

"Today we demand more and more breadth from our workers. Our responsibility as
managers is to provide them access to information they need. The need is particularly
acute when we hold factory workers and other, non-logistics professionals, accountable for managing materials. This book is one of the best ways to introduce your practitioners to the host of logistics related processes they must master. It meets the critical
tests of clarity, practicality and rapid access to specific tools and techniques."

> John J. Bruggeman
> President
> Bruggeman & Associates

Manufacturing
for Survival

Manufacturing for Survival

The How-to Guide for Practioners and Managers

Blair R. Williams

Addison-Wesley Publishing Company
Reading, Massachusetts
Menlo Park, California   New York
Don Mills, Ontario
Wokingham, England   Amsterdam
Bonn   Sydney   Singapore   Tokyo
Madrid   San Juan   Seoul   Milan
Mexico City   Taipei

Many of the designations used by manufacturers and sellers to distinguish their products are claimed as trademarks. Where those designations appear in this book and Addison-Wesley was aware of a trademark claim, the designations have been printed with initial capital letters.

The publisher offers discounts on this book when ordered in quantity for special sales.

For more information, please contact:
Corporate & Professional Publishing Group
Addison-Wesley Publishing Company
One Jacob Way
Reading, Massachusetts   01867

Library of Congress Cataloging-in-Publication Data
Williams, Blair R., 1938–
    Manufacturing for survival : the how to guide for practitioners and managers / Blair R. Williams.
        p.   cm.
    Includes bibliographical references and index.
    ISBN 0-201-63373-6 (alk. paper)
    1. Production management.   I. Title.
TS155.W5386   1995
658.5 — dc20   95-26000
                                                                    CIP

Copyright 1996 by AT&T

AT&T

ISBN: 0-201-63373-6

1 2 3 4 5 6 7 8 9-CRW 98979695
First printing, December 1995

This book is dedicated to all manufacturing
practitioners who, like me,
search for ways to plan and execute effectively.
I hope this book will help.

Contents

Unit II: Manufacturing Planning

Unit III: Manufacturing Execution Overview

Unit IV: Customer Service

Foreword

Hal Mather

New books on the concepts and theories of improved manufacturing appear almost daily. Academics and consultants, me among them, write to explain these concepts and theories, the assumption being that readers can easily apply these ideas to their businesses.

The track record says this assumption is unfounded. Most readers need a clearer road map to implement the current body of knowledge into their companies.

That's why, when Blair asked me to write this foreword, I agreed. He has written a "how-to" book, which is more like a handbook than a textbook. It is a far-ranging book, as any book attempting to discuss a manufacturing company has to be. Among other subjects it covers competitiveness to inventory management, MRP II to JIT, and Benchmarking to Quality Management—all with a focus on customer satisfaction.

Blair has written the book from a practitioner's viewpoint. This is evident in the chapter organization. Chapters start out with executive overviews and dive into details. Each section of a chapter includes the description of a process that guides the reader to the successful application of the section's ideas. Sections end with a list of dos and don'ts that highlight key actions and pitfalls.

I predict this book will find a prominent home on every practitioner's bookcase. It will be well thumbed, because its organization is closer to that of a handbook than to a book you sit down to read from beginning to end. For those who heed its advice, there will be many more successful implementations of the things that make your company more competitive.

Hal Mather
President, Hal Mather Inc.
Management Counseling and Education

Preface

Why This Book?

I have been troubled by the lack of available "how-to" information on manufacturing planning and execution. It seems preposterous that the United States, which has contributed so much to scientific manufacturing, has no current, readily available information on this subject. In an attempt to fill this void, I have written this book. It is a book written by a practitioner, who has experienced—and continues to experience—many of the problems that people face in and around a factory. It seeks to provide a practical guide for all personnel engaged in the manufacturing of discrete products—that is, it seeks to explain how to plan and execute effectively and competitively.

I emigrated to the United States in 1976 and worked for a renowned railcar manufacturer, first in Indiana, and then in its original factory in Chicago. In 1976, the company designed and built Superliner railroad cars. My understanding of U.S. manufacturing was based on my knowledge of American manufacturing pioneers such as Taylor, Gilbreth, Ford, Sloan, and others. I also knew that America had built much of the equipment that had won the Second World War. Imagine my surprise when I found that working for the 150-year-old manufacturing company with a legendary reputation for building railcars was an exercise in frustration. Institutionalized knowledge of building railcars was missing. The design transfer was over a year late and we attempted to design and build simultaneously. This is never easy, and it is a disaster when basic manufacturing management know-how is missing. Of course, having a difficult customer did not help, as they continued to make changes all through production. The Superliners were built about 300 percent over budget and over two years behind schedule. Subsequently the company was taken over and the railcar building operation shut down.

While working for this railcar company I switched from line operations to inventory control. I also became aware of the materials management function, by becoming a member of APICS (American Production and Inventory Control Society) and being certified as a production and inventory control manager (CPIM). In 1981 I joined a renowned pump company in New Jersey as their materials manager. This company also had over a century of manufacturing experience, but again I found a woeful lack of manufacturing and materials management knowledge by the personnel running the factory. Product was made by sheer force of effort—twenty hours of overtime a week, including Saturday, was common. Rework was rampant; there was a continuous shortage of parts; purchased products had to be expedited on an order-by-order basis and customer order shipment had less than a 50 percent on-time delivery record. (Not surprisingly, this company was also taken over a few years later.)

I then was offered and took a job with a small company making standard lighting fixtures in four factories. Here again, customer orders were constantly late, there were large amounts of inventory, and operations seemed to be managed by expediting and constant fire fighting. In this company I was also exposed to marketing and sales management for the first time, and found that they too seemed to operate without much institutionalized functional knowledge.

I have now been in the United States for 19 years, working directly in manufacturing most of the time. I have met hundreds of practitioners and visited a large number of factories.

It appears that there are two major categories of manufacturing companies. There are those companies—General Electric, Hewlett-Packard, Motorola, Ford, and Caterpillar—to name but a few, that have institutionalized manufacturing knowledge. Under competitive pressure from the Japanese, they refined their techniques and practices and became more competitive. There are also those companies—probably a majority—that lack basic knowledge of manufacturing. In these companies, what is applied is common sense, a lot of muscle, and varying degrees of technology—this approach is not bad, but more effort and time is required to achieve national or international competitiveness.

The 1970s and 1980s have witnessed a decline of United States smoke-stack industries and the loss of U.S.-manufactured product to Japanese imports. Manufacturing competitiveness was a much-debated subject. A strong intellectual group, headed by Harvard Business School, proposed the formation of a manufacturing strategy aligned to a corporate strategy as the means of reestablishing manufacturing competitiveness (Skinner 1987). Other reasons given for the decline in competitiveness included excessive government regulations, unfair trade practices, an uneducated workforce, poor design, and poor management. Evidently many reasons contributed, over a period of time, to the erosion of manufacturing competitiveness, not unlike a frog who will *not* jump out of a pan in which the water temperature is slowly raised, so will boil to death! (On the other hand, if the frog is dropped into boiling water, it jumps right out.)

As a person on the front line I looked in vain for mention of the lack of practical manufacturing management knowledge as one of the main reasons for lack of competitiveness. It seemed clear that well-developed strategy, design, technology, or infrastructure cannot compensate for a lack of knowledge on *how to* make a product efficiently. If factory personnel do not know how to maintain accurate data records, or how to reduce lead time or control inventory, production will be costly and delayed. I never understood *why* this connection and its prevalence was not more recognized. Some reasons for this may be:

■ The competitive companies keep such knowledge within their company, and use it as a competitive advantage.

■ Thinkers and developers of manufacturing theory and manufacturing executives are too far removed from the reality of the shop floor.

■ The experts in the field (be it railcars or pumps) do not document their knowledge and it dies with them.

■ Manufacturing has always had the reputation of not needing any specific knowledge or skill to plan and execute—"anyone with intelligence and common sense can manage a factory."

■ Manufacturing is associated with unpleasant working conditions, and few leading university students select manufacturing as a profession.

Still just "off the boat," I was unsure whether my observations and conclusions had merit, even though they were corroborated by many other practitioners! The McKinsey report, *Manufacturing Productivity*, published in 1993, helped to confirm some of my thinking. The report shows that the United States has lost manufacturing competitiveness in the discrete product industries—autos, auto parts, consumer electronics, metalworking, and steel—and that this loss has occurred in the last two decades. It also showed that the "organization of functions and tasks" was one of the principal reasons for Japan's productivity advantage and that this production factor covers the whole continuous improvement approach to manufacturing in Japan.

Having determined that there was a lack of basic manufacturing knowledge in many industries, I asked myself how and where a person could acquire such knowledge. I first asked the question to help myself, and subsequently applied it to the manufacturing community. *There is no simple way and there is no single source.*

The keeper of some of the body of knowledge is APICS. This body focuses on the planning aspect of production and inventory control (Manufacturing Resource Planning II). In addition it offers a theoretical understanding of the topics covered. There are practical how-to techniques included, but these are scattered and have to be dug out and collected. There are few national manufacturing apprenticeships of the type common in Europe or Japan, where most large manufacturing companies (with government support), hire persons with a mechanical aptitude and formally train them in best manufacturing practices. Some companies in the United States, such as GE, have institutionalized "in-house" training programs in manufacturing, but such companies are the exception. I could not find a book focused exclusively on the "how-to" of manufacturing! How then does a practitioner easily learn the basics of manufacturing? How much of the lack of competitiveness of the United States stems from the lack of basic knowledge of effective manufacturing techniques and practices?

I have tried to address these issues in *Manufacturing for Survival.*

Competitiveness

Competitiveness leading to profitability remains the long-term objective for all manufacturing.

An anecdotal story (source unknown), best sums up competitiveness.

> *"Every morning in Africa, a gazelle wakes up. It knows that it must run faster than the fastest lion or it will be killed. Every morning a lion wakes up. It knows it must outrun the slowest gazelle or it will starve to death. It doesn't matter whether you're a lion or a gazelle. When the sun comes up you better be running."*

The book has been structured and written to show how to achieve manufacturing competitiveness and profitability by providing superior customer service. Superior customer service can be achieved by making a high-quality, efficiently designed product, with a motivated and empowered workforce. Further, to ensure profitability, a product has to be planned and then executed effectively, with lead time and inventory being the two primary controls that measure effectiveness. A comprehensive management information system is needed to plan, communicate, and track the progress of manufacturing. Finally, the product must satisfy a customer need and be in demand thus confirming a company's competitiveness and ensuring its profitability. A schematic model, shown below, best illustrates this structure.

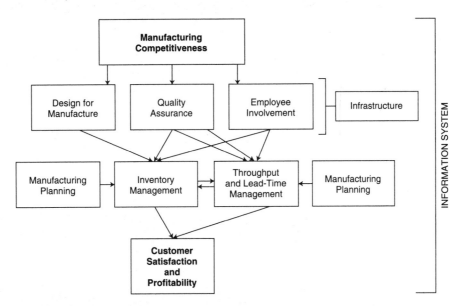

A Schematic View of *Manufacturing for Survival*

What Is Covered

This is a book written by a practitioner, with over thirty years of experience, who is still a practitioner working in a factory. It seeks to provide a practical guide to all personnel engaged in manufacturing discrete products on *how to* plan and execute effectively and competitively. There are few books exclusively covering manufacturing practices. This book partially fills the gap by documenting some of the best manufacturing practices. The book *does not* cover all practices in all industries. It is merely a cross-section of some practices in some industries. It will provide a primer for a person seeking to know what to do in most manufacturing planning and execution activities. It is a response to a need expressed by many of my manufacturing colleagues—"Is there any book that can tell us, simply and briefly, how best to perform this activity?" I hope this book will satisfy that need and that practitioners will be able to identify and implement some of these best manufacturing principles and practices. I wrote the book with this expectation.

Blair R. Williams
Scotch Plains, New Jersey
May 1995

Acknowledgments

I would like to acknowledge George Brandenburg for being a constant source of wisdom and help, Hal Mather for so generously agreeing to write a foreword, and Don Frank for his interesting and challenging remarks. For spending time reviewing various portions of the manuscript thanks also go to Maurice Plourde and Ken Lewis of Interactive, Inc.; Mike Hunter of Hunter Associates; Rick Titone of "Why How Co.," and John Bruggerman. Also thanks are due to many colleagues from AT&T, particularly Rick Frisby, Ron Glass, and Steve Levine. Thanks to Mike Greenstein for recommending me to Addison-Wesley and to my editor Jennifer Joss for her faith in me. A special thank-you to Tamar Gisis for a splendid last-minute effort. Finally, thanks to Pia Viscelli and my son Julian for trying to correct my numerous dangling participles and other such grammatical indiscretions.

No project of the size of this book can be undertaken, let alone completed, without a solid psychological foundation. For this I must acknowledge and thank my wife Ellen, who never doubts me, always supports me, and ministers to my innumerable wants and demands. Many of us have had more than our share of luck in this life, and I am one of those so favored with my wife Ellen.

How to Use This Book

Focus

The practitioner is the focus of this book. The book describes how to set up and run various manufacturing processes, and concentrates on the *process.* The reader will not have to go through a great deal of strategic and theoretical information. The bibliography lists references that provide such information. Examples are provided to illustrate concepts, but it should be noted that the book is one big case study that describes actual techniques and practices.

Scope

The book primarily covers discrete product manufacturing. Within this scope, major manufacturing processes are identified and techniques used to plan and execute these processes are described. The practices described provide the practitioner with a start; they are not meant to be exhaustive. They represent some of the approaches that are used to manufacture effectively.

Organization

Each major manufacturing process has a separate chapter. Chapters are subdivided into sections that cover the most important activities within the process.

Unit I is an overview of the elements of the manufacturing infrastructure. It describes these elements—competitiveness, quality, employee involvement, and design—rather briefly. It is not meant to serve as a how-to guide.

Unit II, Manufacturing Planning, and Unit III, Manufacturing Execution, are detailed how-to descriptions of practices and techniques.

Unit IV, Customer Service, is a summary of some of the present-day approaches and measurements used in the field.

Format

The book is divided into four main units. Each unit is divided into chapters and each chapter has several sections.

Each section is organized into the following areas:

- Purpose and Description;
- Discussion, where some of the strategic issues and approaches are discussed;
- Process, which is a key section that describes how to set up and run the process; and
- Dos and Don'ts, a critical section that highlights some of the most important practical actions and points to common errors.

Structure

Each section is written to be independent. Practitioners should be able to refer to the particular section that they are interested in and get most of the information they need to plan or execute that activity without having to go to another section. This structure leads to some overlapping of concepts and techniques. This is intentional.

Manufacturing Infrastructure

An Overview

In the preface it was stated that the book "describes how to set up and run various manufacturing processes, and the reader will not have to go through a great deal of strategic or theoretical information." In Unit I, however, I do present some strategic information.

In order to set the stage for describing techniques and practices to plan and execute manufacturing (Units II and III), it is necessary to describe certain elements essential to the manufacturing process. These elements are *quality, employee involvement,* and *design for manufacturability.* They make up what is called the manufacturing infrastructure; they are the foundation on which effective manufacturing is built. Each of the elements is a complete function and each has been extensively written about. Their treatment in this book is brief and limited to describing how to implement each particular function.

It is necessary first to provide a brief overview of manufacturing strategy and productivity as a lead in to the infrastructure elements.

Manufacturing strategy and productivity are very popular topics, and are the subject of continuous discussion. Since the United States has lost some of its manufacturing dominance in the 1970s and 1980s, the issue of how to regain industrial leadership is of national importance. A process for developing a manufacturing strategy is described, followed by an examination of the productivity of U.S. manufacturing.

Quality management is described briefly, and a process for implementing a quality program is described—it is stressed that quality is a huge function and has been well addressed in the manufacturing literature. Obtaining an *ISO 9000 certification* is described, as today it is practically a requirement for doing business in Europe.

Employee involvement is covered generally and a process for forming, developing, and growing *teams* is described. Employee involvement,

1

people empowerment, and work through teams are all new approaches to twenty-first century manufacturing. It is expected that the successful use of these approaches will restore the manufacturing edge of the United States. A note on *benchmarking* concludes this chapter. Benchmarking is an essential element in the process of developing competitiveness. It is a common thread throughout the book—effectiveness in planning, executing, and customer service has to be determined by comparing internal performance with the best in the competition.

Design for manufacturing is described with a section on new products and processes and a section on best design practices. It is expected that the next competitive frontier will be the speed at which new products are brought to market and how effectively these products will satisfy customer needs. As such, both these elements are critical to sustaining the competitiveness of the United States. Like quality management and employee involvement, design is a function that has been extensively studied and documented.

Approach and Scope

It is worth repeating that the treatment of these infrastructure elements is brief, as they are outside the primary thrust of this book. This whole unit can be bypassed without affecting the practitioners' knowledge of how to plan and implement. Each of the elements has been extensively studied and written about (the bibliography lists some useful references). These elements have been included in this book because they are the foundation of all manufacturing and provide a backdrop to *how* manufacturing is planned and executed to satisfy a customer.

1

Infrastructure Elements

Summary of Chapter

The chapter covers *competitiveness* from both a strategic and a factory productivity point of view. It draws on a McKinsey report that compares the productivity of the United States, Japan, and Europe. *Quality management* is touched on briefly, with more detail on the substance and requirements of *ISO 9000*. Finally *Employee involvement* and *teams* are described. The chapter ends with a section on benchmarking—a necessary requirement for determining competitiveness.

Contents

- Competitiveness: Manufacturing Strategy
- Competitiveness: Manufacturing Productivity
- Quality Management: Overview
- Quality Management: ISO 9000
- Employee Involvement and Teams
- Teams: Formation, Development, and Growth
- Benchmarking
- Measurements

Relationship of Infrastructure Elements to Other Chapters in the Book

This chapter is intended to set the stage for the core manufacturing activities of planning and execution. The topics covered in the chapter—competitiveness, quality management, and employee involvement—are the foundation on which all manufacturing must be built. These elements are treated briefly, and have been included to remind the reader of their importance. Each of the elements is worthy of a separate book in its own right.

Competitiveness: Manufacturing Strategy

Purpose and Description

For an industrial company to survive, it must plan how to make and sustain consistent profits.

Strategy is a set of plans that seek to satisfy customer needs more effectively than the competition. Manufacturing strategies are the approaches used by manufacturing to support a company's effort to develop and sustain competitiveness. For a manufacturing strategy to be effective, it must be an integral part of the company's overall strategy.

Discussion

Wickham Skinner was probably the first person to develop formally the concept of a manufacturing strategy. (Skinner, 1985). Skinner states that "a company's competitive strategy at a given time places particular demands on its manufacturing function, and, conversely, that the company's manufacturing policies and operations should be specifically designed to fulfill the task demanded by strategic plans." For example, if the corporate strategy calls for a company to be the low-price leader, manufacturing must have a strategy that supports this plan, and manufacturing techniques and practices must be adopted that will produce a low-cost product.

In many cases manufacturing strengths will be leveraged into the formulation of a corporate strategy. A company with a culture of strong technical innovation is likely to develop a corporate strategy of becoming a high-tech market leader.

Developing a Manufacturing Strategy

1. Understand the ways in which manufacturing creates a competitive advantage in the industry. One way of doing this is to identify the customer needs that manufacturing can help satisfy better than the competition does.

   Example: A company's strength is its low-cost, high-volume manufacturing of fasteners.

2. Understand the changing role of manufacturing over time. Initially product performance may be the competitive strategy. As more competitors enter the field, the competitive edge may change to product variety and delivery. Later in the life cycle, the product may become a commodity, and competition may be based on price or service.

Determine if manufacturing can satisfy part of the cycle or the whole cycle.

Example: Initially superior fastener design provides the advantage. Later in the cycle price is dominant.

3. Establish a corporate strategy linked to the identified manufacturing strengths of the company.

Example: A company's business strategy is to be a low-price market leader in fasteners.

4. State a manufacturing strategy and corresponding goals. Ensure that there is understanding of the priorities among the four competitive elements of manufacturing: cost, delivery, quality, and flexibility. Communicate the strategy. A card that can be easily carried by all employees can be an effective reminder of the strategy.

Example: "The company will satisfy its customers by providing them with fasteners at the best price on the market. Manufacturing will support this objective by making fasteners at a cost below that of any other company. Standardized product will support all customer requirements and there will be less than 0.1% product return."

5. Compare the company's performance to that of the competition and identify shortcomings. This is called *benchmarking;* it consists primarily of measuring the company's performance against the best in class.

Example: Compare the cost of the company's line of fasteners with the cost of comparable lines of other fastener manufacturers, nationally and internationally.

6. Determine the actions to be taken to close the gap or to ensure that a competitive position is maintained.

Example: Install special-purpose high-production screw machines.

7. Ensure that goals and the measurements of these goals support and strengthen the chosen objectives.

Example: Measure the cost and price of the fasteners. Set targets to reduce cost and price.

8. Focus on the strategy and the goal that the strategy seeks to attain. Companies frequently deploy their resources across a variety of strategies, resulting in most of them being under-supported and under-managed.

Example: Resist any inclination to produce special types of fasteners that have low sales volume.

9. Safeguard against tactics or approaches becoming goals.

   Example: To improve sales abroad, getting ISO 9000 certification may be adopted as a tactic. Ensure that this does not become an end in itself.

10. Treat manufacturing as an equal partner in planning corporate strategy and ensure that manufacturing contributes to the long-term competitiveness of the company. Periodically review manufacturing's alignment to the corporate strategy.

11. Ensure that manufacturing is linked to all the other functions of the organization and to external entities such as customers, suppliers, and distributors.

12. Stress continuous improvement. Innovate production processes. Leverage technology to help achieve the goal.

13. Learn to live in a world economy and align manufacturing goals to fit conditions in the global economy. Be aware that foreign cultures, practices, and industrial policies can be very different from U.S. ones. It is important to understand foreign companies as well as possible. Enhance international distribution and services.

Dos and Don'ts

DOs

1. Do ensure that manufacturing is an integral part of a company's strategic and business planning processes.
2. Do develop, document, and communicate a corporate strategy and the role of manufacturing within that strategy.
3. Do establish priorities among the competitive tasks of manufacturing (cost, quality, delivery, and flexibility).
4. Do make technical and other resource trade-offs consistent with the objectives of the company.
5. Do ensure that the whole strategic goal-forming process is interactive.

DON'Ts

6. Don't lose focus and be distracted by too many other objectives.
7. Don't plan to use all the capacity of a facility. Allow slack for variations.
8. Don't neglect the company's established strengths—build on strength.
9. Don't think short term in developing strategy.

Competitiveness: Manufacturing Productivity

Purpose and Description

A nation's prosperity depends on its ability to compete. In order to be competitive, a nation must use its resources of labor, capital, and material more effectively than those of other countries. The use of these resources to produce goods and services defines the productivity of the nation. Formally stated, *productivity* is the ratio between the outputs of goods and services and the inputs of the resources that produced them. It is one of the most fundamental measures of a country's prosperity.

Productivity is commonly measured by labor productivity and by multifactor productivity. *Labor productivity* measures output per hour in terms of dollars of revenue produced, and *multifactor productivity* measures output that includes labor and other resources such as equipment, energy, and materials. There is a strong correlation between manufacturing productivity and business success—the more productive a factory, the lower its costs and the higher its profits.

Discussion

Can American factories compete in the global marketplace? There is quite a lot of research that says they can: ". . . with effective management of the total manufacturing system, manufacturing in the United States can be cost competitive with off-shore production and, further, can provide significant advantages in staying abreast of and responding rapidly to changing customer demands." (National Research Council 1992).

The McKinsey Global Institute, in a report entitled *Manufacturing Productivity* (October 1993), attempted to reconcile the dichotomy between the labor calculations that show U.S. productivity to be higher than that of Japan and Germany and the common perception that the United States has fallen behind the productivity of Japan and Germany. The study covered nine industries. The labor productivity in Japan compared to that of the United States may be seen in Figure 1.1. The width of the bar represents the number of workers employed in the industry.

Clearly Japan is more productive in the discrete-product industries (steel, auto parts, metalworking, autos, and consumer electronics), whereas the United States is more productive in the process-type industries.

The McKinsey report investigated why productivity differences exist in countries that have comparable technical know-how, capital, and labor

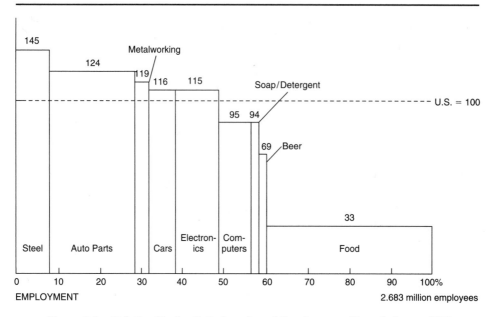

Figure 1.1.   Relative Productivity Levels and Employment Share in Japan, 1990
Source: McKinsey Analysis. Reprinted with permission.

(see Figure 1.2). The study determined that the physical output of a worker depends on three areas:

1. **Output.** What the output is in terms of mix, variety and quality
2. **Production factors** that include:
   - the technology, plant, and equipment in use
   - the scale of production
   - the ease of manufacture of the product
   - the skill and motivation of the workers
   - the parts and raw material
3. **Operations** that include:
   - capacity
   - how the many tasks of production are organized in detail.

Observations on the Findings

The following findings have been summarized from the McKinsey report.

- Design for manufacturing and organization of functions and tasks have a major influence on productivity in the auto, auto parts, con-

● High (Major effect on productivity)
○ Low
X Undifferentiated

	Processed food	Beer	Steel	Metalworking	Cars	Parts	Consumer electronics	Computers	Soap and detergent	"AVERAGE"
Output										
• Mix, variety, quality	●	○	○	○	X*	X	X**	○	●	○
Production factors										
• Machinery, equipment, buildings (technology, intensity, age)	●	●	●	●	○	○	○	X	X	●
• Scale	●	●	●	●	X	X	X	X	○	●
• Design for manufacturing	X	X	X	●	●	●	●	○	X	●
• Basic labor skills, intrinsic motivation	X	X	X	X	○	X	○	X	X	X
• Raw materials, parts	X	X	X	X	○	X	X	X	X	X
Operations										
• Capacity utilization	X	X	X	X	X	X	○	X	X	X
• Organization of functions and tasks	○	○	●	●	●	●	●	○	○	●

*   Adjustment for quality made to the productivity data
** Influence is high in the opposite direction

Figure 1.2.   Summary of Causes of Labor Productivity Differences at the Production Process
Source: McKinsey Global Institute, *Manufacturing Productivity* (October 1993).

sumer electronics and metalworking industries (all discrete product type industries). In both of these areas the Japanese are more productive than the U.S.

Design for manufacturing. An important source of productivity advantage for many companies has been their ability to create product designs that are less complex, use fewer parts and are easier to assemble, without producing products that are different from the customer's perspective. Innovations in this area have usually been introduced first in manufacturing operations in Japan.

Organization of functions and tasks. Somewhat surprisingly, there are very large differences in productivity among plants that look similar and that produce similar products. These productivity gaps result from the way the product is organized. The most productive operations in Japan have refined and re-refined their production methods in order to pull as much labor out of the process as possible. There is no one large change that has been implemented, so that the nature of the improvement seems mysterious. It is the accumulation of thousands of small changes, to the point where the placement of every small part and every machine and the movements of every worker are optimized for productivity. It not only includes the optimization of time and motion, but also the management structure. For instance, the delegation of responsibilities, such as production worker empowerment and suggestion systems where improvements are directly implemented, played a large role in the way operations in Japan were able to achieve high productivity. (McKinsey Global Institute, October 1993, Chapter 3, pages 7 and 9.)

- Technology, machinery and plant, and scale are important in the metal-working, steel, beer, and processed-food industries (all, except metal-working, process type of industries).

- Differences in the basic skills of workers do not appear to play an important role in explaining productivity differences in any of the nine industries!

The report goes on to state that "most, if not all of the things that cause significant differences in productivity are subject to choice by managers. That is certainly the case when it comes to the manufacturability of the product (design for manufacturing) and how the workplace is organized. Yet great differences persist."

Other interesting findings of the report are that Japan took over the productivity leadership in the case of autos, auto parts, metalworking, steel, and consumer electronics in the last two decades (1970s and 1980s). The report also states that becoming productivity leaders in manufacturing is a function of the intensity of competition to which the managers are exposed and, more specifically, the degree to which they are faced by direct competition from producers on the leading edge.

Implications for Companies

Once again, much of the information for this section has been summarized from *Manufacturing Productivity* (McKinsey 1993).

The most important implication for companies is that management can transfer best practices in technology, product design, and the organization of functions and tasks from one geographical location to another.

1. Determine the best practices (benchmarking). Implement best practices by ensuring that the appropriate staff/workers are properly trained. (Remember the basic skills of workers and their capacity to be trained is similar in the United States and Japan.)

2. Establish a competitive advantage with the productive best practices. Transfer the best practices to all facilities nationally and internationally.

3. Import innovations, new technologies, and operating processes to close productivity gaps.

4. Standardize parts, components, and modules to enable them to be made through automation.

5. Develop a strong capability of design for manufacturing. This is especially useful for discrete assembled-product industries—autos, auto parts, metalworking, consumer electronics, and computers. Modular designed and standardized parts translate into greater flexibility and lower costs.

6. Grow an organization that is flexible with cross-trained workers and an emphasis on quality, speed, and responsiveness to the customer.

7. Develop new organization forms emphasizing teamwork, front-line empowerment, communication, and bottoms-up continuous improvement.

8. Ensure that efficient and integrated information systems are in place to record and monitor manufacturing products and processes.

9. Drive for speed. Reduce total lead time from design through manufacture to delivery to the market.

10. Develop strategic partnerships with key suppliers, as these are a significant cause of productivity differences. Successful supplier relationships are characterized by:
    - A balance between the need to reduce cost and a sharing in improvement gains.
    - No duplication of effort in quality, delivery, or other production factors.

- A two-way flow of technical advice and engineering support.
- Early involvement of both suppliers and customers in product development and design.

11. Establish close communication, and, if possible, become a partner with key customers. Such relationships help to identify customer needs and to establish product demand.

12. Acquire new best practices. This can be done by benchmarking, transferring management, acquiring a company or employee who knows the best practice, and licensing. Companies do not spend as much time in this activity as they do in acquiring capital equipment. There is opportunity to redress this balance.

Dos and Don'ts

DOs

1. Do ensure that best practices are identified, established, and communicated.
2. Do train the organization in the specific skills needed to execute the best practices.
3. Do focus on product design, since it is a critical driver of manufacturing productivity.
4. Do focus on cultivating a new relationship with the workforce that increases workers' involvement and accountability, and gives them more decision-making authority (empowerment).

5. Do organize for improving productivity through empowered teams.
6. Do design measurements to reinforce productivity improvement.
7. Do recognize and reward improvement.

DON'Ts

8. Don't neglect information systems. They are the glue that holds the whole infrastructure together and allows optimum decision making.
9. Don't lose focus by having too many objectives.
10. Don't focus on results only. Look at the processes behind the results.

Quality Management: Overview

Purpose and Description

Product quality is an indispensable condition of competitiveness. In fact, it is an assumed condition without which a company cannot survive.

At its simplest, *quality* is defect-free performance that meets or exceeds a customer's expectation.

Discussion

Quality is probably the most written about subject in business today. Total Quality Management (TQM) is a comprehensive approach to managing the entire business to ensure customer satisfaction through employee involvement. It includes the achievement of goals in product quality, delivery, price, and performance. Describing the TQM approach is beyond the scope of this book (see the bibliography for further reading in this area). This section seeks to provide a brief, general overview of the practice of quality management.

The U.S. Government GAO (General Accounting Office), in a study commissioned by Congressman Don Ritter in May 1991, reviewed 20 companies that were among the highest-scoring applicants for the Malcolm Baldrige National Quality Award in 1988 and 1989. It found that "Companies that adopted quality management practices experienced an overall improvement in corporate performance. In nearly all cases, companies that used Total Quality Management practices achieved better employee relations, higher productivity, greater customer satisfaction, increased market share, all resulting in improved profitability" (see Appendix 1.1). Today ISO 9000—a quality certification—is fast becoming a world requirement for manufacturing companies. (See the next section on ISO 9000.)

A comprehensive definition of *quality* is difficult to decide upon because there are so many valid approaches to defining it. There are three commonly accepted definitions, any one of which, or any combination of which may be used. These are:

1. Conformance to requirement with no defects. This is the definition of Phillip Crosby. In this context, products need well-defined specifications. A product that is consistently manufactured to meet specifications is a quality product.

2. Fitness for use. This is the definition of Joseph Juran, founder of the Juran Quality Institute, and is user oriented. Fitness for use has five dimensions:
   - Quality of design. This distinguishes one product from another—a Chevrolet from a Cadillac. Quality is designed into the product.
   - Quality of conformance. Does a product deviate from its design specification?
   - Availability. Is the product reliable and maintainable?

- Safety. This covers the possibility of injury—is the product safe?
- Field use. Is the product used effectively by the customer?

3. The eight elements of quality. This is the approach of David Garvin of Harvard Business School. The eight elements of quality are: performance, features, reliability, conformance, durability, serviceability, aesthetics, and perceived quality. Quality is measured independently against each of these elements.

This section focuses on the more traditional definition of quality—making a defect-free product that performs per requirements and satisfies a customer.

The Japanese must be given credit for having raised the practice of quality to the first and most important objective of manufacturing. It is ironic that they restored practices developed in the United States and used Americans (particularly Deming) to teach them the practices. Given all the discussion and literature covering quality, the essence of quality management is often obscured: quality is achieved by persistent, day-to-day attention to detail and by continuous improvement.

Finally, *quality* is a culture of doing things right. A story from the 1980s, told to me by a colleague, illustrates this point:

> American manufacturers resent their unfortunate reputation regarding quality control. Many have intensive programs to deal with the issue. "Let's make it tough on them," said the midwestern purchasing agent, writing out the specification for the company's first order from a Japanese subcontractor. "On ball bearings, let's accept no more than three defectives in every ten thousand." Tough it was. Far more stringent than the rates allowed to American companies. And so it was with great excitement that the firm opened the Japanese shipment when it first arrived. In each crate of ten thousand they found a letter.
>
> Dear Sirs,
>   Enclosed please find the ball bearings you ordered. We do not know why you wished to receive three defective bearings with every ten thousand, but we have enclosed them, wrapped separately and identified with cross-hatching, so you will not mistake them for good ones. Sincerely,

Process for Implementing Total Quality Management

1. Communicate the need for superior quality from top management down. Top management must be visibly and constantly involved in

quality management. The company should have a quality policy statement that clearly establishes the importance of quality.

2. Develop customer-related measurements. These measurements must include performance and reliability, and should be extended to customer satisfaction.

3. Benchmark the process (by comparing company's performance with that of the best companies), and then set aggressive goals for quality. (See "Benchmarking" in Chapter 1.)

4. Determine what is to be measured and when. Standardize quality measurement and reporting methods across the facility to ensure a common language and understanding. Purchased parts and components are usually inspected at the supplier's factory in the supplier's process or by an inspection agency. Manufactured parts are inspected in the process—preferably through statistical process control by the operator. Finished goods are tested for conformance to specifications and useability.

5. Decide the precision to which quality is to be measured. Defect rates are the commonest means. Defects used to be measured as a percentage of defective (parts per hundred that are defective, or percent defects), until the 1980s, when the Japanese raised quality to a new level by measuring defects in terms of parts per million (ppm) or 0.000001 or one ten-thousandth of 1 percent! Unfortunately, percent defects is still a common level of measurement.

6. Involve all employees in teams, in the goal-setting process (see the section on employee involvement and teams), and in the achievement of those goals. Include customers and the suppliers in the process (this is covered in Chapter 10 on "Supplier Partnerships" and in Chapter 11 on "Customer Relations").

7. Measure and review. This is a critical step. Look at trends and take corrective action if necessary.

8. Develop a few key quality approaches:

   ■ Quality cannot be inspected into a product. The process itself must have built-in quality checks.

   ■ The source of quality is the operator, the machine, the material, and the process. Attention must be paid to controlling these inputs and the control should be exercised by the operator in a continuous process similar to statistical process control (SPC). (This is discussed in detail in Chapter 10.)

- Operators should inspect their own work or the work of the preceding workstation. Inspectors should only be used in exceptional cases.

- Operators must be permitted, even encouraged, to stop the line if they detect a defective part or process.

- There should be no circumstances when defective product will be accepted.

9. Provide operator training on quality both at a generic level and at a specific "how-to" measure level, similar to the provision in SPC. There should be a quality-training curriculum that is a mandatory requirement for every employee. The curriculum should include, at a minimum:

- Quality awareness—the need for and the competitive effect of quality

- Problem-solving methodology, including the common quality tools—one of the common methodologies, such as the Quality Improvement Story (QIS)

- The use of tools such as flowcharts, line graphs, bar charts, scatter diagrams, Pareto diagrams, and fishbone diagrams (see Chapter 10, "Continuous Improvement," for descriptions of the tools)

- Statistical process control (SPC) for all operators

- Root cause analysis

- Team training (see the Employee Involvement section in Chapter 1)

10. All defects must be investigated to determine the root cause of the defect. The easiest and one of the most effective means of performing this analysis is the *5 W technique,* which "asks 'why' 5 times" to get to the real cause of the defect. Establish an open investigative environment, with a focus on causes and not individuals.

11. Focus on the process. Set up process controls to be preventative rather than reactive. The SPC method of control is a good example.

Error proofing or poka-yoke should be used where possible. (This technique is described in Chapter 10 under Poka-yoke)

12. Understand the real cost of quality. It is estimated that most companies spend up to 20 percent of sales revenue on quality. There are generally four types of costs. These are:

- Appraisal. These are inspection and testing costs and include test set equipment maintenance.

- Preventative. These are built-in checks such as SPC and preventative maintenance, training, and helping suppliers.

- Internal failures. These include process defects, scrap, rework, and redesign.

- External failures. These are warranty failures, returned product, and complaint reconciliation. These are usually the most costly, both in terms of direct cost and in terms of loss of goodwill.

13. Set up a culture of and an attitude to quality improvement. Quality management requires discipline, persistence, and meticulous attention to detail.

14. Celebrate success. Showcase achievements. Reward quality role models.

Dos and Don'ts

DOs

1. Do start the quality management process from the chief executive.
2. Do understand that quality management is an attitude of doing the right things the right way and should involve everyone in the organization.
3. Do make sure that proper training is provided.
4. Do devote more time and attention to prevention than to correction.
5. Do understand that execution is in the details that must be performed daily.
6. Do focus on the customers and verify that they are being satisfied.
7. Do make certain that the quality approach is designed around processes.
8. Do constantly improve and evaluate.

DON'Ts

9. Don't expect to improve the product by increasing the amount of inspection conducted on it.
10. Don't underestimate the resistance to change, and be prepared to accept setbacks.
11. Don't expect quick or easy results. Culture is being changed and this is a time-consuming process.
12. Don't expect quality to be improved "when time permits," as a special, one-time project.

Quality Management: ISO 9000

An Overview

The International Organization of Standardization (ISO) is made up of national standards boards from 91 countries. ISO 9000 became the European Economic Community (EEC) norm; it was approved by the European standard-setting organization in 1987 and revised in 1994. *Certification to ISO 9000 is becoming a requirement for all companies doing business in Europe.*

ISO 9000 has a five-part quality standard:

- ISO 9000 is a guide to selection and use of the remaining standards.
- ISO 9001 is a specification for design/development, production, installation, and service.
- ISO 9002 is a specification for production and installation.
- ISO 9003 is a specification for final inspection and testing.
- ISO 9004 describes the principal concepts and applications for a complete quality management system.

"**Quality System**—The organizational structure, responsibility, procedures, processes, and resources for implementing quality management" (ISO 8402-1986, paragraph 3.8).

Table 1.1 Comparison of ISO Certification Levels

	9001	*9002*	*9003*
Design	*		
Development	*		
Service	*		
Production	*	*	
Installation	*	*	
Final Inspection	*	*	*
Testing	*	*	*

ISO 9001 Elements

ISO 9001 is a complete quality system model, since it covers quality assurance from design to product servicing. Section 4 of ISO 9001 describes 20 elements for a quality system. Each company seeking registration under this section is evaluated for compliance to the requirements of each element, and all the elements require documentation of procedures and maintenance of results. The elements are listed here by their ISO 9001 section numbers.

4.1. **Management Responsibility**

This element defines: the quality policy—the policy, objectives, and commitment to quality with executive responsibility; the organization—the responsibility and authority of personnel who manage, perform, and verify the quality of the product; and the management review—assessment at defined intervals to ensure the suitability and effectiveness of the quality system.

4.2. **Quality System**

This element describes a documented quality system that meets the ISO criteria and ensures effective implementation of quality procedures.

4.3. **Contract Review**

This element defines the customer's requirements and ensures that the supplier's quality system is capable of meeting the customer's requirements.

4.4. **Design Control**

This element defines design and development planning activities, design input and output requirements, design verification of final product, and design change control.

4.5. **Document and Data Control**

This element establishes procedures for document and data approval, issue and control, and review and control of document changes and modifications.

4.6. **Purchasing**

Subcontractors must be selected and controlled on the basis of meeting the quality requirements. Purchasing data must describe the product and its requirements clearly. Purchased product must be verified to ensure that it conforms to specifications.

4.7. **Purchaser-Supplied Product**

The supplier must ensure that purchaser-supplied product is verified, stored, and maintained at quality standards.

4.8. **Product Identification and Traceability**

Where appropriate, product must be identified and tracked from receipt through production, delivery, and installation.

4.9. **Process Control**

This element must contain documented work instructions, monitoring and control of process and product characteristics, preventive maintenance of equipment, and approval of processes and equipment used, including software and criteria for workmanship.

4.10. Inspection and Testing

This includes receiving inspection and testing, in-process inspection and testing, final inspection and testing, and maintenance of inspection and test records.

4.11. Inspection, Measuring, and Test Equipment

This element describes the need to control, calibrate, and maintain inspection, measuring, and test equipment (including test software).

4.12. Inspection and Test Status

This element establishes procedures to determine conforming and nonconforming product.

4.13. Control of Nonconforming Product

This element establishes procedures to ensure that nonconforming product is not inadvertently used or installed. The authority for reviewing and disposing nonconforming product must be defined.

4.14. Corrective and Preventive Action

This element establishes, documents, and maintains procedures for investigating the cause of nonconforming product and the corrective action needed to prevent recurrence. It also ensures that procedures required to deal with customer complaints and preventive actions to ensure controls are effective.

4.15. Handling, Storage, Packaging Preservation, and Delivery

This element provides methods and means of handling, storing, packaging, and delivery that prevent product damage and deterioration.

4.16. Quality Records

This element establishes and maintains procedures for identification, collection, indexing, filing, storage, maintenance, and disposition of quality records.

4.17. Internal Quality Audits

This element establishes the need for conducting independent, comprehensive, planned, and documented audits to verify the working of the quality system.

4.18. **Training**
This element establishes procedures for identifying training needs and providing training.

4.19. **Servicing**
Where required, this element verifies that servicing requirements are met.

4.20. **Statistical Techniques**
Where appropriate, this element establishes statistical procedures for verifying process capability and product characteristics.

Process for Obtaining ISO 9000 Registration

The following process was compiled from a presentation made by Sandford Liebesman from AT&T Quest on "A Roadmap for ISO 9000 Implementation."

1. Evaluate the existing quality management system. Communicate to all employees the company's intention of seeking registration. Provide a brief orientation of what this entails and the advantages that will accrue from the certification. Build initial involvement and commitment. (Time requirement: 1 month)

2. Plan the compliance effort. Form a core coordination team, establish process management teams to address specific elements, define the scope of the registration, and select a registrar. Conduct information workshops for all management and team members. The final activity of this step is to develop a project plan. (Time requirement: 1 month)

3. Develop and document the quality management system. At this point, new processes—such as document control and corrective action—are being developed along with existing ones. Processes should be documented as they *are, not* the way teams would ideally like them to be. (Time requirement: 2 to 4 months)

4. Roll out the quality management system. This phase involves two or three internal audits and a preassessment by the registrar. The preassessment is very important because the organization gains an understanding of the auditors and the auditors learn about the organization. (Time requirement: 4 to 6 months). The chances of passing a registration in the first audit increase dramatically when a preassessment is performed.

5. Conduct the registration process. This step consists of a period of preparation for the audit and for the actual on-site registration audit itself. Preparation should include training, conducting mock audits, and creating a team to support the site visit. A detailed plan for the days of the audit is extremely useful. (Time requirement: 2 months)

6. Maintain and monitor the quality system (after registration). Here internal audits, management reviews, and corrective and preventive action plans play a key role in maintaining certification.

Survey of Successful AT&T Organizations

The information in this section is taken from 35 AT&T and NCR organizations. The time and effort needed to prepare for and successfully complete an ISO certification was determined to be a median of one year (see Table 1.2). Staff years represents the number of staff times the number of years (e.g., six staff years = six staff used for one year or three staff used for two years).

Nine Lessons: These are the most important lessons learned in the course of obtaining ISO 9000 certification in the experience of 35 AT&T and NCR organizations in 1992 and 1993.

1. Management must take a strong role in leading the effort. Management must show commitment and support for quality and a depth of understanding of the policy. Management must also be trained in process management and improvement.

2. The basic philosophy should be to go well beyond minimum compliance. The successful companies did many things because they were the right things to do and not merely to comply with the standard.

3. The registrars view internal quality audits as very important. One of the first things registrars look at during an audit are the results of

Table 1.2 Benchmark of Becoming ISO 9001 Certified

Company	Small	Medium	Large
Size	Less than 500	500 to 2000	Over 2,000
Staff used in years	0.4 to 4.7 staff years	1.8 to 12.5 staff years	6 to 22.5 staff years
Median of staff used	1.9 staff years	5.0 staff years	17.5 staff years

internal audits and what action management has taken as a result of the audit findings.

4. The registrars will spend a great deal of time tracing requirements through the design and development process. The audit trail is developed through records of results and the documentation of process reviews. In ISO 9001 there are 18 specific statements requiring quality records.

5. The auditors view corrective action as the means of ensuring continual quality improvement. Evidence of corrective action helps to demonstrate that the quality system has been and will continue to be effective.

6. It is advantageous to format the quality manual according to the structure of the ISO standard and to prepare a matrix identifying the ISO clauses that each internal process addresses. This format ensures that all the ISO requirements are covered.

7. ISO training should be provided to everyone in the organization who affects the quality of the product or service. The elements of a training program should include an understanding of quality management principles and policy, ISO standards and definitions, process controls, corrective action, and root cause analysis. Training should be highly customized to the environment of the organization.

8. Preassessment by an outside agency should be performed during the compliance process. A high value comes from an "independent set of eyes," which provides insights into gaps in the system.

9. One very important factor in a successful registration is a comprehensive logistics plan for support during the site visit. The recommendation is for two escorts per auditor. One stays with the auditor and the other communicates with the rest of the organization. The escorts must be the most knowledgeable persons in the field. Work out a process for corrective action during the visit.

Most Difficult Elements with Which to Comply

Some elements encountered during ISO 9000 certification are more difficult to comply with than others. The eight most difficult ones are ranked in order of difficulty in the following table; number 1 is the most difficult.

Rank	Element Number & Description
1.	4.5 Document control
2.	4.11 Inspection, measuring, and test equipment
3.	4.2 Quality system
4.	4.17 Internal quality audits
4.	4.18 Testing
6.	4.1 Management responsibility
6.	4.9 Process control

Element compliance: ranked in order of difficulty

Employee Involvement and Teams

Purpose and Description

The purpose of employee involvement through teams is to create an environment that will sustain a competitive level of productivity.

Employee involvement means inclusion of the employee in the design and operation of the system. It recognizes that individual employees are best equipped to know and understand the problems of their own work area. Employee involvement provides employees with a strong motivation for wanting to improve their performance continuously, and gives them the responsibility and the authority (empowerment) to resolve their process problems. Further, when involved and empowered employees are organized into effective teams, their collective performance can reach even higher levels of productivity.

Discussion

A key driver of superior performance is an organization where *all* employees, individually or in teams, are committed to continuous improvement. Employee involvement in problem identification, and employee empowerment that allows employees to take actions to resolve these problems, are essential to achieving world-class productivity.

The concept is simple and attractive: create a vision of the company's goals, involve all employees in identifying with this vision, allow them to organize into teams, support the teams by giving them responsibility and authority for undertaking their tasks, and a high level of performance will be achieved! Alas, the reality of involvement and empowerment is vastly different from this simple concept. At the root of the concept is change. Traditionally manufacturing has been managed by hierarchical organizational structures, with rigid levels of authority and control. Transforming the culture of factory management to relinquishing control is probably the most difficult part of the change. The employee change required is almost as difficult. Employees have to overcome a mindset that expects to be told what to do, and they have to develop a self-confidence that will permit them to exercise initiative and make decisions.

Most examples of successful employee, management, and team integration have occurred in cases where a factory is faced with survival. What is needed then is a knowledge and understanding of *how* to organize and train employees so that they become involved and motivated to high and sustained levels of performance. Today, *empowerment* and *self-managing teams* have become buzz words, and expectations are being created that this process can be implemented and maintained with some basic simple activities. The result is lip service to the concept of empowerment, which does more damage than good.

At the root of all employee involvement are two assumptions:

- Employees can be motivated to want to perform at a high level of productivity.
- Team performance is superior to individual performance.

Both assumptions are reasonable, and there is considerable evidence to confirm them. What is not as clear are the circumstances and the stimuli required to realize and sustain employee involvement and team productivity. Neither assumption is a natural part of organizational growth or maturity. Different companies have different requirements for developing employee involvement. These needs depend on a company's existing culture, on its economic urgency, and on the vision of its leadership. As already noted, most of the publicized examples of superior employee involvement have occurred in companies facing survival. Employee involvement still remains principally an act of faith!

One of the critical factors affecting competitiveness in the twenty-first century will be a company's ability to develop and use the potential of its employees. Call it *people power.*

In the 1990s employee involvement has been and is one of *the* most discussed, written about, and worked on subjects in manufacturing—indeed, in all organizational development. By necessity, this book can cover this subject only briefly. An effort will be made to outline a process for setting up employee involvement that concentrates on *how to.* This is followed by a section on forming, building, and sustaining effective teams.

Process for Employee Involvement

1. Create a "vision" for the company. The vision must be perceived to be worth striving for by all employees. Capture the vision in a simple slogan.

   Examples: Ford—"Quality is Job #1" or Avis—"We try harder."

2. Operationalize the vision by translating it into a business focus— such as quality, performance, or price—so that all employees can relate to it and so that it can be easily identified, measured, and tracked. The vision and its business focus must also be identified with and aligned to a strong employee need. This alignment of a company's vision and objectives with an employee's needs ensures employee commitment, the cornerstone of involvement.

   Example: Ford's quality vision was tied to competitive survival and the employee's need for job security.

   The intention of involving the employees directly in achieving the vision should also be articulated. Upper management must be seen to support and "walk the talk."

3. Involve the union and gain its support in the development of the vision and in the deployment of the vision.

4. Market the vision to *all* employees. Senior managers and union officials must communicate the vision, its business focus, and the role of the employee in its achievement. The business focus should be translated into a meaningful work activity at *every* level of the organization, and its achievement should also be linked to the competitiveness of the company. An effective means of communication is third-party teaching, where every manager deploys the company's vision and business goals to the staff in terms of the department's responsibilities.

   Example: A quality goal for the number of acceptable defective parts per million can be developed and deployed to every level. Furthermore, the quality goal, as it is being improved, must be related to an increase in the company's sales or profitability.

5. Ensure that debate and discussion take place and feedback is received and responded to, while communicating the employee involvement process. The process must be conducted in an open and

truthful manner; doing so will create and strengthen an environment of trust.

6. One of the best ways to solicit employee input for ways to improve their work process is a suggestion program. Such a program should be simple to use, and boundary conditions should be spelled out (such as all union contractual subjects are off limits). It should contain a description of what an employee can expect (reward and response time). There should be a clear understanding that suggestions can be rejected, but that when this happens reasons will be provided.

7. Provide an envelope or scope of what the employees will be allowed to do through their own efforts, and, since this is at the crux of empowerment, it should be extended to most functions in the employees' area. The principle is to move decision making to the actual working level of the organization.

8. Select pilot areas for implementation. Initial successes are crucial. If the means of employee involvement is through a suggestion scheme, be prepared to meet a flood of ideas and ensure that they are dealt with efficiently.

9. Conduct regular meetings of all employees. Communicate and debate progress and improvement. The employees must feel that they are contributing to the development and functioning of the process.

10. Ensure that ideas and improvements are communicated to other areas and parts of the organization, so that the whole organization learns.

11. Trace the involvement process to the external business success. No process can really succeed unless it actually contributes to the success of the organization. This success should be visible to all employees.

12. Be prepared to repeat the procedure several times as the process requires constant repetition and reinforcement. Culture is being changed and trust is being developed. Both these activities require time, consistency, and patience.

13. Develop trust through personal contact and personal relationships. It is imperative that management work with production associates to develop this trust.

Barriers to Successful Employee Involvement Activities

The U.S. GAO conducted a survey of companies that had or were implementing some form of employee involvement. These companies were asked to identify barriers that were a "great" or "very great" hindrance to the implementation effort. The following table summarizes the results.

Barrier	Companies classifying barrier as significant (%)
Short-Term Performance Pressure	43
No "Champion"	26
No Long-Term Strategy	25
No Skills Training	23
Unclear Objective	21
Lack of Tangible Improvements	20
Lack of Feedback	18

Data Source: U.S. General Accounting Office, May 1988

Teams: Formation, Development, and Growth

Discussion

The most effective and widely used form of employee involvement is the use of *work teams*. A *team* is a group of individuals working together to reach a common goal. *Teamwork* implies a sharing of responsibility and decision making and an acceptance of joint accountability. The difference between a *team* and a *work group* is that the former develops synergy. The collective performance of the team exceeds the individual performance of the members, through interdependence and sharing. The responsibility and autonomy of a team will depend on the type of task it is commissioned to perform.

Teams are not an end in themselves; they are a means of increasing employee productivity, to improve a company's competitive performance. The achievement of this objective must be measured, compared to a benchmark, and appropriate actions taken.

Process for Implementing Work Teams

1. Establish a vision and business goals. Like individuals, teams must understand and buy into the vision and goals. (See the earlier section on Employee Involvement.) Deploy these goals to the level of a specific organization (a factory).

Example: Vision: To provide world-class customer service.
One of the goals: To ship 99 percent of all catalog items within 24 hours.

2. Decide how to structure the teams. Teams may be structured in many ways, and which way is most appropriate depends on the type of product and process, the working environment of the organization, and the health of the business. Involve members at all levels of the organization in this process. In the planning of work teams, ensure that union personnel are involved.

Example: Environment—A low-volume, high-tech, costly engineered product with considerable process variability, dominated by design, product and process engineers. Teams in this environment should have 10 to 12 persons.

■ Cross-functional *product teams* set up for every major assembly. Teams consist of engineering—design, product, process and testing, shop supervision, quality, materials management, project management, and accounting.

■ Cross-functional *platform teams* set up for common processes such as manufacturing systems, supplier material management, quality control, program management, and common operations such as fiber splicing or circuit-board manufacturing.

■ *Work teams* consisting of shop floor operators, the shop supervisor, the supporting engineers, and a facilitator. There may be one or several work teams supporting each of the product teams.

3. Develop a team charter. In this charter, spell out the role, responsibility, and authority of the teams. Specify key goals that will be measured and for which the team will be accountable. The charter should contain specific limits of what the team is expected to do and spell out what the team is empowered to do. For an example of a team charter, see Figure 1.3.

Example: Teams will be expected to meet the delivery commitments as defined by the schedule. Teams are authorized to spend up to $25,000 for resolving a specific problem.

Initially, team members may be selected to represent all facets of the product or process. Later on the team should be allowed to determine its own membership.

Management's Commitment to Process Management Teams

The undersigned, representing the management team for design and manufacture, empower several process management teams to improve the processes used to define requirements, select technology, qualify, build, and certify product. The performance of the team will be strongly considered in assessing individual performance and in allowing individual performance awards for the assigned team members.

Each team will be constituted and charged to:

1. Understand the needs of its customer.
2. Keep a focus on improving its products and processes.
3. Develop key indicators of the performance of the areas assigned to the team.
4. Track and report key indicators.
5. Develop initiatives to improve the process continually.
6. Use the indicators to demonstrate the effectiveness of each initiative.
7. Involve all appropriate resources of the division in the improvements.

In return, we the management state:

1. Each Process Management Team (PMT) will be empowered to implement, through standard change control procedures, all improvement recommendations that they have identified, that involve purchase expenditures, not exceeding $25,000.

2. Any recommendations that require expenditures of more than $25,000 to implement will be considered, and a response containing an answer or a plan to achieve an answer will be provided in a timely fashion (objective is five working days).

3. Written explanations or directions will be provided for all rejected suggestions.

4. Satisfying the needs of the customer through continuous quality improvement of products and processes is the highest priority.

5. Team members will receive the training necessary to accomplish their mission.

6. All employees will be encouraged to identify issues and opportunities without fear of reprisal.

7. Adequate facilities for team meetings will be provided.

Approved
Sd. All members of the senior management team.

Figure 1.3. An Example of a Team Charter

4. Provide training sessions where the charter and the roles and responsibilities of the team and its members are discussed. Members of the team must understand and accept the purpose and the value of teams. Training should include guidance on interpersonal interaction dynamics and team behavior. Training is an ongoing process, and frequent formal and informal sessions must be conducted. Training must seek to obtain some level of commitment from the team members.

5. Deploy the organization's goals to each of the teams. Ensure that the teams are consulted and that there is open discussion and debate

before the goals are deployed. The team must be part of the goal-setting process, and goals must almost always be agreed upon. In addition to having deployed goals, teams should be encouraged to set their own goals commensurate with improving their process. A team should have 6 to 10 goals. Goals should be weighted, as they will have different priorities.

Example: Goal: Total product lead-time = 10 weeks.
Each product team will be assigned a lead-time goal consistent with the overall goal. Furthermore, this lead-time goal may be considered to have a weight of 30 percent of a product team's goals—indicating that it is a priority.

6. Provide a process for the teams to follow. This process may consist of general rules on the frequency and length of meetings, procedures for resolving conflict within the team and between teams, ways to request functional resources, and so on. The process should include a problem-solving methodology, such as the Quality Improvement Story, so that there is consistency in the approach of all teams. Training should be provided in the use of the technique selected.

7. Ensure that there is a role for management. Teams take over the managing and decision-making roles of department managers, either leaving them without a role and frustrated, or causing them to hang on to controlling functions, thereby impeding the empowerment of teams. A useful and productive role for managers is that of a team champion. A team champion's role is to establish team priorities, to provide resources, and to act as a spokesperson and advocate for the team. Such a role is beneficial to both the process and the manager.

8. Provide a trained facilitator for each team (one person can facilitate many teams). This is critical, and its absence could abort the whole process. Each team should have a coordinator or a leader. Initially this person may be appointed, but later, as teams mature, team members can elect their own coordinator.

9. Adopt different tactics for different teams depending on their stage of development. Typically most teams go through four stages:

   ■ Forming. In this stage there is a great deal of uncertainty and cynicism. Members are unsure of their roles and the scope of their responsibility. They are usually unwilling to exert the authority delegated to them. In this stage teams have to be encouraged and cheered. Some shortcomings and mistakes may be overlooked. The champion must be a strong supporter of the team and work closely with them until they establish confidence in the process.

- Storming. This stage is characterized by interpersonal conflicts among the team members, as team members start to exercise their initiative and authority. The champion must keep contention constructive, and see that it is issues and not personalities that are addressed. This is the stage in which the problem-solving methodology should be stressed, as it will focus on fact finding and disciplined decision making. Usually little progress will be made toward the team's goals.

- Norming. Team members develop respect for one other and start to work together, directing their energy toward problem solving. The champion must now remove barriers from the path of the team and act as a resource provider. The team may function with a great deal of autonomy.

- Performing. The team members are cooperating and collaborating and are performing productively to achieve or exceed their goals. The team starts to direct itself and monitor and correct its performance. The champion may only need to coach the team occasionally, to help establish their goals or to encourage the team to develop innovative solutions.

10. Provide a regular scheduled review—usually monthly—for all teams to present their achievements and the issues with which they are confronted. This is an occasion for the teams to "show off" their performance and also to ask for help if it is needed. Management involvement in this meeting is crucial, and management must come to the meeting to listen, asking questions only if there is need of clarification. An attempt to check or correct performance at this meeting is likely to have an adverse effect on the process. Mistakes must be openly admitted and attention focused on the remedy and preventive action—*not* on the person who has made the mistake. Deficiencies in team performance should be noted and discussed by the champion with the team at a subsequent meeting.

11. Hold teams accountable. Some teams may not perform. Changing the team composition or the coordinator may help. As a last re-

sort, a manager may temporarily be a member and coordinator of the team.

12. Sustain the team process. This is like winding a ball of wool—progress is slow and mostly tedious, but if the process slips, a large portion can come undone very quickly. The champion must constantly observe the team's performance and plan how to challenge and enhance this performance.

13. At the core of the team and employee involvement process is *trust*. There must be an honest and open environment. Mistakes are acceptable. Everyone "walks the talk."

14. Teams consisting of operators require more attention and their progress may be slower. There is need for more facilitation and organization.

Dos and Don'ts

DOs

1. Do remember that measurements drive behavior and so measurements should be carefully selected, should be easily understood by the team, and should be easy to obtain.
2. Do ensure that teams have **SMART** goals (Specific, Measurable, Agreed upon, Realistic, Time bound).
3. Do ensure that teams are responsible and accountable for achieving their goals.
4. Do make sure that top management are visibly and consistently involved.
5. Do ensure that managers have a role, preferably as champions, and are trained to allow their teams to operate independently.
6. Do confirm that the team goals are in alignment with the company's goals.
7. Do encourage a bias for action.
8. Do see that good productive ideas are implemented. Publicize successes.
9. Do recognize and reward superior performance—when it happens.
10. Do insist on collecting and working with facts.

DON'Ts

11. Don't expect initial buy in or acceptance. Expect initial cynicism.
12. Don't expect quick results. Expect up to a year and more for sustained results.
13. Don't fail to test reality. Wishing and willing success is short-lived.
14. Don't expect that all teams will function effectively.
15. Don't expect easy or pat answers. There are no prescribed solutions.
16. Don't eliminate jobs as a result of the productivity of teams.
17. Don't fail to communicate—there can never be too much communication.

Benchmarking

Purpose and Description

Benchmarking is a technique used to compare a company's performance with that of the best in the industry and to drive the company to improving its performance to equal or exceed the best or benchmark.

Benchmarking was defined as "the continuous process of measuring our products, services and practices against our toughest competitors or those companies renowned as leaders" (Kearns 1979).

> Example: Xerox compared the cost of U.S.-manufactured copy machines with the cost of their Japanese associate (FUJI-Xerox) and found that FUJI-Xerox could sell their copier at what it cost to be manufactured in the U.S. New manufacturing processes and components were adopted and U.S. manufacturing costs were reduced.

Discussion

The credit for developing the practice of benchmarking must go to Xerox who, in the late 1970s, with their Xerox machine market being taken over by Japanese competitors, tried to determine how their costs compared with the Japanese costs.

Benchmarking against the competition's performance is difficult, as detailed information is not easily accessible or available. There is doubt about whether comparisons are being made of like processes—apples to apples. Thus it is usually more effective to benchmark a noncompetitor's performance on similar functions. Investigating a noncompetitor will provide detailed information on best practices that can be adopted and used to improve performance.

Benchmarking may also be conducted against internal operations and/or generic processes. Figure 1.4 illustrates the steps of the benchmarking process, and each step is detailed in the following list.

Process

1. Identify what functions or processes are to be benchmarked.

2. Determine what measurements and criteria will be used to benchmark the function or process. Collect internal information.

3. Determine the method of collecting information. Prepare a detailed questionnaire on information to be collected. There can be one or several sources of data and information on practices used. Sources

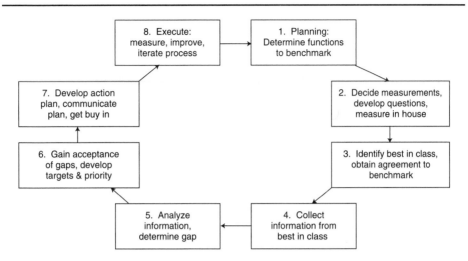

Figure 1.4. Benchmarking Process

of information include internal information, company information, combined studies with other companies, public domain information, library searches, professional and trade data, consultant information, and external experts and studies.

4. Determine who (which company) will be used for the comparison. Ensure that the company chosen is the industry or functional leader. Obtain agreement from the company selected.

5. Collect information from the companies selected. This may involve site visits and/or meetings.

6. Analyze the information and determine the "gap" between company's current performance and that of the best in class.

7. Gain acceptance of the benchmarking results. Develop priorities of gaps to be closed.

8. Decide on what targets to set to close the gap. Targets must be measurable goals, must have a time reference, and should be developed with the major stakeholders.

9. Communicate the targets to all involved and obtain a "buy in."

10. Develop action plans that have clear lines of responsibility and authority to reach the targets with the major players.

11. Execute action plans. Measure. Improve. Revise action plans if necessary, until the target is achieved.

12. Iterate the entire process, at least once a year.

Dos and Don'ts

DOs

1. Do set a proper project scope—not too large and not too small.
2. Do involve the principal stakeholders in the analysis, target setting, and action plans.
3. Do have a champion (a leader) who will carry the message across the company.
4. Do ensure that the benchmark focus is on the target work process as well as on the measurement.
5. Do understand that benchmarking processes involve attitude and behavior change.

DON'Ts

6. Don't neglect to define specifically the project plan and its goals with responsibility and dates for completion.
7. Don't keep benchmarking as a staff exercise—conduct frequent feedback sessions with all stakeholders.
8. Don't encourage or be satisfied with imitation. Seek new and innovative solutions.

Measurements

Typical set of questions is used to benchmark some manufacturing functions. These questions address the areas of materials management, supplier management, and information systems management which are all part of manufacturing.

Materials Management

Materials management covers making and/or stocking an optimum quantity of materials so all customer orders can be fulfilled.

Scale of Operation

- Number of active product families
- Number of product options/features
- Number of active stock keeping units (SKUs)
- Commonalty index: average number of parents per SKU
- Number of active bill of materials (BOMs)
- Depth (number of levels) of BOMs by product

- Transactions per day, categorized by type (e.g., receipt, disbursement, and so on)
- Number of material personnel directly planning with MRP
- Material cost as a percent of total cost

Inventory Scale

- Number of active SKUs in the warehouse or storeroom
- Number of stocking locations by type (e.g., small parts, bulk fill, and so on)
- Warehouse/storeroom inventory accuracy—by count and by location

Turn Rates

$$\text{total inventory turns} = \frac{\text{cost of sales}}{\text{average \$ inventory cost}}$$

$$\text{raw material turns} = \frac{\text{cost of direct material in sales}}{\text{average raw material cost}}$$

$$\text{finished goods turns} = \frac{\text{cost of sales}}{\text{average finished goods cost}}$$

Material and Delivery

Context is extremely important. All the following indicators should be measured, at the very least, in terms of

- The absolute number of occurrences
- The percentage of total SKUs or total orders

The measurement of the elements below should be considered in the context of the number of items being worked with. For example, shortages per month should be measured in absolute numbers, such as "10 shortages" and as a percentage of total SKUs, such as "0.1%" (there being 10,000 SKUs).

- Shortages by month
- Number of line stoppers
- Occasions of line stoppers
- Number of active items at 0 balance
- Number of short interval orders
- Customer orders rescheduled per month
- Engineering changes made to product on order and to product in progress

- On-time delivery =

$$\frac{\text{Number of orders delivered on time by required date}}{\text{Number of orders scheduled to be shipped in same period}}$$

Supplier Management

Supplier management seeks to compare the performance of a company's suppliers with the best in class.

Quality and Delivery

- Percent of supplier's total business generated by benchmarking company
- Orders or pieces rejected by key supplier
- Number of key suppliers on a ship to stock program with self-inspection
- Number of key suppliers on a ship to stock program with external inspection
- Number of key suppliers on demand pull
- On-time delivery by key supplier =

$$\frac{\text{Number of orders shipped}}{\text{Number of orders scheduled}} \quad \text{(by required date, by promise date, or by lead time)}$$

Improvement

Types of innovative supplier partnership arrangements for reducing cost and improving delivery

- Point-of-sale data supplied by the customer and used by the supplier to manage flow of inventory to point of use (e.g., P&G manages the supply of diapers to Wal-Mart stores based on sales)

- Back-flush system used to pay supplier upon system of final product

- Lead-time compression combined with manufacturing flexibility, leading to supplier making to order instead of making to stock, with no impact on service levels

- Supplier's personnel located at customer's premises to replenish stock, eliminating the need for a buyer

- Number of continuous improvement suggestions received per supplier to improve cost, delivery or quality

Cost Trends

(Note: It is important to base all cost measurements on the total cost divided by life-cycle costs rather than on the latest "price".)

- Reduction of cost for the same product
- Material substitutions/product redesign suggested by suppliers to improve costs

Information Systems Management (Hardware, Network, and Application Software)

The cost and reliability of an information system is benchmarked against the scale of the operation.

Scale of Operation

- Data disk space for one quarter's information
- Transaction volume by type
- Maximum number of concurrent users
- Average number of users
- Full-time operating personnel: hardware, software, and application (MRP)

Cost

- Cost per transaction =

$$\frac{\text{Salary of dedicated personnel} + \text{hardware and software expense}}{\text{Number of transactions}}$$

- Cost of system maintenance

Service

- Downtime in percent and actual time (hardware and network)
- Innovative ways of using the system to reduce cost and improve performance

Manufacturing Engineering

Manufacturing engineering compares how well the company is reducing its lead times and making improvements. The comparison is made by scale of operation.

Scale of Operation

- Number of personnel working on shop floor
- Total material, labor, and overhead cost of product

Lead Time

- Lead time for "A" type products and major assemblies
- Times butt-to-butt for "A" type products and major assemblies
- Level of work in process inventory as a multiple of the weekly delivery rate
  - Average
  - Major assemblies
- Average lot size used and basis of that lot size
- Lead-time reduction-time and percent over last three years on a year-to-year basis
- Are bottlenecks identified? How are bottlenecks managed?
- Are buffers used? How are buffers determined?

Continuous Improvement

- Number of personnel with more than two job skills
- Number of personnel with more than three job skills
- Number of suggestions received per employee
- Number of received suggestions implemented
- Innovative programs for improving productivity

Appendix 1.1: A Case Study of Attitudes Toward TQM (Total Quality Management)

A survey was conducted in 1992 on the perceptions and attitudes toward TQM of a large group of management personnel—managers, engineers, and other manufacturing professionals. The study revealed that:

- The majority reported that they can identify the customers for their organizations (88.2%) and customers for their functions (91.7%).

- Over 40 percent of the engineers and 50 percent of the general management responses indicated that there were *no* methods in place for preventing problems.

- Forty-four percent of the respondents stated that there were no mea-surements in place to track improvements, and 50 percent say that measurements are *not* used to set objectives.

- Sixty-three percent stated that competitive benchmarking is *not* used to evaluate quality progress.

- Thirty-five point four percent disagree with the statement that "my manager walks the talk" regarding quality improvement, and 38.3 per-cent state that higher-level managers do not practice quality precepts.

- Seventy-five percent of management and 81.2 percent of technical per-sonnel saw quality improvement as everyone's responsibility.

- On the unit's working environment, personnel agreed that there is respect (82%), open communication (77.6%), team work (76.9%), and trust (72.4%).

The results of the survey were analyzed by top management and the following conclusions were drawn:

What is getting in the way of people buying into the quality thrust?

1. Lack of a systematic management process to deploy goals. Causes of this lack include:
   - Inadequate quality measurements
   - Inadequate quality plans
   - Designers are perceived as not contributing to quality values
   - Unawareness of best practices
   - Success is at best spotty

2. Employees experience a leadership vacuum. Causes of this include:
   - Visibly demonstrated leadership is lacking
   - Wanting to see leaders "walk the talk"
   - A need for top management to be more directive
   - A lack of clear policy directives, explicit expectations, and meaning-ful measurements consistently applied, and a lack of feedback on results

3. Culture and beliefs about quality impact and acceptance of the quality thrust. Causes include:
   - A reluctance on the part of many employees to change behavior in support of quality process

- A lack of the strong belief in quality as performance at individual and group levels
- A resistance to "Quality Programs"
- An attitude of "we are the world's best" in quality
- Compromising quality to meet schedule

4. Lack of balance between results and process focus. Causes include:
   - Resistance that is intensified by managers driving for results
   - The feeling of compulsion to pass work that did not meet "quality standards"
   - The misunderstanding that quality approach is to work harder and to patch fixes
   - The view that being on a team is additional work, not as "real work"

5. Poor communication. Causes include:
   - Barriers to resolving conflicts among quality, cost, and schedule
   - Process improvements that conflict with schedules
   - Little input on schedules and thus lack of commitment to those schedules

6. Lack of customer contact and focus
   - Company driven by internally focused measurements
   - Managers who behave like the customer
   - Difficulty in obtaining accurate, unfiltered, direct customer requirements
   - Need to assure that right things being done per the needs and priorities of the customer.

Appendix 1.2: Findings of the U.S. GAO (Government Accounting Office), May 1991

Overview

From the results of the diverse companies studied, total quality management appears to be useful for small companies (less than 500 employees) and for large companies (over 500 employees), and for companies that sell a service as well as companies that produce and sell manufactured products. It took an average of 2.5 years (the range was 1 to 5 years) for the 20 companies of the study to realize initial benefits. Allowing sufficient time for results to be achieved was as important as initiating a total quality management system.

Important Features of TQM

- Customer satisfaction is critical to remaining competitive in the marketplace. Ultimately, customer satisfaction, both internal and external, drives quality efforts. Organizations, therefore, need to determine what customers want and they must have processes in place to meet those customer needs.

- Top executives must provide active leadership to establish quality as a fundamental value to be incorporated into the company's management philosophy.

- Quality concepts need to be clearly articulated and thoroughly integrated throughout all activities of the company.

- Top executives need to establish a corporate culture that involves all employees in contributing to quality improvements.

- Companies need to focus on employee involvement, teamwork, and training at all levels. This focus should strengthen employee commitment to continuous quality improvement.

- To succeed, total quality management systems must be based on a continuous and systematic approach to gathering, evaluating, and acting on facts and data.

- Suppliers should be made full partners in the quality management process. A close working relationship between suppliers and producers can be mutually beneficial.

The results of the companies we studied indicate that TQM systems are promising ways to strengthen a company's competitiveness in both domestic and world markets.

2 *Design for Manufacture*

Summary of Chapter

In the section on manufacturing productivity (Chapter 1) design was established as one of the significant factors affecting manufacturing competitiveness. This chapter describes some of the best practices used to ensure that new product and process design is completed quickly and effectively. To this end it describes concurrent engineering in detail. The process for developing effective designs is reviewed with a focus on manufacturability.

Contents

- Overview of Design for Manufacture
- New Product and Processes
- Concurrent Engineering
- Best Design Practices

Relationship of Design for Manufacture to Other Chapters in the Book

Manufacturing effectiveness is dependent on effective design. It is estimated that over 80 percent of the product cost is determined by the design. Even though this chapter on design is brief, it is the foundation of the core of this book, namely manufacturing planning and execution. Effective design is also one of the strongest influencers of customer satisfaction.

Overview of Design for Manufacture

Purpose and Description

Design must not only provide the concept of a product that functions effectively—it must also create a competitive advantage by allowing the product to be manufactured quickly, at a reduced cost, and with improved quality.

As competition becomes more global, the time to make and market is being reduced, product options are being increased, and all the while the price of the product is being decreased. Design for manufacturing is now a sophisticated, computer-based process, where an effective design is carefully planned for ease of manufacturability. Furthermore, well-designed products are easy to use, reliable, and please the customer.

Discussion

Design for manufacture (DFM) is today usually referred to as *Design for X*, the DFX representing manufacturing, testing, installing, servicing, and other requirements in the product's life cycle. For purposes of this book, DFM will be described, but it should be understood that the principles apply to all forms of "design for."

Manufacturing competitiveness is being increasingly determined by a company's ability to rapidly specify, design, produce, and deliver a quality product at a low price. It is estimated that 70 to 80 percent of a product's cost is determined by its design, and once a product is designed only marginal cost improvements can be made. Poor design thus has a multiplier effect—it makes products more difficult to make, increases their manufacturing cost, and limits efforts to improve them. Today significant effort is being spent on ensuring that products are designed to enhance their manufacturability. Design is now being incorporated into a concurrent process, involving all the principal stakeholders in marketing, costing, and manufacturing a product (see the section on concurrent engineering later in this chapter).

Design also determines if a product can be used easily and provides reliable and effective service. Good design pleases customers. Psychologist Donald A. Norman, in *Psychology of Everyday Things*, states that well-designed objects should have what he calls *visibility*: that is, they should contain visual clues to their operation. A key to visibility is the *natural mapping* that relies on the mind's ability to grasp and remember naturally appropriate actions. An example of natural mapping is that turning a steering wheel to the right steers a car to the right. In another example,

Norman arranged his light switches so that their position corresponded to the positions of the lights being controlled—the leftmost light is controlled by the leftmost switch.

Good designs delight and sell.

New Product and Processes

Purpose and Description

Companies that do not stay in the forefront of product design are unlikely to survive. New products are the lifeline of a business.

Customers are constantly demanding more sophisticated, more efficient, and/or more elegant products. Product designs that are successfully introduced first become the market leader and capture the lion's share of the profits. Unfortunately, not all new-product introductions are successful, either from an innovation point of view or from a manufacturability point of view.

Discussion

In order to succeed, product design must be focused on the customer and must be manufacturable. Speed to market is considered one of the most critical aspects of new-product development. Perhaps the ability to introduce new products and processes effectively will determine the next competitive frontier of manufacturing.

One of the fundamental issues confronting decision makers is whether to develop a new product or to improve the existing product—radical change or incremental change? Obviously incremental change will increase the speed to market dramatically, but it will leave a company susceptible to being overtaken by a competitive design. On the other hand, radical change is more likely to fail. The decision of which approach to follow depends on the company's exposure to risk and the status of the competition's products. There are several types of risks: market—will the product sell; technological—will the technology work; competitive—has the competition a superior product; and financial—how much money will be required to be invested and what is the probability of return. Companies develop partners to help share the risk.

The track record on the introduction of new product over a broad range of industries is poor, with 30 to 40 percent of new products failing to be profitable (Abrikian 1981). Further studies have shown that new-product introduction usually takes two-and-a-half times longer than estimated (Leenders and Henderson 1980).

Despite the risk and the record, most companies have to continue to develop and introduce new products. It is a competitive imperative, necessary for survival, particularly in view of increasing global competitiveness. The new-product introduction process tests a company's ability to handle change. New products also test a company's functional integration, flexibility, and responsiveness.

Process for Introducing New Products

1. Identify the need for the new product from company research, customer surveys, or competitive offerings. Allocate adequate resources to address the technical feasibility and develop a timetable of introduction.

2. Secure early management commitment for the project by preparing a business justification and obtaining approval of funding. Ensure that human resources are allocated to the project. Senior management must be convinced of the need for the product and make sure that it fits the company's strategy. The project approval should constitute an empowerment of the project personnel to execute the development of the design.

3. Develop cross-functional teams. Ensure that all needed functional expertise is represented—design, marketing, manufacturing engineering, manufacturing operations, materials, quality, costing, and project management. Suppliers, whose equipment or material is used extensively, should also participate. If there are dominant customers, their input should be obtained. This is a concurrent engineering team.

4. Ensure that all personnel involved in design understand that poorly designed product damages the company's competitiveness—manufacturing costs are higher, quality is suspect, and the product is unreliable.

5. Set targets for product performance and quality, cost, and time. Use benchmarked competitive standards. Define team goals clearly in terms of *what* and *when* separately for the development and manufacturing stages. Ensure that the concurrent engineering team is responsible for the product until it is proven on the manufacturing floor.

6. Use commercially available parts and material. Also use standardized and commonly manufactured parts. This requires having a database of standardized and commercially available parts and insisting that engineers specify them first if they fulfill the design requirements.

7. Apply the principles of Computer-Aided Design (CAD) and Computer-Aided Manufacture (CAM). There are excellent software packages available that will guide the designer to use the most efficient practices.

8. Establish design rules for the manufacturing processes that have been chosen in the form of a check list of items for the designers to choose. Also ensure that testing requirements are specified. Designers must relate product design with product testing and specify products whose functions can be easily tested. Rules for printed circuit design are the most common example.

9. Consult with suppliers and design parts to suit their expertise and capability.

10. Formulate plans to develop the product and the process in conjunction with each other. Wherever possible they should be worked on in parallel. This will reduce the development time cycle. Set up the process to turn out designs quickly. There is evidence that the first to market captures most of the profit of the product.

11. Set up formal design reviews at appropriate stages of the design process: initial requirements, initial design concept, design transfer, final design for prototype, testing completion, and final design for manufacture. Ensure that all involved functions are included in the review.

12. Develop an approach to transfer technology from R&D to Manufacturing. This is one of the critical, if not *the most critical* step in new-product introduction. It is a complex task, fraught with political overtones and "turf" issues.
    - Determine when a product is ready to be transferred from design to manufacturing, and formally document the design transfer process. This is a balancing act between having all the bugs ironed out and reducing the time to market. It should be the goal of the cross-functional team to agree to and document the handover process.
    - Ensure that Design and R&D conduct sufficient tests on the prototype. (The building and testing of a prototype is the most common and recommended means of certifying a design.) Verify that the prototype gives a realistic performance.
    - Provide overlapping responsibility until the process is stable and can be performed independently by an operator. Ensure that the process capability comfortably exceeds the product specification requirements.

- Train individuals and teams responsible for the product and process. Build in a means of qualifying the operators and reviewing their skills at periodic intervals.

13. Determine how knowledge will be transferred. This may be done with classes and on-the-job training, and reinforced by well-prepared documentation.

14. Provide strong project management to understand the critical path and the time-phased requirement of resources. This is a must for the development of large and complex products. Project management should also coordinate the activities of the teams and monitor and control performance.

15. Attempt to minimize changes to the design or specification after release to manufacturing. This may require the understanding and active support of marketing and sales. Changes are disruptive and costly to a manufacturing process (see Table 2.3).

16. Track the manufacturability of the product, recording the lead time and the number of products that have to be reworked or are scrapped. Determine the root cause of the defect and resolve it.

17. Ensure that all customer returns of defective product are carefully investigated. Again, the root cause of the defects must be determined and fixed.

Dos and Don'ts

DOs

1. Do adopt a concurrent team approach to product design, manufacture, and testing. Authorize (empower) the team to make decisions.
2. Do ensure that design is driven by customer needs.
3. Do set goals for and measure performance, time, and cost. Set intermediate milestones for review of progress.
4. Do take corrective action to close gaps.
5. Do structure and formalize the transfer of the design to manufacturing, with a clear understanding of when the design is ready for transfer.
6. Do encourage the establishment of design through models prior to introduction to manufacturing.

DON'Ts

7. Don't neglect to use the expertise of key suppliers. Establish an environment of partnership with them in the design phase.
8. Don't allow changes to be made when the product is in manufacturing unless there is strong reason for the change. Measure the changes made and review this metric frequently.
9. Don't delay forming the cross-functional team. Establish it at the start of the design.

Concurrent Engineering

Purpose and Definition

The purpose of concurrent engineering is to design a manufacturable product in the least possible time at the lowest cost.

The APICS dictionary (1992) defines *concurrent engineering* as "A concept that refers to the participation of all the functional areas of the firm in product design and activity. Suppliers and customers are often included. The intent is to enhance the design with the inputs of all the key stakeholders. Such a process should ensure a product that can be quickly brought to the marketplace while maximizing quality and minimizing costs."

Discussion

Concurrent engineering (CE) is a time-based strategy in which functions, traditionally performed in series, are performed in parallel or concurrently. The principal functions that concurrent engineering integrates include conceptual design, detail design, prototyping, process design, planning of make-buy piece-part decisions, manufacturing, software,

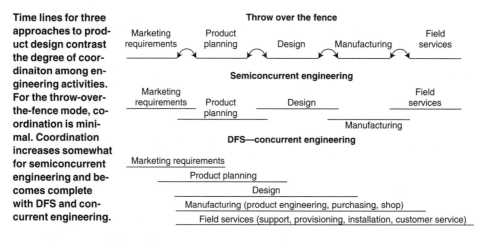

Time lines for three approaches to product design contrast the degree of coordinaiton among engineering activities. For the throw-over-the-fence mode, coordination is minimal. Coordination increases somewhat for semiconcurrent engineering and becomes complete with DFS and concurrent engineering.

Figure 2.1. Effect of Concurrent Engineering on Various Business Functions
Note: DFS Design for Simplicity
Source: Copyright © 1990 AT&T. All rights reserved. Reprinted with permission.

Table 2.1 Qualitative Comparison of Design Methods

Method	Develop time/effort	Product change activity	Product reliability	Overall product cost	Product quality	Time to market	Customer delight
Throw over the fence	Short	High	Low	High	Low	Unpredictable	Low
Semiconcurrent engineering	Medium	Medium	Medium	Medium	Medium	Predictable	Medium
DFS— concurrent engineering	Medium	Low	High	Low	High	Predictable	High

Source: Copyright © 1990 AT&T. All rights reserved. Reprinted with permission.

hardware, field support, purchasing, suppliers . . . in short, all players involved in a product. Representatives from all these functions participate in the design of the product from the inception or conceptual stage. Concurrent engineering is most often associated with time to market, with the overall objective of reducing the entire life cycle of a product from design through manufacturing to sales and distribution. Along with the cycle-time reduction, the cost of development and manufacturing are also significantly reduced.

The concurrent engineering process involves working in teams across departments, and accepting joint responsibility and accountability. Like all processes involving change, it is difficult, takes a long time to implement, and has to be persistently sustained until it is institutionalized.

Concurrent engineering may be contrasted with the "hand-grenade-over-the-fence" approach (design throws the design grenade over the fence into manufacturing and lets it explode, unmindful of the destruction it causes). A simple schematic (Figure 2.1) will illustrate the difference in the approaches.

With Design Decisions So Critical . . .

Design is a tiny piece of the development pie, but it locks in the bulk of later spending. It may be observed that up to 80 percent of the cost is committed at the design engineering phase, even though as little as only 8 percent of the cost may have been incurred (Table 2.2).

Table 2.2 Costs Committed at Various Stages of Manufacture

Development Phase	*Percent of Total Costs Cumulative*	
	Incurred	*Committed*
Conception	3 to 5	40 to 60
Design Engineering	5 to 8	60 to 80
Testing	8 to 10	80 to 90
Process Planning	10 to 15	90 to 95
Production	15 to 100	95 to 100

. . . And Traditional Engineering So Expensive . . .

The typical cost for each change made during the development of a major electronics product (DataQuest Inc.). Observe the exponential increase in the cost to change in the production phases on the design (Table 2.3).

Table 2.3 Ratio of Costs Incurred Due to Change at Various Stages of Manufacture

When design changes are made	Cost
During Design	$1,000
During Design Testing	$10,000
During Process Planning	$100,000
During Test Production	$1,000,000
During Final Production	$10,000,000

. . . Concurrent Engineering Pays Big Dividends

Benefits from designing manufacturability, quality, and ease of maintenance are seen in the product at the start. (National Institute of Standards & Technology as seen in *Business Week*, April 30, 1990.) It may be seen that the time and productivity gains are significant (Table 2.4).

Keys to a Successful Concurrent Engineering Process

1. Support of top management.
2. Use of a common design database and standardized parts and components.
3. Provision of adequate education and training.

Table 2.4 Benefits from Concurrent Engineering Improvements

Activity in Concurrent Engineering	Percent Improvement
Development Time	30 to 70 less
Engineering Changes	65 to 90 less
Time to Market	20 to 90 less
Overall Quality	200 to 600 higher
White-Collar Productivity	20 to 110 higher
Dollar Sales	5 to 50 higher
Return on Assets	20 to 120 higher

Source: Tables 2.2, 2.3 and 2.4 are reprinted with permission from *Business Week,* April 30, 1990

4. Designation of an effective and willing team leader and members.
5. Co-location of team members.
6. Regularly scheduled review meetings.
7. Empowerment of concurrent engineering team.
8. Acceptance by team of responsibility and accountability for success.
9. Recognition of team success.

Best Design Practices: Design for Manufacture and Assembly (DFMA)

Purpose and Description

Best design practices enable products to be manufactured with the highest quality, in the shortest time, and at a competitive cost.

Design for manufacturability (DFM) is a process for matching design requirements, determined through identifying customer needs, with manufacturing capabilities. The process includes design for manufacture and assembly (DFMA), design for simplicity (DFS) and design for X (testability, installability, reliability, and other process demands). The essence of all DFM is to reduce the number of parts and simplify the means by which they are assembled. Results have shown that DFM is most successful when developed with cross-functional teams that are established early in the process and that represent all important functions of the design and manufacturing processes. The process of working in parallel on the elements of design and manufacturing is also referred to as *concurrent engineering.*

This section relates to only mechanical and electro-mechanical assemblies.

Discussion

In the previous section the importance of design was stressed, as the design determined 70 to 80 percent of the total life-cycle cost of a product. The cost of the design phase itself is probably only 10 percent of the total life-cycle expenditure. The major life-cycle costs are incurred in the manufacturing phase through changes and rework, and in the service phase in the form of warranty repairs and returns. The costs can be conceived in the form of a cost curve (Figure 2.2).

The effect of starting a design late on increased costs in production and service is illustrated in Figure 2.3. When design is started with concurrent engineering early in the product life cycle, the design is completed at the start of production, allowing manufacturing to make product with few changes and resulting in highly reliable product with few warranty repairs or returns. When design is started late, it continues into production, and manufacturing has to deal with the confusion of numerous changes. The probability of extensive warranty repair and returns is increased. The curves in Figure 2.3 are often used to illustrate the difference between Japanese and U.S. approaches to manufacturing in the 1970s.

Using best manufacturing and assembly practices, it has been dem-

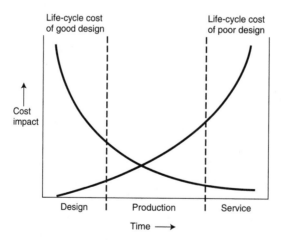

Figure 2.2.   Design Costs and Life-Cycle Costs

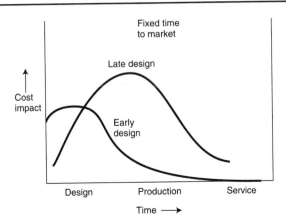

Figure 2.3.   Effect of Design Timing on Costs

onstrated that the manufacturing cost can be reduced by 40 to 50 percent while doubling or trebling the productivity.

In a recent study, the National Research Council identified five areas of research priorities for U.S. manufacturing (NRC Report 1991). One of these areas was the product realization process, which combines the activities of design, concurrent engineering, and customer need satisfaction. The NRC study observes that "A weak approach to the development of robust, reliable, manufacturable products is a major bottleneck in the production realization process," and "Research is greatly needed in both robust design and product realization process."

Process for Developing Best Design Practices

1. Use research and design, marketing, and product management expertise to define and specify the conceptual design of a product; make an estimate of its market life and its likely sales volume.

2. Form a design team with representatives from design and manufacturing. The team should have clearly defined goals of time and cost. The initial product requirements should be tentatively specified at this point.

3. Determine the *optimum design zone,* which is a balance between the cost of the project and the number of parts needed (see Figures 2.4 and 2.5). Figure 2.4 represents low-end electrical and mechanical systems containing 25 to 50 parts. The initial cost of parts will decrease as the

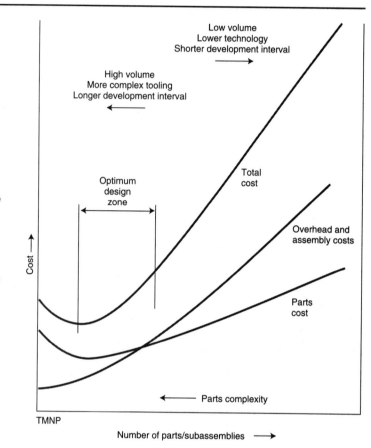

Cost/complexity chart of low-end electronic and mechanical systems (25 to 50 parts) in high-volume production. In the optimum design zone, the larger tooling cost is offset by the higher volume, the increase in development interval is small because of the simple system design, and no major technology breakthrough is required to approach the theoretical minimum number of parts.

Figure 2.4. Cost Complexity Chart for a Small Assembly
Source: Copyright © 1990 AT&T. All rights reserved. Reprinted with permission.

parts are increased (less complicated parts), and then increase for additional numbers of parts. The assembly costs, including inventory costs, will increase as the number of parts increase. Figure 2.5 represents a similar cost/complexity relationship for a larger system, and, as can be seen, the optimum design zone increases.

4. Simplify the design of the parts used. Adopt a design-for-assembly process (Boothroyd and Dewhurst 1987). This approach computes a design efficiency that measures its manufacturability. The four primary criteria are:

1. The orientation and handling of the component—complexity of movement
2. The difficulty of the actual assembly process
3. The type of process—manual, semi-automatic, or robotic
4. The handling and palletizing of parts—how easily parts can be picked up and transported; how easily parts can be supplied.

There are tables that rate the difficulty of the design. The technique also relates efficiency of design with the number of parts. (See Appendix 2.1 for design for simplicity rules.) A general approach to design for simplicity (DFS) is:

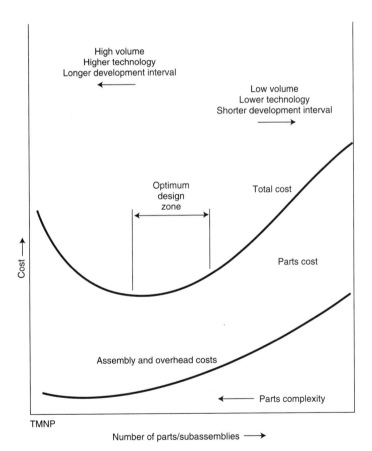

Cost/complexity chart for larger electronic and mechanical systems (about 500 parts) in medium-volume production (5000 to 10,000 units per year). The optimum design zone is characterized by moderate tooling cost, strategic technology, and a moderate development interval.

Figure 2.5. Cost Complexity Chart for a Large Assembly
Source: Copyright © 1990 AT&T. All rights reserved. Reprinted with permission.

Step 1. Analyze assembly requirements.
- Develop an assembly sequence: Obtain a part and subassembly list and a structure tree (see bill of material, Chapter 4), and develop an assembly-and-process sequence.

Step 2. Challenge the existence of every part.
- Examine the function of each part
- Minimize the number of parts used. The lowest cost part in a design is the one that is eliminated!

Boothroyd and Dewhurst used three criteria for evaluating the necessity for a part. Find candidates for elimination by asking:

1. Must this part be separate because it moves with respect to all other parts already assembled? (Can a flexible material be used instead?) Motion.

2. Must this part be made of a different material or be isolated from all other parts already assembled? (Is there a property of the material required?) Material.

3. Must this part be separate from all other parts already assembled because necessary assembly or disassembly would otherwise be impossible? (Fasteners are seldom considered essential, as they can be part of the assembly.) Service.

   If the answer to all of these questions is No, a separate part is not required (See Table 2.5).

Table 2.5 Parts Reduction Worksheet   **Product No.**

Part	Quantity	Motion	Material	Service	CFE	Quantity
Top plastic	2	N	N	N	Yes	1
Message light	1	N	Y	N	No	0
Feet	4	N	N	N	Yes	4
Dial buttons	12	Y	N	N	No	0
Dome assembly	1	Y	Y	N	No	0
	Total Parts = 20				Candidates for elimination CFE	Sum of CFE = 5

Theoretical Minimum No. of Parts (TMNP) = Total Parts − Sum of CFE = 20 − 5 = 15

$$\text{Parts Count Design Efficiency} = \frac{\text{Theoretical Minimum}}{\text{Total Parts}} = \frac{15}{20} = 75\%$$

Step 3. Improve design:
- Simplify assembly requirements
- Eliminate and/or combine parts
- Study trade-offs
- Modify the design for higher design efficiency

The expected parts reduction with the application of design for simplicity technique may be seen in the graph in Figure 2.6. The figure shows parts count before and after application of DFS principles. The typical improvement with DFS is a reduction of from 25 to 45 percent in number of parts. The theoretical minimum number of parts was 65 percent of the pre-DFS parts count.

Studies have also shown that for larger systems (having 500 or more parts), assembly time is usually 20 to 60 percent less than the pre-DFS assembly time, with an average of 20 percent.

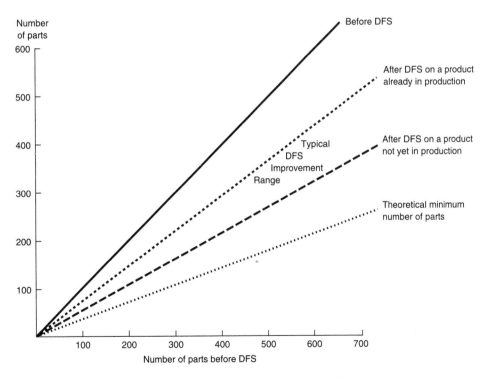

Figure 2.6. Typical Improvement with Design for Simplicity
Source: Copyright © 1990 AT&T. All rights reserved. Reprinted with permission.

For simple products (having about 50 parts), the reduction in assembly time is even more dramatic, averaging about 80 percent (see Figure 2.7).

5. Reduce the complexity of the parts by:
   - Reusing existing parts, processes, and technology
   - Standardizing parts and processes
   - Working with suppliers and using their expertise

6. Iterate the design and continue to simplify. Measure the progress in terms of assembly time or parts count design efficiency (above).

   Example: The IBM Selectric typewriter reduced the number of parts from 2,300 to 190 in four years.

   An example of a product development process may be seen in Figure 2.8.

7. Invite principal suppliers to help design the parts that will be made by them and form supplier partnerships. If possible, plan for the supplier to perform functional tests at his or her facility.

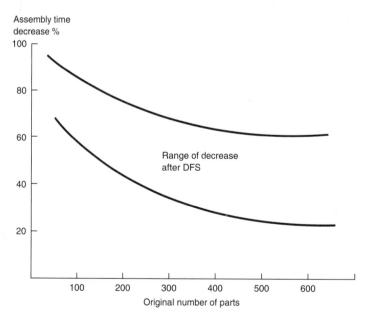

Figure 2.7. Relationship Between Assembly Time and Number of Parts
Source: Copyright © 1990 AT&T. All rights reserved. Reprinted with permission.

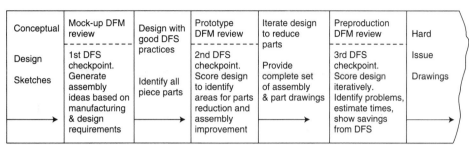

| Conceptual

Design

Sketches | Mock-up DFM review
- - - - - - - - - -
1st DFS checkpoint. Generate assembly ideas based on manufacturing & design requirements | Design with good DFS practices

Identify all piece parts | Prototype DFM review
- - - - - - - - - -
2nd DFS checkpoint. Score design to identify areas for parts reduction and assembly improvement | Iterate design to reduce parts

Provide complete set of assembly & part drawings | Preproduction DFM review
- - - - - - - - - -
3rd DFS checkpoint. Score design iteratively. Identify problems, estimate times, show savings from DFS | Hard

Issue

Drawings |

Hardware Development   ⟶

Figure 2.8.   Product Development Process
Source: Copyright © 1990 AT&T. All rights reserved. Reprinted with permission.

Dos and Don'ts

DOs

1. Do work the stages of design, development, and manufacturability concurrently.

2. Do minimize hand-offs from one department or function to another.

3. Do plan to spend more time in the design phase and less in the manufacturing phase.

4. Do reward individuals on results (does it meet a need and sell).

5. Do place a major emphasis on common parts (a database of common parts is a must).

6. Do set goals for performance, quality, cost, and time. Measure and take corrective measurement when necessary.

DON'Ts

7. Don't allow changes easily once the product is in manufacturing. The confusion, cost, and probability of defective work increases dramatically.

8. Don't build dead-end designs to satisfy deadlines.

9. Don't attempt to maximize profits by avoiding "continuous improvement."

Appendix 2.1: Rules for Mechanical Design for Simplicity

1. Reduce the number of parts.
2. Don't fight gravity.
3. Eliminate the need for fasteners.
4. Increase modularity and subassemblies.
5. Use standard parts.
6. Enhance self-nesting and self-alignment capability of parts.
7. Avoid tangling parts.
8. Increase symmetry.
9. Avoid wires and flexible parts.

Manufacturing Planning

Overview

Unit II begins with a section on *manufacturing information systems*. Today, to manufacture effectively, a company must have a computer-based information system. Many companies have had bad experiences with information systems—partly because the systems they used were inappropriate to their products and processes, and partly because they have not implemented the systems effectively. Manufacturing Resource Planning (MRP II) is a computer-based planning and scheduling system, designed to improve management's control of manufacturing and its support functions. MRP II seeks to integrate all functions of a company—financial, marketing and sales, and engineering—with manufacturing. Using a common *manufacturing database* it provides a single set of configurations and numbers that the entire company can work with and aim toward. At the core of MRP II is the group of manufacturing planning functions.

Manufacturing planning determines what, how many, and when to manufacture. It starts with a *forecast* of product requirements at an aggregate or product-family level, and converts these estimates into a high-level plan called a *production plan* or a *sales and operations plan.* All areas of a company collaborate to develop this plan, and the company's commitment to the plan is seen in the execution of its strategy. For example, if a company's strategy is to provide superior service, its production plan might provide more inventory to cover sales.

The *master schedule* plans the products that need to be made by a factory to meet the demand generated by customer orders and forecasts. The master production schedule is the factory's master plan. Before implementation, the master production schedule is checked against critical resources—a process called *rough-cut capacity planning*—to determine whether the factory has the capacity to meet the schedule. The original master schedule is then modified as required. Once this condition has been satisfied, the master production schedule is input to a computerized technique called *materials requirement planning,* which determines what, when, and how many components and assemblies need to be made or bought. The manufacturing plan is checked to confirm that *constraints* have been identified and sufficient *kanbans* are allowed (both these activities are part of manufacturing execution). This is a major departure from the conventional approach, known as *capacity requirements planning,* where a system check is made of available capacity to make MRP's planned components and assemblies. The whole process of manufacturing planning is interactive and iterative.

The objective of the entire process is to ensure that a company is able to translate its business strategy into an effective manufacturing plan that satisfies customer orders, while using the company's resources optimally.

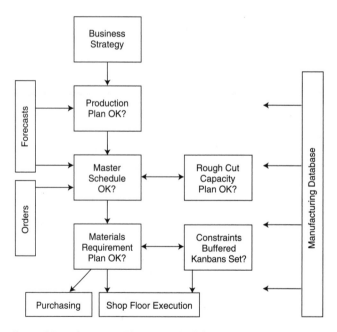

Schematic 2.   Manufacturing Planning Model

The modules above constitute the essential elements of the manufacturing planning model (see Schematic 2), which in turn is part of a larger *Manufacturing Resource Planning II* (MRP II) model. A software package integrates the modules of the MRP II system. Although most packages have a common overall approach and follow similar calculation techniques, modules are designed and used differently at the working level. A system requires different characteristics to service different types of manufacturing environments. The four main types of manufacturing environments are make to stock, assemble to order, make to demand, and make to order (these environments are covered in the master scheduling overview section).

It is essential that the practitioner understand the product-manufacturing environment and purchase a system that meets the needs of the product-manufacturing process. Since companies typically spend over $1.0 million and take two years to implement an MRP II system, an appropriate system must be selected.

The manufacturing planning process depends heavily on accurate and timely data. Even though this requirement appears to be the easiest to meet, it is most often the most difficult to attain and maintain. The principal data requirements of an effective planning system are:

- Inventory data: on-hand balances of storeroom material, work in process, and finished goods
- On-order information: material on order, both manufactured and purchased
- Item masters, bills of material, and shop routings
- Lead times needed for manufacturing and purchasing items

The practitioner will do well to remember that data accuracy remains a cornerstone requirement, even though there are many other functions to understand and manage in the modules of MRP II.

In the 1980s and 1990s, the MRP II system has been complemented with execution practices that seek to reduce lead times, minimize inventory, initiate manufacturing only after receiving a customer demand, and "pull" product through the shop and from assembly through to the supplier. These approaches are part of a just-in-time (JIT) manufacturing process. Today the MRP system is used to plan and order material, while the JIT practices are being used to operate the shop floor. The two systems are complementary and should be used to build on each other's strengths.

Finally, implementing a manufacturing system and maintaining it effectively requires trained and motivated personnel. The system invariably represents a new way of planning and tracking product. This means that practitioners have to learn a new way of performing their work, and this requires change. The difficulty of effecting this change should not be underestimated. It requires persistence, patience, and immediate hands-on help at all times, particularly on the shop floor, for the first year. It requires constant and meticulous attention to detail. Training and retraining are an essential element of the process. It is estimated that the people element is as important as the technical part of the planning process, if not more important. JIT emphasizes this aspect of manufacturing with worker empowerment and the use of teams (covered in Chapter 1 of this book).

3 Manufacturing Information Systems

Summary of Chapter

The first section of this chapter provides an overview of an effective manufacturing system and details a generic process for deciding how to select an effective information system, based on a company's product and process characteristics. The second section describes a detailed implementation process. This process is particularly important as the implementation of manufacturing information systems has not had a good record.

Contents

- Overview and Selection of a Manufacturing Information System
- Implementation of a Manufacturing Information System

Relationship of a Manufacturing Information System to Manufacturing Planning

Today, to be competitive, manufacturing must have an effective information system in order to plan and execute its functions. Such a system is loosely called *MRP II* (*Manufacturing Resource Planning*), but a better description is *Manufacturing Information System* (*MIS*), as this term covers planning, execution, and customer service. The MRP part of the information should cover the planning function of manufacturing.

Overview and Selection of a Manufacturing Information System

Purpose and Description

A manufacturing information system is needed to record, track, and control the manufacturing operation, so as to improve the profitability and competitiveness of the company.

A manufacturing system is built on an information base (*database*) describing all parts that are used (the *item masters*), how they are put together to form assemblies (*bills of material*), how they are made (*routings*), and factory capacity (*work centers*).

Discussion

Manufacturing information systems are usually referred to generically as *MRP II* (*Manufacturing Resource Planning II*) systems. There are several hundreds of MRP II software packages available, and their cost can run from a few thousand dollars (for PC-based systems), to a few million dollars, depending on size and sophistication. The record of companies that have been able to improve their productivity and profitability through the use of MRP II is mixed, with 50 percent stating that the MRP II equaled expectations. The balance is about equally split: 25 percent exceeding and 25 percent being less than their expectations (Oliver Wright Companies—Survey 1990).

In view of the cost and performance, the selection of an appropriate system is extremely important. The term *appropriate* covers usability (it can be easily learned and used) and functionality (it has the required logic for listing, tracking, and controlling the type of products being manufactured by the factory). All too often an MRP II software program is purchased and then extensive reprogramming is done to make it suit the current mode or a preconceived way of tracking product. Modifications to software can cost several times more than the original program! More importantly, modifications can alter the performance of the software, because conflicts in its logic may be created that cause other standard features to malfunction. There is also a tendency to incorporate too much functionality or sophistication into a manufacturing system. Functionality is important, but ease of use is even more important. Apart from the technical nature of MRP II, it represents a new way of doing business and is subjected to a resistance to change on the part of its users.

This section focuses on the selection of an appropriate MRP II system; the next section addresses its implementation.

Selecting an Appropriate Manufacturing System

Generic Considerations: User Friendly The system should have the following demonstrated characteristics:

1. Simple sign on and easy movement between screens and modules (point and click is preferable). This capability may be a function of the host software and its configuration.

2. Add, change, and delete transactions should be performed (if authorized) with the ease and flexibility of a manual system.

3. The system should have the capability of being able to design screens and reports to correspond to existing forms of work. A fourth-generation language allows these functions to be performed easily.

4. Bar-code capability—wanding information into the system to avoid using the type keys.

5. Availability of special-purpose key definition. The key (usually these are function keys) should be linked to frequently used information screens. An example is the F1 key for the Stock Status Inquiry screen.

6. There should be a "windowing" capability so that the user can work on several screens at a time and switch easily between them.

7. There should be on-line help. Help should also be intuitive: if the user is doing something wrong, the system should prompt the user on how to correct it. Help should also act as a reference, where searches based on a description or part of a description are used to find a part number.

8. There must be flexibility to change superficial aspects of the system—reports, screens, inquiries, and field arrangements—without changing the base code.

9. Ensure that there is a simple report writer so that everybody in the organization can use it to write customized reports. It should be as easy to use as popular PC-based commercial software.

10. There must be strong system support from the software developer. A hotline and capable technical project managers are some of the requirements. The attitude of the software company to service is critical: will they consistently walk an additional mile?

11. The software manufacturer's product must have a quality reputation and should be free from bugs. Bugs hurt the credibility of the system and its implementation.

12. All the modules of the system must be integrated.

13. There should be easily understandable user documentation and training manuals.

Functional Considerations: Appropriate to the Manufacturing Environment

1. Understand the product's manufacturing strategy. Generally this means make to stock, assemble to order, or make to order (these strategies are covered in the Manufacturing Planning section). Each of these strategies requires different master scheduling and shop-floor capabilities.

2. Determine the product's technical requirements. Will these requirements be subject to frequent change? If they are subject to frequent change, the system, particularly the change control and routing modules, must have the flexibility to handle the changes.

3. Understand the predictability of the manufacturing process—also known as *variability*. Determine how much rework is likely to occur and whether this is likely to be a part of the process. The flexibility of the system is an important consideration in dealing with rework and variability.

4. Determine the level of traceability required. Will all components or only some components or subassemblies require serial numbers to be traceable? Is lot traceability required?

5. Decide on the data requirements of the process. Will data, such as testing data, have to be collected at every operation step? Will verification that all parts for an assembly have been selected be required?

6. Decide on how the shop floor is to be controlled. Will the MRP II shop floor module be used or will a form of just-in-time be used? What will be the system requirements of using a form of JIT?

7. Appraise the working culture of the organization's personnel. Are factory operators comfortable with computers? Is the environment progressive? Is management comfortable with computer-screen information as distinct from written reports? A working culture that is

not comfortable with computers will reinforce the need for the ease-of-use capability of the software package.

8. Assess the level of validation required. Can inputs be made to the system without verifying the accuracy and integrity of the information? Is an operator required to do his or her own entry for each process, and does the operator's qualification for performing this step have to be verified?

9. Determine what in-house or other systems have to be sustained, and ensure that the system can be integrated with these existing systems. Costs of system integration can be substantial (if indeed integration is possible at all!).

10. Ensure that the system is capable of providing customer order delivery dates based on availability of material and capacity. The system should also be able to track the progress of a customer order during manufacture.

11. Determine the need for EDI (electronic data interchange) for communication with suppliers and customers.

12. There should be an understanding of the total cost required to implement the system. Implementation should be distinguished from installation, and should mean that the user can successfully work on the system.

Dos and Don'ts

DOs

1. Do ensure that the system fits the product and process requirements of the company.
2. Do check that the system is functioning in another company that has a *similar* product and process environment.
3. Do check out the software company's reputation for customer support.
4. Do expect to take about a year to implement the system.

DON'Ts

5. Don't buy a general system if your product or process has special requirements.
6. Don't plan to modify a system to suit anything but simple and/or superficial working requirements (such as screens, formats, or terminology).
7. Don't buy all the modules of the system. Buy only what is required and what will be used initially.
8. Don't get pushed into a popular system. Remember that the principal value of the system is in meeting the specific needs of the product and process.

Implementation of a Manufacturing Information System

Discussion

Manufacturing information systems (MRP II) have now been around for almost 40 years. A survey conducted by APICS shows that a system can take more than two years to implement. There are many horror stories of companies spending large amounts of money and jeopardizing their business while the information system is being implemented.

Implementation of a manufacturing information system is a major undertaking. It needs knowledge of software, hardware, manufacturing and materials management—and, above all, an ability to effect change in people. A good benchmark for a factory having about 300 operators, and implementing a complete manufacturing system with the main modules, including: order processing, forecasting, master scheduling, materials requirement planning, shop floor control, purchasing and account ing, is 12 months with 3 full-time technical personnel and a designated software-company-supplied project manager. Such a system (hardware and software) should cost around $400,000 (in 1993 dollars) all inclusive—installation, training, software-company consulting services, and common interfaces. If extensive interfaces are required with other systems, or if the software is to be extensively modified, the cost will increase proportionately.

Finally, the approach and attitude to implementing a manufacturing information system should be that the system is required to help the user (shop operator, material planner, etc.) to function more productively. The system must serve the user. It is doubtful if sustained productivity can be obtained unless the users feel that the system is a help and not an imposition.

The process described below is based on a very successful case study. The approach used contained many conventional steps, but also had many unconventional ones.

Process for Implementing a Manufacturing Information System

1. Develop a business case to install a new system or replace an existing inappropriate information system. The proposal must be based on need and must project future profitability and competitiveness. Obtain management approval for funding and management's commitment for implementation. The project should have a champion—a knowledgeable, dedicated, and passionate believer in the need for an information system. This person may be a functional manager or a department head.

Note: It has been the experience of most implementations that upper management's active commitment is needed for a successful implementation. Another approach is for top management to empower and support the functional management and not be directly involved in the implementation, other than for periodic status reviews. In the latter case, top management must accept the recommendations of the implementation team.

2. Select an in-house project manager to a full-time position. The person must have superior manufacturing and software knowledge. The person should be from within the organization and should have the confidence of the factory personnel at all levels.

3. The project manager should lead the effort to determine the product, process, and work culture requirements of the system. All functional groups should be involved: manufacturing, engineering, materials, quality, human resources, and accounting. Where direct marketing and sales are involved, they should also help in the determination of these system requirements. There should be a set of questions developed on each major function of MRP II (see overview)—order entry, forecasting, production planning, master scheduling, and MRP—pertaining to the specific product and process requirements of the company. (There is software available with formulated questions that can be used as a starting point.) Requirements and answers should be prioritized (for instance, they should be rated as essential, nice to have, and not required). This step should provide a learning and an involving experience for all the major stakeholders. The output of this step is a specification document, listing the requirements of the system. (See Appendix 3.1.) This step will typically take longer than the software evaluation process itself.

    Steps 1, 2, and 3 are critical. Neglect any of them and the implementation will be jeopardized.

4. Typically the requirements document developed in Step 3 is sent to about 15 to 20 software houses that claim to satisfy all the system requirements. Clarifications and demonstrations should reduce the number to three to five finalists. (See Appendix 3.2, for example.) Document the capabilities of each of these systems against the specification developed in the earlier step. Develop a rating system that lists all the major requirements and assign points to every system capability that meets the specification. The whole idea is to develop as objec-

tive a rating as possible. Every effort must be made to select software that meets the system requirements in its "vanilla" or original form.

> Note: There has been a tendency to engage outside consultants for the software selection and the system implementation. If there is in-house expertise, it should be used in preference to going outside. If in-house expertise is not available, an administrative in-house project manager should still provide single-point accountability, but expertise should be obtained from outside.
>
> The vanilla version of the system selected must meet the company's specifications. Modifying the code and logic of the purchased software to meet custom requirements is strongly discouraged. This should be one of the ground rules—not to be violated unless there are compelling reasons. Modifying software increases the cost substantially, delays implementation, and usually makes the user ineligible for software upgrades. This is not to be confused with making superficial changes, such as modifying screens or reports. The latter is to be encouraged.

5. Before final selection, ensure that demonstrations are conducted in the factory and all the principal stakeholders (whose input was solicited in Step 3) are exposed to the final software candidates and their reactions noted and satisfied. Like Step 3, this should be a learning and an involving experience. Before making the final decision, visits to organizations that have similar products and processes, and that use the system being investigated, *are a must.* Visits to the software supplier's development center is equally important. Consider the strength of the vendor's organization, its growth, its years in business, its user base, and its reputation in the market (from users). Again, the specification document should be used as the basis of comparisons, and findings against key functional requirements must be carefully written down.

> Note: Every attempt must be made to maintain objectivity. Often, however, there is little to choose between the final two or three software programs. The final decision is then made on the level of comfort the project manager and the champion establish with representatives of the software organization. The chemistry has to be good initially, as this will help to sustain a positive relationship in a long and sometimes problematic implementation. Reasonable checks on the quality of the supplier's software and their support and reputation must be conducted, but ultimately a level of trust has to be es-

tablished that will provide the basis for a long-term working relationship.

6. The objective is to select a vanilla software package that meets the system requirements. If modifications have to be made, these should be identified up front and included in the quotation.

7. Decide on the most appropriate system from a functional and a budgetary point of view.

> Note: This is an extremely difficult step and there is a tendency to drag this step out with long negotiations and the need to obtain approvals at many levels. If Steps 1 through 6 are done thoroughly, the final selection should be quick. If there is clearly one best system the decision is easy; if there is more than one system, an approach is to ask each vendor to provide a best quote and then decide. Do not horse trade—it will create mistrust at the start of the partnership.

8. Develop a realistic implementation schedule. After fulfilling the data requirements, focus on the shop-floor modules rather than the master scheduling and materials requirement planning modules. The latter are self-contained and can be successful only if the shop-floor modules (data accuracy) are working properly. Break the implementation plan into phases with measurable milestones and avoid scheduling phases that are too far out into the future.

> Note: Developing a realistic schedule is extremely difficult. Typically schedules are overambitious. Estimates of both time and cost should be doubled! Generally implementation starts with the planning modules and the shop floor is implemented after these are working. This is *not* recommended.

9. A cross-functional team is the best vehicle to use to oversee the implementation. Members should include shop supervision, engineering, materials, systems, and product management. Ensure that the team, along with their champion, has the authority to decide on all technical and cost issues within the scope of the project. Implementation is often bogged down by a long decision-making procedure and this should be avoided.

10. Decide on what training should be provided, to whom and when. Train accordingly.

> Note: Typically most implementations stress the need for training top management and all personnel in the working of MRP and its

benefits. There is another approach that trains only on a need-to-know basis and as much as possible "on the job." Under the latter approach, the shop operators will be taught what, how, and why system transactions must be performed. Other typical types of training are concept training (for management), system training (for the implementation team), application training (for staff supporting the line), and report writer training (for all persons directly using the system).

11. Provide ongoing and continuous front line shop floor support and training. A full-time system training specialist is strongly recommended. This person will provide "on-the-job" daily help, will document and issue local operating instructions, and will develop special reports and screens required by the shop floor.

    Note: Along with the project manager and champion, a full-time specialist (system administrator) is one of the three key personnel needed for a successful implementation. The person will act as a hands-on consultant. This is not a common feature in most implementations.

12. Ensure that the software supplier's organization is used during implementation and until the project is functioning effectively. One of the criteria of a system selection should be the supplier's demonstrated capability in implementing the software—on the basis of both time and cost.

13. Provide constant hand-holding during the early phases of the implementation to overcome genuine fear and hence opposition or neglect. The best ways of doing this is through the system administrator/trainer (covered in Step 9 above). Be sensitive to user requests and strive to address them and inform the user that they have been addressed.

14. Provide customized screens and reports to suit the specific requests of the user. These superficial changes have a strong influence on obtaining the user's "buy in."

15. Publish periodic progress reports, highlighting successes. Stress the specific role of the information system in improving productivity: for example, reducing lead time.

16. Keep management in the loop. The champion must guide management, and informing them is the best means of relieving anxiety. Ultimately, a system functioning effectively is the surest way to win sup-

port. A nervous and overactive management (commonly the case) can hinder or stop the progress of an implementation.

17. Encourage all professionals, including management, to use the system to obtain information. Provide training on how to use the report writer for obtaining simple custom reports. Make available simple instruction sheets for accessing and using the system. As in the case of the direct users, special screens and reports should be developed to help management obtain specific information. The elimination of paper records, a little at a time, will force the use of the system.

18. Ensure that there is frequent and comprehensive backup capability and that it is performed. Provide for a manual system of information control in case the system crashes and cannot recover quickly.

19. Implement a formal problem-reporting and change-control process through the system. This will provide system integrity and control.

Dos and Don'ts

DOs

1. Do have a full-time project manager to provide single-point authority and accountability.

2. Do arrange for a software presentation and demonstration from the final two or three suppliers to *all* the stakeholders.

3. Do be prepared to deal with opposition, cynicism, and lack of interest. Be prepared to iterate instruction and training constantly.

4. Do develop a realistic schedule with major milestones and track progress against these milestones.

5. Do involve the implementation team. Buy-in at their level will prevent many problems down the road. They should provide positive support all through the implementation.

6. Do recognize and celebrate successes. The process is an uphill task and may appear never to be succeeding, so positive identification of success is essential.

7. Do encourage discontinuance of manual paper backups.

DON'Ts

8. Don't horse trade between vendors in the final package selection.

9. Don't plan to modify the software selected. Stay close to the vanilla version.

10. Don't neglect the software supplier as a source of expertise. Insist on a capable supplier project manager to help the implementation.

11. Don't engage an outside consultant or expert to lead an implementation. They are best suited to be technical advisors.

12. Don't ever lose sight of the fact that one of the main reasons for a system is to help the user.

13. Don't underestimate the huge change required in culture and learning. It never ends.

Appendix 3.1: Characteristics of an Ideal System

Shop Data Collection Characteristics

- Common screen format for all shop collection data
- Use of point-and-click, menu-driven screens
- Use of bar-code technology for increased accuracy and efficiency
- Screens can be customized by area
- Transaction input response time should be no longer than 2 seconds
- Capability to track WIP and piece parts by serial number
- Test set data loaded into database without operator intervention
- Data are entered only once
- System must control sequence of data entry, i.e., Step 2 cannot be entered until Step 1 has been entered
- Rework is handled properly, with capability to de-allot and allot parts
- Variance product is handled properly

Shop Data Retrieval Characteristics

- Capability to view all data on-line for the entire project
- Standard on-line report menu for commonly requested data queries
- One day maximum turnaround for special data queries
- Capability to retrieve entire build history, including all piece part information both on-line and in report form
- Capability to perform on-line queries given one or more search parameters
- Capability to retrieve shop data in graphical form, both on-line and in report form
- Paper storage of data can be minimized because it is easy to retrieve information from database
- Ability to create ad hoc reports
- Availability of on-line engineering analysis tools

Database Characteristics

- Stand-alone system
- Single database or multiple databases logically linked
- Relational database, SQL compatible
- 4 GL language feature for editing and report writing

MRP II Characteristics

- Single storage of item master data
- Data "dictionary" queries
- Capability to track and plan activity at the project level
- Ease of route building
- Routing construction and shop floor use of routing on the same platform
- Ability to interface with corporate systems, such as purchasing
- All inventory (WIP and storeroom) stored on same platform
- Ability to create ad hoc reports and on-line queries
- Use of point-and-click technology
- Ability to track project by several versions, such as current plan, recovery plan
- Material requirements and availability analysis capability
- Capacity requirements and availability analysis capability
- Exception reporting capability

Appendix 3.2: Example of Special Ability Vendor Comparison

Y = Currently can accommodate
N = Cannot accommodate
M = Can accommodate with a modification
W = Forthcoming release will accommodate
? = Not sure, need further clarification

> All of the above systems have the following modules: Rough cut capacity, Pegging, Import/Export capability, Copy capability, Change control, On-line inventory, Single point entry

	V1	*V2*	*V3*	*V4*	*V5*
General					
"Search for match" capability on fields by description	Y	Y	Y	Y	Y
Print bar-code on lot traveler's (routings)	?	Y	Y	Y	Y
Screen builder can easily create screen in a few minutes	Y	Y	?	Y	W
Full-function menus for modifications	Y	?	?	Y	?
Cross-reference tables	?	?	Y	Y	?
Flexible field lengths	Y	Y	?	Y	?
On-line help for fields	Y	?	?	Y	?
Hot key definition	Y	Y	Y	Y	Y
Word processor for text	Y	?	?	W	?
Auto CAD interface	Y	?	?	Y	?
Routings					
BOM's and routings built on same screen	Y	?	?	?	?
Change routings on fly	Y	?	?	?	?
Shop Floor Transactions					
Data entry—expandable fields for extra info, easily edited	Y	?	?	Y	?
Multiple reason codes	Y	Y	M	Y	N
Easy to split lot to any step	?	?	?	Y	N
Time-stamp solution	Y	Y	Y	Y	Y
Pre-assignment of lot numbers	W	?	?	M	?
Single-screen entry of all components when used	N	Y	?	Y	?
Reports					
Multiple levels of report writer	Y	Y	Y	Y	Y
Component availability by routing	Y	Y	Y	Y	Y
Can send reports by e-mail	?	?	Y	M	M
Report templates (like lower query menus)	?	?	?	Y	?
Scheduling					
Schedule totally based on demand	Y	?	?	M	?
Schedule by project	?	N	N	M	?

Source: Appendixes 3.1 and 3.2 were developed with the help of Steve Levine, AT&T.

4 *Manufacturing System Database*

Summary of Chapter

This chapter describes the basic building blocks of a manufacturing information system. The features and functions of the item master, bill of materials, routing, and work centers are described. The chapter also describes how each of these data modules is set up and maintained. Finally, the process of ensuring the integrity of the database through engineering change control is discussed.

Contents

- The manufacturing system database
- Item master
- Product structures (bills of materials)
- Routings
- Work centers
- Engineering change management

Relationship of the Manufacturing System Database to Other Major Planning Processes

The manufacturing system database is the foundation of all manufacturing planning. Every item processed on a manufacturing system must have a unique *item master.* Master planning and material requirements planning are directly dependent on the system database for *bill of materials, routing,* and *work centers.* All data must be valid and accurate, and must be available when needed. *Engineering change* must be controlled so that data has integrity and traceability.

The product data module that defines the system database elements is the first module that is installed when implementing a manufacturing system. All the activities of an MRP II are programmed on these database elements.

The Manufacturing System Database

Purpose and Description

A manufacturing system database is an organized assembly of computerized information that is used to define products and processes. It is the foundation of an integrated manufacturing planning and control system. An effective manufacturing database defines and describes how items are structured, ordered, manufactured, stored, and sold. A manufacturing database consists of the following primary information files:

- **Item Master File.** This primary file stores data records describing the item, the policies concerning its ordering and inventory, planning data, engineering data, accounting data, and sales data—all information that a manufacturing company needs to control an item.
- **Product Structure File.** This file contains records defining how a product is constructed. It lists all the raw materials and components, and the quantities of each, that go to make up a subassembly. It establishes a relationship between a subassembly (parent) and its parts and/or components (child) for every subassembly and assembly, until the whole product is covered. The structure of this relationship of parts is called a *bill of material* (BOM).
- **Routing File.** This is a set of information detailing the method of manufacture of an item. It includes the operations to be performed, their sequence, where they are performed, and the times required to set up and run them.
- **Work Center File.** This file contains records detailing the capacity of a production facility comprising identical machines or groups of people. It includes the productive capacity of personnel or machines in the work center.

Discussion

The principal requirement for an efficiently run manufacturing operation is an integrated computer system. *Integrated* is the key word, and it implies that all information comes from a common database, so that the same up-to-date information is shared by all users. Primary requirements of a manufacturing system include simple and quick access to information, effective storage, and easy input and output. These requirements are provided by the hardware and software used.

The importance of maintaining the accuracy and integrity of the manufacturing database is often overlooked. Insufficient time and resources are applied to ensuring that records are accurate, even though the concept is accepted intellectually. A manufacturing system depends totally on the information provided to it. The basic product and process records of item masters, product structures, routing, and work centers must be accurate and must be continually audited to ensure that they are *kept* accurate. Later on, this book describes the requirement of record accuracy in planning for materials, and the same warnings will be sounded. Item master, bill of material, routing, and work center data are the foundation of all manufacturing planning and execution, and must be kept valid, accurate, and integrated.

Item Master

Purpose and Description

An *item master* is a record that describes a part, a component, a subassembly, or an assembly. It is also called a *part master.* The item master contains all the data about a part, its ordering and inventory policies, planning data, and accounting and sales data.

Discussion

As a general rule, a unique part number should be assigned to items that move in and out of inventory and that need to be valued separately. If an item is part of a subassembly, it is said to be the *child* of that *parent assembly.* The same item can own other parts and be a parent. An item can thus be both a child and a parent. The parent is said to be at a higher level (0 is the highest), than the child.

Example: A Desk drawer is a child of the parent assembly Desk. The Desk drawer in turn is a parent of the drawer sides, ends and handle.

Items can be classified into:

■ Final assembly or end item. This item owns other parts, but is not owned by any item.
■ Subassemblies. These are items that are owned by a higher-level parent item and in turn own a child item or items (parts).

- Purchased parts. These are raw material, piece parts, and/or components. These parts are owned by other parts, but do not own any parts. They are at the lowest level of a product assembly.

Process for Setting Up and Maintaining the Item Master Records

1. Assign a *part number* (Part No.) or *item number.* This can be alpha, numeric, or alpha-numeric and can be of any length (seven to eleven characters are most common). Whenever possible use numeric characters only, as field results show that this helps to reduce transactional errors.

   Advantages of a significant numbering system (that is, a system where the part numbers themselves mean something) are found in the ease of item identification and in generally fewer gross errors of wrong usage of parts. However, a significant part numbering system requires maintenance and effort to set up. For most systems with a large number of part numbers, a nonsignificant system of part numbering is recommended.

2. Assign a *unit of measure* (UOM). These are measurements, such as *each* for pieces, *pounds* for weight, *feet* for length, and so on, by which an item is stocked and managed. This stocking unit of measure may be different from the unit of measure used to purchase the item.

   Example: Steel is usually bought in pounds or hundredweight but stocked and consumed in feet.

   The system should convert the purchased unit of measure to the stocking unit of measure when an item is received.

   As will be seen later, Sales very often use *dollars* as a unit of measure, whereas the factory uses *pieces* or *units.* Sales should measure their progress in sales and units, and should be strongly encouraged to provide their forecasts and results to manufacturing in units.

3. Assign a *revision status indicator.* This field contains a letter indicating the status of an item and an alpha-numeric field that indicates the revision level of the item.

   Example: A letter indicates the status of the item—A approved, U unapproved, R released, and an alpha-numeric indicating the revision level—A1, A2 and so on.

4. Assign a *source code*. This is usually a letter describing the source of a part and how it should be handled by inventory.

> Example: M-Manufactured; P-Purchased; Z-Manufactured or Purchased; S-Subcontracted; C-Customer supplied; X-Phantom or blow through; F-Free or floor stock; and so on.

> *All of the above four fields are mandatory*—that is, they must be entered to have a valid item master.

5. Provide a *description*. This may or may not be a mandatory field. It is not required for any systemic reasons, but it helps the user to sort a listing of items. Descriptions should be complete.

> Example: The description "Screw" is incomplete. A proper description is "Screw machine Csk $1/4 \times 1-1/2 \times 12$ tpi."
> Also commonly used but inadequate descriptions are "unknown," "blank ( )," and "To be specified."

6. Set up *nonmandatory elements*. There are virtually an unlimited number of nonmandatory fields that provide information about an item. Usually these nonmandatory elements are organized by function or department, such as Engineering, Purchasing, Material Control, Accounting, etc.

> Example: material control elements: ABC code; quantity on hand; buyer or planner; order policy; safety stock; fixed lead time; per unit lead time; scrap factor; cycle count frequency, and so on.
> Purchasing elements: vendor code; vendor name; address; number of orders placed; dollar amount of orders placed, and so on.

7. Conduct regular checks on the validity of the item master records.

> Example: Run a "childless" report that lists manufactured (M) parts that have no dependent parts.
> Run a "nowhere used" report to look at parts that have no assembly links.
> Run a "usage report " year-to-date and last-used to establish candidates for removal.

8. Maintain control over the size of the part population by running a report, sorted by description, that lists duplicates and "look alikes." Eliminate or standardize these parts.

Dos and Don'ts: Item Master

DOs

1. Do ensure that all material that is stored or costed is identified by a valid and *unique* part number.
2. Do try to use numeric part numbers only, with as few digits as possible.
3. Do provide controlled access to modifying item-master records by providing password security. Allow change to be made only after approval from a change management committee. Assign a specific person to be responsible for creating and deleting all item masters.
4. Do designate a person from each function to be responsible for maintaining the accuracy and integrity of function specific records.
5. Do set up tables of acceptable values for units of measure, ABC codes, lead times, order quantities, safety stocks, and so on, to be used to verify the validity of the item master functional fields.
6. Do recognize that packaging or marking an item may require giving the item a different part number.
7. Do standardize the parts to reduce the part population.
8. Do measure the accuracy and integrity of the item-master database.

DON'Ts

9. Don't let the engineering department have total control over part-number generation or revision. Use a team approach.
10. Don't get caught in the trap of "it's really the same, except . . ." instead of using standardized parts.

Product Structures (Bills of Materials)

Purpose and Description

A bill of material (BOM) defines how a product is made. It does this by linking every item (parent) with the parts that go into it (children). It does this for successive levels of subassembly until the final assembly of the end item is reached. Material requirements planning uses a bill of material to determine the quantity and type of parts required to build a particular product.

Discussion

An example of a partial bill of material is shown below. The final assembly is given a level code of 0, and every succeeding level down is given a level code in increments of 1. These are called *low-level codes* and are used by MRP to decide how BOMs are exploded.

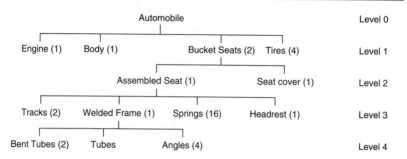

Figure 4.1. Bill of Material Structure Levels and Parent-Child Relationship (number in parentheses is the quantity of an item used)

In the bill of material shown in Figure 4.1, the automobile is the end item, or Level 0. It is the parent of many Level 1 assemblies (children) including 1 engine, 1 body, 4 tires, 2 bucket seats and so on. The bucket seat is the parent for Level 2 assemblies (children), including 1 assembled seat, 1 seat cover, and so on. The assembled seat is parent to Level 3 subassemblies including 1 welded frame, 16 springs, and so on. An item that does not have any children under it is a purchased item.

Process for Setting Up a Product Bill of Material

1. Determine how the product will be made and assembled.

2. Identify all the parts that go into the product and ensure that they have item masters.

3. Set up the parent-child relationship based on how the product will be built. Attempt to build as few levels as possible, while following the building process.

4. Verify that each parent is complete at every level by entering the BOM mandatory *data elements:* parent part number, child (component) part number(s), child quantity per parent assembly, and effectivity dates. (See the example of BOM in Table 4.1.)

5. Ensure that a parent identified by a unique part number has only one set of children (a unique combination of materials and components). If the type or number of the children change (in their design, for example), the parent must be given a new part number, or a new revision level.

6. Conduct regular checks on the validity of BOM records.

> Example: Run a "childless" report that lists manufactured (M) parts that have no children.
> Run an "orphan" report that lists purchased (P) parts that are not attached to a BOM. These will be valid only in the case of a distributor.
> Run a BOM chain report that verifies the integrity of the BOM.

7. Conduct regular shop floor audits by comparing the actual assembly of a product with its BOM, determining differences and resolving them.

8. Assign single-point responsibility for maintaining the BOM data. This should be either design or manufacturing engineering.

BOM Structures

BOMs have several types of structures or displays. Some commonly used are:

1. A Single-Level Bill of Material. This report lists all the direct components for an assembly. For Table 4.1 below, the parent assembly "Assembled seat sporting" contains four subassemblies—items 1, 2, 3, and 4.

> Example: Part No. AS001 Desc. Assembled seat sporting Rev. B2
> Lead time 2 days Lot Size 100 Date 12.22.94

Table 4.1 Single-Level Bill of Material

Item	Description	Part No.	Rev.	Quantity per Assembly (QPA)	UOM
1	Track —C type— feet	TK001	A1	2	Feet
2	Welded frame bucket seat	FR 010	A3	1	ea(each)
3	Spring coil 12pi NiCr	SC100	B1	16	ea
4	Headrest bucket seat	HR002	A2	1	ea

2. An Indented Bill of Material lists all an assembly's components, with the number of the level indicated by the amount of indentation of the component Part no. See Table 4.2.

> Example: Part No. BS001 Desc.: Bucket seat sporting Rev. B2

Table 4.2 Indented Bill of Material

Level			Description	Part No.	Rev.	QPA	UOM
1			Bucket seat sporting	BS002	B2	2	ea.
	2		Seat cover, cloth gray	CV020	A1	1	ea.
	2		Assembled seat sporting	AS001	B2	1	ea.
		3	Track —C type— 3 feet	TK001	A1	2	feet
		3	Spring coil 12pi—NiCr	SC100	B1	16	ea.
		3	Headrest bucket seat	HR002	A2	1	ea.
		3	Welded frame bucket seat	FR010	A3	1	ea.
		4	Bent Tubes 1in.dia. NiCr	BT004	A2	2	ea.
		4	Tube 1in.dia.2ft. NiCr	TT002	A1	2	feet
		4	Angles 2x2x3 NiCr	AA010	A3	4	feet

3. A Summarized Bill of Material. This lists all components for a given assembly with total quantities required across all levels needed to manufacture the component's parent.

> Example: For a screw S#100, if 4 are used in Level 2, 8 in Level 3, and 8 in Level 4, then the summarized BOM will call out 40 S#100 screws.

4. A Single-Level Where Used BOM. This lists all the parents where the components are used. It is the reverse of the single-level BOM. The component is listed on top and all the parents that use the component are listed below it, using the same format as the single-level BOM.

5. An Indented Where Used BOM. This traces usage of a given component in its parent and, in turn, the parent's parent, until the end item is reached. It is the opposite of the regular indented BOM.

Effectivity

Parts and assemblies are subject to change. It is essential to capture planned and actual change on the BOM, so that obsolete parts are not ordered and replacement parts are ordered in time to be available when needed. The BOM has an "In Effectivity" and "Out Effectivity" associated with the parts that are changed.

> Example: Part No. AS001 Desc.: Assembled seat sporting Rev. B2
> Lead time 2 days Lot Size 100 Date 12.22.94

In the BOM shown in Table 4.3, the spring coil drawing is due to be revised on the 1st of January 1995. It has an out-effectivity of 12/31/94 and

Table 4.3 Bill of Material Showing Effectivity

Item	Description	Part No.	Rev.	QPA	UOM	In-Date	Out-Date
1	Track -C type- 3 feet	TK001	A1	2	ea.		999999
2	Welded frame bucket seat	FR 010	A3	1	ea.	060194	
3	Spring coil 12pi—NiCr	SC100	B1	16	ea.		123194
3	Spring coil 12pi—NiCr	SC100	B2	16	ea.	010195	
4	Headrest bucket seat	HR002	A2	1	ea.		999999

an in-effectivity of 01/01/95. The revision level changes from B1 to B2. The track and headrest have no plans to be changed, and so they have an out-effectivity of 999999. The welded frame bucket seat was changed on 06/01/94.

Alternate Parts

The BOM effectivity can be used for setting up alternate parts. In this application the alternate part is included in the BOM with QPA of 0 and no effectivity. When an alternate is needed, before the order is released the BOM is modified with the alternate given the required QPA and in-effectivity and the regular part the out-effectivity. When the regular part is available, the process is reversed.

Phantoms

A phantom is an item, not a bill of material. A phantom is usually a subassembly that is actually built, but not stocked—it is merely an assembly stage of work in process. MRP will drive requirements through the phantom part to its components. Phantoms are a convenient means of identifying a group of items (subassembly), without having the system plan supply for the subassembly. Phantoms can exist at any level of the bill of materials.

A phantom must have the following attributes:

1. Lead time must be set at zero.
2. Order policy should be set at lot for lot.
3. The item master should be coded, to enable MRP to recognize it.

Example: A welded automobile body may be classified as a phantom. Say it is built by welding an underframe, two sides, a roof, and a trunk (all subassemblies). The body or phantom level is seldom stocked, as after being welded, the body immediately goes on to other operations. This phantom classification may be to account for the number of bodies built.

Planning Bills (aka Pseudo Bills)

A planning bill is a grouping of items, assemblies, and components to form an artificial parent item that is never built, only planned for. The parent of a planning bill has a large number of configurations, and to create an item master and a regular BOM for each would be impractical and burdensome because it would call for setting up a large number of parents. In creating the planning BOM, the percentage of parts and options used historically are expressed in the quantity-per-assembly field. The percentages can be over 100 if there is a need to overplan for some options. Planning bills are used to help forecast or master schedule product that has many options. Planning bills are also used in planning rough-cut capacity.

Example: Consider the automobile BOM. The four assemblies listed are engine, body, bucket seat, and tires.

Each of these assemblies can have many options. Engines can be 4, 6, or 8 cylinder; the body can be coupe, four-door or hatchback; bucket seats can be leather, linoleum, or cloth; and tires can be regular, radial, or steel-belted. Given the above options, that auto can have:

3 (engines) $\times$ 3 (bodies) $\times$ 3 (seats) $\times$ 3 (tires) = 81 combinations

If one more type of tire is introduced, the number of combinations will double!

There are many more such assemblies and options, and the possible combinations will run into thousands. Clearly these combinations cannot all be built. But there is a possibility that any one of the combinations can be sold. A planning bill helps to plan for dealing with this situation by planning for and building a few parts only that can be assembled into many combinations.

Applications of a planning bill are modular bills and super bills.

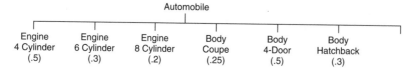

Figure 4.2. Example of a Planning Bill (numbers in parentheses show historical percentages of options built)

In the planning bill above, the automobile is an artificial part and will never be built. Historically, 50 percent of the autos are ordered with 4-cylinder engines, 30 percent with 6 cylinders, and 20 percent with 8 cylinders. If sales plans to sell 10,000 automobiles, the planning bill will advise the factory to make 5,000 4-cylinder engines, 3,000 hatchbacks, and so on. When a customer wants a 6-cylinder engine in a hatchback, with cloth seats and radial tires, the specific auto can be assembled from the assemblies available.

Planning Bills have the following attributes:

1. The artificial parent defines a family of many products.
2. The planning bill can combine common parts (100 percent) with features (add up to 100 percent or over) and options (can have any percent that is likely to be ordered).
3. The planning bill can be combined into modules of options available to the customer.
4. The planning bill helps to forecast and master schedule when the exact configurations are not known.

Dos and Don'ts (for BOMs)

DOs

1. Do assign a person(s) (from design or manufacturing engineering), a specific cradle-to-grave responsibility for maintaining the accuracy and integrity of the bill of material files.
2. Do verify the completeness and accuracy of the BOM by checking the actual quantities used in assembling the product. Set an accuracy target of 100 percent.
3. Do confirm that the product is built in the same sequence as the BOM level structure.
4. Do use phantoms to represent transition assemblies that are built. They help to structure the BOM in line with the manufacturing process.
5. Do ensure that a BOM cannot be deleted while there is an open order for the product it represents.
6. Do limit access to the add, change, and delete features of the BOM file, by providing password security.
7. Do make engineering change implementation a joint engineering/manufacturing activity, preferably a change board.

DON'Ts

8. Don't let more than one bill of material system exist in the company. There should not be separate engineering and manufacturing bills.
9. Don't neglect to ensure that the unit of measure used is consistent. If an item uses 10 feet and has a unit of measure of each, there is danger that requirements will be generated for 10 units each requiring 10 feet, instead of one unit requiring 10 feet.

Routings

Purpose and Description

Routings inform the factory personnel how a part is to be made. A *routing* is a document that details the sequence of operations and methods used to manufacture a product. It also includes information on the time the operation is expected to take and the machines and tools it will use. Routings allow the manufacturing system to estimate the capacity available. Routings are the basis of a shop traveler instructing the shop on how to process a part and provide for work in process tracking.

Discussion

Routings are a basic information requirement for manufacturing product. Today paper routings are being replaced by electronic (on the manufacturing system) routings. Good control can be exercised over electronic routings, and the errors of using outdated versions are eliminated. Many information systems allow process engineers to develop custom routings, including rework, as the product is being made. This is a particularly useful feature in high-tech product where the exact manufacturing process is not yet determined.

Process for Setting Up a Routing

1. Develop a draft routing or sequence of operations that matches the BOM for the product being considered.
2. Trace the *actual* sequence of operations a product goes through.
3. Estimate or determine realistic operation and setup times.
4. Compile the *mandatory* data elements into the routing form. These are:
   - Operations sequence number: the chronological sequence in which the product is built.
   - Operations description: brief description of work to be done.
   - Work center number: every operation is performed in a work center—a machine or a person(s).
   - Standard job setup and standard run time per unit.
5. Develop a combined *routing and process sheet* (see Table 4.4). A process sheet includes scheduling information. This serves as a routing sheet and a standards guide.
6. Develop an alternate routing for items that go through work centers that are likely to be process capacity constraints.

Table 4.4 Format of a Combined Routing and Process Sheet

Part No.	Rev. Desc.	UM	QPA	Next Assy.	WO #	Qnty	Date
W.C.	Oper.# 01	Description	Setup hrs	Run hrs/pc	Tool #	Dt. Sched	Dt. Comp	Qnty Complete
	23							

Dos and Don'ts

DOs

1. Do assign specific responsibility to a person in product or manufacturing engineering for the accuracy and integrity of the routing.

2. Do control access to the routing file by password security.

3. Do verify that the routing data are consistent with the actual sequence of operations.

Work Centers

Description and Purpose

A *work center* is a specific production facility consisting of one or more people and/or machines that perform similar tasks, and through which work is processed. Work centers are needed to establish capacity available, so that work can be scheduled and job start and completion dates determined.

Discussion

A work center contains information on:

- Capacity: number of workers, number and types of machines, number and length of shifts, and how much the work center is utilized.
- Performance: productivity or efficiency factors for labor and machines.
- Costs: labor and burden rates for each work center.

A work center may be a single worker or machine and can extend to a cell of similar machines.

Process for Establishing a Work Center on the System

1. Determine and identify all unique work centers through which product must flow.

2. Collect the data elements required to define the work center's capability. Included are:
   - Work center identifier and description
   - Hours per shift, number of shifts per day, and days per week
   - Number of machines and how used per shift and per day
   - Number of people and how used per shift and per day

3. Observe and collect data on how many hours in a shift/day a resource is used. This is utilization. If individuals consistently work 6 hours in a 8 hour shift, the utilization is 75 percent.

4. Determine how well a resource is used. This is efficiency. If the standard output for a machine is 16 pieces per shift and 20 pieces are consistently produced, the efficiency is $20/16 = 1.25$, or 125%.

5. Verify from historical records that the utilization and efficiency factors are accurate.

Dos and Don'ts

DOs

1. Do assign specific responsibility to a person in product or manufacturing engineering, for the accuracy and integrity of the routing.

2. Do control access to the work center file by password security.

3. Do verify that the work center data are consistent with the actual operation of the work center.

DON'Ts

4. Don't consider a person to be available until he or she is trained and productive (improved on a learning curve).

Engineering Change Management

Description and Purpose

Product and processes are subject to change, thus requiring corresponding changes in the database records. Change has to be managed or it can result in unapproved product being used, and/or obsolete product being ordered and as a consequence unnecessary cost being incurred.

Discussion

Engineering change control is an entire subject in itself; in this book only the highlights will be captured. A major advantage of an integrated system is its ability to plan and control engineering change. There must be a company policy on how changes are made and who is authorized to make them. Generally there are two types of changes:

1. Mandatory or immediate changes. Change must be made immediately when products fail to function, when safety is endangered, or to comply with the law.

2. Phased-in or optional changes. Here the change may be introduced to minimize costs. Such changes include quality improvements, cost reductions, improving practices, accommodating customer requests, and so on. Essentially an attempt is made to phase out the old product or to modify existing product. These changes require an effectivity date and a schedule. At times the exact date that a change should become effective is not known, as the change is driven by when a part is no longer in inventory, or when a given serial number is processed. In such cases the system should allow effectivity to be expressed by a serial number or a per quantity processed.

Change control is evoked when the *form, fit,* or *function* of a part or a product is affected. Changes are made to a product or part, a drawing, a BOM, and an item master. It is important that the advantage gained by a change be clearly established, as change is almost always disruptive and costly.

Change control is usually maintained through *Engineering Change Notices* (ECN), which detail what is changed and why, how and when to change it, and who is responsible for making the change. Change can be initiated from any source, but generally originates from engineering. An example of an ECN can be seen in Figure 4.3.

Process

1. Set up a change review board. It should include representatives from materials, design, manufacturing, manufacturing engineering, and quality.

```
ENGINEERING CHANGE NOTICE            No. _____ Dt. _____
┌─────────────────────────────────────────────────────────────┐
│  Type of Change:      ☐ IMMEDIATE       ☐ Phased in           │
├─────────────────────────────────┬───────────────────────────┤
│  Orig. By      Dept.        Dt. │  Part No. _____ │
│  Appd. By      Change Board. Dt.│  Drgs affected _____ │
├─────────────────────────────────┴───────────────────────────┤
│   Reason for Change:                                          │
│   ☐ Correct drg error    ☐ Improve reliability   ☐ Reduce cost│
│   ☐ Improve quality      ☐ Improve serviceability ☐ Other     │
├───────────────────────────────────────────────────────────────┤
│  Impact on:      Cost:        Schedule:     Performance/Reliabilty:│
├───────────────────────────────────────────────────────────────┤
│  Effectivity:                                                 │
│  ☐ Immediate. Stop all current work. Replace with new part.   │
│  ☐ Rework all units. Change to new part after Ser. No. ___ or Dt.│
│  ☐ Use all existing product before change.                    │
├───────────────────────────────────────────────────────────────┤
│  CHANGE: FROM:      TO:              On Date:                  │
│  (Write details of Old and New Part Nos., Quantities, where new supply│
│  will be received and when, and any other instructions)       │
├───────────────────────────────────────────────────────────────┤
│  DISTRIBUTION: Mats    Pur.    Design    Mfg. Eng.   Prod. Eng.│
│  Acctng.      Quality.  DataAdm.    Prod Suptd.   Shop Sup.    │
└───────────────────────────────────────────────────────────────┘
```

Figure 4.3.   Sample of an Engineering Change Notice

2. Determine the procedure to be followed for instituting changes and issue a formal control document. (See discussion above.)

3. Train all personnel who are likely to be affected by changes, so that they understand the change procedure and follow it.

4. Ensure that all changes have an effectivity date, a revision level (both captured in the BOM or item master), and are associated with an ECN number.

5. Ensure that the change process deals with multiple items. A single ECN should be capable of affecting the parent-child and/or the where used relationships.

6. Record an effectivity type for each revision. This helps to classify why changes are generated, and it can be used for reducing the number of changes.

   Example: "1" for data driven, "2" for inventory supply driven, or "3" for serial number driven.

7. Maintain a status level for each revision: *U* for unapproved, *A* for approved, and *R* for released. Usually MRP plans requirements for an unapproved part, so it should be monitored.

Dos and Don'ts

DOs

1. Do ensure that the ECN process is an integral part of the information system and information on changes is available in real time.

2. Do ensure that training is provided to all personnel, including shop operators and store attendants, on how the change control process works and on the importance of following the change.

3. Do include the importance of an evaluation of the cost and schedule impact of a change, before accepting a change.

4. Do establish the culture of only making changes when there is a real advantage to be gained!

DON'Ts

5. Don't have a process that is too involved or too lengthy. Such processes are usually circumvented by users.

6. Don't allow engineering changes to remain on the database indefinitely. If they become permanent, they should be replaced by approved drawings, if temporary they should be purged after the approved period of time.

Forecasting for Manufacturing

Summary of Chapter

The overview provides a general administrative view of the forecasting process. Forecasting principles describe basic terms and rules and detail a specific forecasting procedure. Techniques used to develop accurate forecasts are described next, and the chapter ends with a section on managing the forecasting processes to either minimize forecasts or improve forecast accuracy.

Contents

- Overview and management considerations
- Forecasting principles
- Forecasting techniques
- Managing forecasts in manufacturing

Relationship of Forecasting to Other Major Manufacturing Planning Processes

Forecasting predicts future demand (orders) that will be placed by customers. It sends this forecast demand at an aggregate or product family level to the *production planning* process. It also sends the forecast demand at an individual item level to the *master schedule*. Where a master schedule is not used, the forecast is sent to the *materials requirements plan* (although this is not recommended practice). The master schedule and the materials requirement plan schedule what, how much, and when a factory must produce.

Overview and Management Considerations

Purpose and Description

> *Forecasting* enables delivery of customer orders when needed, while op-
> timizing production resources. Manufacturing product requires time,
> and customers are often unwilling to wait for companies to make or build
> a product. Forecasts are used to bridge this gap. With a forecast a com-
> pany can get a head start at buying material and building product, mak-
> ing it available when the customer needs it. This approach contrasts to
> the just-in-time (JIT) process of making product only after the customer
> order is received. In JIT, a customer order is satisfied by rapid response,
> necessitating short lead times and carefully planned buffer inventory of
> raw materials and/or assemblies.
>
> In manufacturing, forecasting is an estimate of future customer de-
> mand. It signals operations to buy material and built products. Forecasts
> can be qualitative, using subjective and informal methods, or quantita-
> tive, using mathematical means based on a predictor (usually past data).
> Generally a combination of the two techniques is used, resulting in a cal-
> culated projection that is modified by market intelligence. Since forecast-
> ing predicts the future, it is subject to error.
>
> Forecasts drive the production plans at the family or product level,
> the master schedule at the end item level, and the materials requirement
> plan at the assembly level.
>
> Forecasts are also required for long-term planning of resources—
> plants, space, equipment, and people. This requirement of forecasting
> will not be covered in this book. See the bibliography for references that
> address this topic.

Executive View

> Since forecasts of product requirements are invariably incorrect and lead
> to schedule and cost risk, the best forecast is no forecast. A forecast is not
> required when the product can be made within the time the customer is
> willing to wait. Further, by manufacturing to customer orders, inventory
> costs are minimized. Today the principle of building to demand is being
> applied together with a focus on reducing manufacturing lead time and
> operation cycle time. Many products are built only when a hard order is
> received. In such cases, however, forecasts may be required to buy long
> lead time material.

Many products must be stocked (off-the-shelf consumer items); others have to be started ahead of the expected order so as to build the product in time to satisfy a customer. Products such as these require forecasts. Shigeo Shingo (1988) developed a P:D ratio, where P is the manufacturing lead time and D is the customer's required lead time (the customer wants delivery of the order in D time). If the P:D ratio exceeds 1.00 (in other words, if the production time exceeds the customer's required time), either a customer's order will be delayed or a forecast is required to start making the product. A high P:D ratio also highlights specific products that *must* have their lead times reduced. A high P:D ratio exposes an order to more risk, as it has to be forecast further out making it more susceptible to inaccuracy. Inaccurate forecasts result in poor service or unsold inventory.

Approaches for improving the P:D ratio are:

1. to plan for and provide contingency capacity,
2. to simplify the product line, and
3. to standardize the product and process.

All of these techniques reduce variability and shorten lead times and are the essence of just-in-time manufacturing. JIT also seeks to eliminate forecasts by making product to a customer demand.

Since forecasting is an attempt to predict future demand, the sales and marketing groups must be the main drivers of providing timely and accurate forecasts. They should also establish a culture that commits to sell what is forecast. Practitioners need to be sensitive to an inherent conflict between the objectives of sales, marketing, and operations. Sales and marketing want ample product available for sale and they like to change forecasts often to react to the market; operations do not like frequent and disruptive change and can only make as much as their capacity will allow.

One of the biggest problems with the forecasting process is that unreasonable expectations are developed, driven by the claims of sophisticated software solutions. Forecasts are inherently inaccurate. Accuracy will be improved by disciplined collecting of data, meticulous analysis, correction of error, and applying known market intelligence to new forecasts. Forecast accuracy also depends on the repetitiveness and volume of the product being ordered. It is reasonable to expect repetitively ordered, high-volume products to have a forecast accuracy of 80 to 90 percent, while infrequently ordered products with variable order quantities may have a forecast accuracy of 50 to 70 percent. Serious consideration should be given to the value of incurring high costs on sophisticated forecasting packages with many options. Such packages may only marginally improve forecast accuracy. They will provide a good information base and attractive graphical representations of the data.

Process for Establishing a Forecasting Cycle

1. Establish a monthly sales and operations meeting at which the marketing, materials, and factory managers review the previous month's business, customer service, and inventory, and then forecast future business. Attempt to get the functional heads (vice presidents) and the president to attend.

2. Ensure that there is a means of translating the forecasts made at the product level (usually in dollars), to the item level in units. The master schedule and the factory are operated in units or pieces.

3. Measure forecast deviations and ensure that the reasons for large discrepancies are understood and corrected. Use the deviation to calculate safety stocks, as detailed later. Watch out for and identify error trends. Determine root causes and take corrective action as a priority.

4. Identify products with predictable sales and products with uncertain demand. Different approaches need to be used to minimize risk for each of these types.

5. Analyze the forecast and separate business trends from seasonal business. Identify sales made due to promotions or other induced-sales activities. Special market or other uncontrollable events that affect the business should also be identified.

6. Measure inventory cost and inventory turns to determine if the forecasting process is reducing inventory, particularly safety stock.

7. Ensure that forecasts are consistent with the business plan and the company strategy.

Dos and Don'ts

DOs

1. Do ensure that marketing, sales, and operations are equal partners in all forecasting activities.

2. Do ensure that the forecast is tied to inventory and develop an inventory plan that accounts for forecast error.

3. Do establish a formal policy of determining how much additional inventory will be allowed to support a company's focus on customer service.

4. Do readjust the forecast based on actual orders being booked.

5. Do remember that a hard customer order is better than any forecast.

6. Do try and get a forecast range: this helps to evaluate risk and plan better.

7. Do seek to eliminate forecasts whenever and wherever possible.

DON'Ts

8. Don't use forecasting accuracy as an excuse for poor delivery performance.

9. Don't expect accurate forecasts. Expect and allow for error based on the sales stability of a product.

10. Don't use sales history "as is." Filter and purge the data until the information represents what can be expected on a normal basis.

11. Don't second-guess marketing and sales' forecasts.

Forecasting Principles

Purpose and Description

Forecasting principles help practitioners understand forecasting and enable them to adopt the right approach to generating accurate forecasts.

There are a few broad empirical principles that should be followed when forecasting. Forecasts have basic elements and all forecasts can be broken down into these elements. Finally, forecasts have error and there is need to measure forecasts and reduce error.

Discussion:

The general principles of forecasting are:

1. Forecasts are more accurate for larger groups of items than for smaller groups. Whenever possible get a forecast for the family or product group, and use it to verify the sum of individual item forecasts within the group.

2. Forecasts are more accurate for shorter periods of time, closer to the present. Postpone making a forecast until it falls within the manufacturing lead time of the product.

Long-range forecasts of two to five years should be used for strategic planning issues such as business policy direction. Intermediate range forecasts of one to one and a half years are used for commitments and expenditure, and should correspond to the budget cycle. Planning of materials and capacity should be covered by forecasts that extend from three months to a year. Finally forecasts that commit a factory to scheduling and building product should be limited to one to three months.

3. Forecasts will always be wrong. Do not develop any expectation that it can be otherwise.

4. Forecasts are a set of numbers to work from, not to work toward.

Quantitative forecasts are a projection of past data and have certain basic patterns. These are:

- An average level of demand. This is the starting point.
- A trend—upward or downward
- Cyclical or seasonal variations
- Randomness or noise

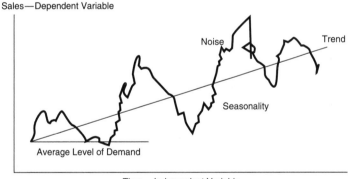

Figure 5.1. Elements of a Forecast

The forecasting requirements for data are:

Data collection elements:

1. Availability—How easily can it be collected?
2. Consistency—Can it be compared over time?
3. History required—Are enough data available? How much history is enough?
4. Units of measure—What UOM is used? Units or currency?

Forecast elements:

5. Independent data—Are they valid predictors?
6. Element being forecast—What is being used? Orders or shipments?
7. Forecast frequency—How often is it necessary?

The quality of data has a direct effect on forecast accuracy. The type of data determines the cost of forecasting.

Most software packets have provision for a management forecast and the system itself will calculate a system forecast based on its projections of historical data.

Process for Forecasting

1. Determine the purpose of the forecast. For business planning, a high-level forecast by the business unit in dollars may be adequate, whereas for master scheduling a forecast is required by saleable item in units.

2. Ensure that the data requirements of availability, consistency, history, and unit of measure (items 1 to 4 above) are all examined and resolved.

3. Determine the forecast predictor (orders or shipments) and the forecast frequency (items 5 to 7 above).

4. Select the forecasting technique to be used (see the following section).

5. Prepare the data. This includes ensuring that the historical record contains only those numbers that are consistent with today's sales. Purge all one-time sales events or aberrations.

6. Set quantitative technique parameters to reflect the type of product being forecast (see the following section).

7. Make the forecast.

8. Measure error. (Error = Actual − Forecast).

9. Analyze and document the reasons for error and develop a means of incorporating lessons learned into the next forecasting cycle.

All of these steps are covered in much greater detail in the next two sections.

Dos and Don'ts

DOs

1. Do ensure that forecasts have a range or an estimate of error.
2. Do compare forecasts with actual results and analyze large deviations.
3. Do forecast as late as possible (not before the start of manufacturing lead time or purchasing lead time).

DON'Ts

4. Don't commit large resources to any sales forecast beyond about six months.
5. Don't presume that the users know or are confident about the whole forecasting process. There is need for constant guidance.
6. Don't let forecast error become a self-fulfilling prophecy; keep looking for and minimizing causes of deviation.

Forecasting Techniques

Purpose and Description

The purpose of a forecasting technique is to produce an accurate and timely estimate of customer demand.

There are many techniques for forecasting product demand. The technique used depends on the type of product and the environment in which it is used. It also depends on whether the forecast covers a short or long time period. The technique should be cost effective, and this depends on the data required and the effort needed to obtain the data. Finally, the technique must be understood and applied using market and sales intelligence.

Discussion

Forecasting techniques can be divided into two broad categories: *qualitative* and *quantitative. Qualitative forecasts* are subjective and depend on judgment, intuition, and personal knowledge. Among the major techniques in this category are market research (surveys), Delphi (specialist consensus), historical analogy, and management review (business estimates). *Quantitative forecasts* use a predictor to estimate future demand. Intrinsic techniques use a time series or historical data as a predictor, while extrinsic techniques attempt to relate and use an external element, such as business conditions, to predict demand.

Qualitative Techniques

Market research: This technique, usually conducted through surveys, is generally used to support product development and promotions. It can be used to provide insights into the likely sales of a product, but should be used with other techniques. It is accurate over the short term and caters to a specific set of conditions. It is costly and not responsive to change.

Delphi Method: This method is based on the knowledge and judgment of a small group of experts. A specific question(s) is asked, such as, "In the next ten years, will wireless communication exceed all other forms of telecommunication? By how much?" Each panel member is asked the question separately, and the answers are circulated until consensus is reached. The technique is useful for long-range forecasts at a product family or generic level.

Historical Analogy: This method attempts to correlate a knowledge of the past to a future event. It is best used when there is similarity between the products being compared. The method is useful for establishing an initial growth trend of a new product.

Management Review: This is similar to the Delphi method, except it is more product-specific. It is used to qualify quantitative methods—a sanity check—and depends heavily on the knowledge and quality of judgment of the managers. It is extensively used to forecast, particularly when dealing with new technologies. It suffers from over-conformity, possible bias, and management ego.

Quantitative Techniques

Moving Average: This is perhaps the simplest of all forecasting techniques. It consists in finding the average of a preset number of periods of actual demand and using the average as the next periods forecast.

Example: Three-period average

$$\text{Forecast for April} = \frac{\text{(actual sales for January + February + March)}}{3}$$

$$\text{Forecast for May} = \frac{\text{(actual sales for February + March + April)}}{3}$$

In using moving averages, the forecast lags behind the actual demand. By increasing or decreasing the number of periods used in the average, the forecast can be made less or more responsive to the actual demand. (A five-period moving average is less responsive than a two-period average. A one-period average lags behind the actual occurrence by a period.) The moving average can be biased by multiplying each period with a weighting factor (providing more weight to the latest period); this technique is known as the *weighted moving average* method.

Example: Three-period weighted moving average, with weighting factors of 1, 2, and 3 applied from the oldest period to the most recent:

Forecast for April =
$$\frac{[\text{Actual sales for (January} \times 1) + (\text{February} \times 2) + (\text{March} \times 3)]}{(1 + 2 + 3)}$$

Forecast for May =
$$\frac{[\text{Actual sales for (February} \times 1) + (\text{March} \times 2) + (\text{April} \times 3)]}{(1 + 2 + 3)}$$

Weighted moving averages require the storage of a lot of data and need many computations.

Exponential Smoothing: This is a form of weighted moving average with fewer calculations and less stored data. The determination of the α (alpha) factor is very important. Generally low α factors (0.2 and less) lead to a forecast that responds less to the immediate past actual results, whereas high α factors (0.5 and above) result in forecasts that chase or are closer to the immediate actual results (see following example). For first order smoothing the formula used is:

$$\text{new forecast} = \text{old forecast} + \text{smoothing factor } (\alpha)$$
$$\times (\text{actual sales} - \text{old forecast})$$

Example: old forecast = 160, actual = 200,
smoothing factor $\alpha = 0.1$
$$\text{new forecast} = 160 + 0.1 \times (200 - 160)$$
$$= 160 + 0.1 \times 40 = 164$$

Example: old forecast = 160, actual = 200,
smoothing factor $\alpha = 0.8$
$$\text{new forecast} = 160 + 0.8 \times (200 - 160)$$
$$= 160 + 0.8 \times 40 = 192$$

It may be seen in the first example, using an α of 0.1, a forecast of 164 follows an actual of 200; in the second example with an α of 0.8, a forecast of 192 follows an actual of 200.

Exponential smoothing has the advantage of not requiring retention of the amount of data that the weighted moving average does. Exponential smoothing is useful for short-range forecasts and is effective when products are fairly stable. It has the advantage of being able to apply market product intelligence by making the forecasts more or less reactive. It is less useful for seasonal products, unless a more complex, seasonally adjusted method is used. It is less useful for lumpy demand, since the latest demand gets the most weight.

Regression Analysis: This is a technique of finding a relationship from historical data (usually straight line, or linear), between two variables, that is, time and the product being forecast. The formulas are part of every software packet and the calculations are made by the computer. The regression analysis determines a constant (A) and a slope (B) for a particular set of sales over a given period of time.

The formula for a straight line is of the form:

$Y = A + (B \times X)$
where Y is product forecast and X is time

If the values of A and B are calculated, they can be applied for any other period (X) and the product forecast Y can be calculated.

Example: If $A = 10$ and $B = 3$, calculated for the last 24 months of tractor sales, then for month 30:
Formula for forecast sales of tractors $Y = 10 + (3 \times X)$
Forecast sales of tractors month 30 $= 10 + (3 \times 30) = 100$

The major assumption of this technique is that the sales will continue to follow the same straight-line projection that it has been following in the past.

To determine that the relationship of A and B established is valid, a *correlation factor* must be calculated. The correlation factor has a maximum value of 1.0, with values above 0.6 indicating a good match between the regression formula and the actual results. A regression with a high correlation has the expectation of producing a good forecast. Factors below 0.5 represent poor correlation and unreliable forecasts. Like the regression analysis, the formula for correlation is contained in any software packet. Regression analysis and correlation can be useful for long-range forecasts. Seasonality can be decomposed and applied on top of a regression analysis, making the technique particularly suitable for seasonal products.

Extrinsic Techniques: Here an attempt is made to relate an external factor to the product being forecast, and to use the values of this factor to predict the sales of the product. For example, if a relationship is developed between housing starts and lumber sales, then housing starts may be used to forecast lumber requirements. Extrinsic forecasts may be extended to more than one external variable. There is a whole body of statistical knowledge devoted to using extrinsic factors to forecast. This method is useful at the product family level and can be valuable when forecasting consumption of large dollar volume items, particularly when there is a strong correlation between the factor(s) and the product. Of course, a reliable predictor (factor) has to first be selected and validated.

Pyramid Forecasting: The essence of this technique is that a product is forecast at several levels, with the lowest level representing the individual stock keeping unit (SKU) or item and the highest level covering the product family. Levels can be forecast independently, with each level being aggregated to the next level. The highest level acts as a check to ensure that too much or too little is not forecast at the lower levels, as the forecast at the highest level is "forced" or imposed on the lower levels.

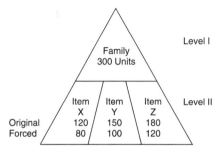

Figure 5.2. Pyramid Forecasting

Example: Level 1 represents family and the forecast is 300 units.

Level 2 represents item forecasts. X, Y, and Z have original forecasts of 120, 150, and 180 respectively, for a total of 450.

The ratio between Level 1 and Level 2 is $\dfrac{300}{450} = \dfrac{2}{3}$.

Therefore forecasts of X, Y, and Z are factored by $\dfrac{2}{3}$ and become 80, 100, and 120 respectively.

Process for Selecting a Forecasting Technique

1. Understand the stage of the product in its life cycle. New products, new technologies, and new markets all need qualitative inputs from management and other experts as quantitative data are scarce or not available.

2. Separate products with stable sales from those with highly variable sales.

3. Exponential smoothing should be used for stable products. Use a low smoothing (α) factor of 0.2 or 0.1

4. Conduct a regression analysis and obtain the A and B factors. If the correlation factor is above 0.5, use the formula for forecasting.

5. Identify products with highly variable sales. Since there is no good forecasting technique for these products, the best strategy may be to stock subassemblies at a lower level. This is an area where qualitative forecasts may have to be used.

6. Identify those products with large volume and/or dollar sales and monitor them carefully. For these products ensure that the past data are purified of all one-time sales quantities. Add into the forecast any market intelligence on promotions, or other factors likely to affect sales.

7. Most forecasting software develops a model for each item based on its history. The model selected is the one with least deviation. If that model is still too inaccurate, try to tune it (see the following section), and until an acceptable level of accuracy is achieved, use a management forecast.

Dos and Don'ts

DOs

1. Do provide adequate training in the concepts of forecasting and how to use the software to forecast accurately. This knowledge has to be constantly reinforced.
2. Do develop a realistic expectation of the accuracy of forecasts.
3. Do use regression analysis for longer range and seasonal forecasts, after first determining that there is a good correlation factor (above 0.6).
4. Do use exponential smoothing for shorter range forecasts. Make sure the "α" factor is set based on the volatility of the product.
5. Do be wary of bias in both quantitative and qualitative forecasts.

Managing Forecasts in Manufacturing

Purpose and Description

Practitioners accept that most forecasts will be wrong. In spite of this they have to use forecasts to plan and drive their business. Forecasts have to be managed. A company is able to provide better customer service at a lower cost by eliminating forecasts or making them more accurate.

There is a need to adopt planning and manufacturing techniques that minimize or eliminate the need for forecasting. The focus of this approach is to book customer orders and deliver them in the timeframe required, instead of forecasting them. This is also the essence of just-in-time manufacturing. Where forecasts cannot be eliminated, forecast accuracy must be improved.

Accurate forecasts increase the probability of on-time order shipments and reduce the likelihood of surplus inventory.

Discussion—Approaches to Managing Forecasts

Forecast management in manufacturing needs to be approached on three fronts. The *first front* is to attempt to reduce or eliminate the need for forecasting. In order to do this, the reason for forecasting needs to be restated. A forecast is required only when the customer needs an item in less time than it takes to make the item. In other words the time to procure materials and manufacture (P) exceeds the time to deliver (D). This problem can be resolved in one of two ways. First it can be solved by obtaining a customer order in advance with sufficient time to make the product. Building to demand entails getting the customer to commit to an order outside the manufacturing lead time. Second, it can be solved by reducing the manufacturing time down to the customer's required time. Reducing manufacturing lead time means reducing P until it is less than delivery time D.

Both of these approaches are being used extensively. They are the preferred way to resolve the forecasting challenge, as the best forecast is no forecast. As mentioned before, this approach is the essence of just-in-time manufacturing.

The *second front* in managing forecasts is more conventional. It is by improving forecast accuracy. This approach must start by ensuring that there is joint responsibility between marketing and operations for both the accuracy of the forecast and the level of the inventory. Most forecasting techniques are based on projecting the past, with the presumption that the future will be an extension of the past. However, statistical forecasts should be modified by management intelligence and judgment. Since the forecast is based on past results, it is essential that the historical record is purged of all one-time or unique events. Along with purging past data, known or expected future events should be added to the forecast projection. There are numerous software programs that will do the mathematical calculations, and all of them have numerous controls for adjusting and guiding the projection. The objective in operating the software is to select a set of parameters that will produce a forecast that will accurately predict the sales of the product.

The *third front* is used where forecasts cannot be eliminated, forecast accuracy cannot be obtained, and a high level of customer service is required. The approach requires demand uncertainty to be hedged with inventory (buffer or safety stock). This is the least desirable of the three approaches, but is used fairly extensively.

All three approaches or some combination of them may be used to manage forecasts.

Special Considerations in Forecasting

One of the problems in forecasting is the conversion of sales dollars to unit quantities. Companies establish forecasting requirements in dollars or units; the more progressive companies require a forecast in both. Since the master schedule has to be loaded in units, every effort should be made to get a forecast in units. Forecasting in units has the following advantages:

- It avoids conversion of dollars to units using an average price per unit.
- Operations and sales speak the same language.

- It avoids having to forecast an average selling price—thus requiring two forecasts, one for quantity and another for average price.
- It eliminates price inflation from misrepresenting sales performance.

Second, there is a decision to be made about whether to use orders taken or orders shipped as the forecast predictor. Usually orders taken or sales booked are the best predictors of meeting the customer needs, as they are independent of internal factors that may affect the actual revenue generated. Shipments, on the other hand, depend on the capability of a factory to ship or on the level of inventory, and thus they may not represent the level of actual demand. Orders taken is a better indication of the rate at which the customer needs the products, and should be reflected as such in the forecast history.

Finally, there is a major cultural change element involved in the introduction of a forecasting discipline. Forecast accuracy is generally not monitored and measured. There is a strong element of defensiveness and this can take the form of not supporting the implementation of the forecasting system. The system will be used successfully only if the fears of all groups are allayed by stressing the reasons for measuring forecast accuracy. The primary reason for measuring forecast accuracy is to understand cause of error(s) and to take corrective action. This aspect cannot be overstressed.

Elements of a Good Forecast

A good forecast:

1. Has an equal chance of being overstated or understated
2. Includes known future events, such as promotions, price increases, incentives, and so on
3. Has a range of maximum and minimum, or has a forecast and an estimated error
4. Is measured, reviewed and revised in a timely (monthly) manner
5. Does not reflect bias

Eliminating or Minimizing Forecasts by Obtaining Hard Orders (Make to Demand)

Process

1. Identify the main customers for each item or product family.
2. Establish an EDI link with the customer. They should be able to place orders directly and know the production status of their orders.
3. Establish a blanket contract with a supplier for an annual quantity with estimated monthly deliveries.

4. Attempt to enter into a partnership with customers to obtain their requirements well in advance of their need date.
5. Establish an agreed-to order lead time with key customers.
6. Manage demand when taking orders by trying to persuade the customer to place orders that require consistent deliveries.

Making to demand is one of the key objectives of just-in-time.

Minimizing or Eliminating Forecasts by Reducing the Manufacturing Time

Reducing the manufacturing lead time will be reviewed in detail in Unit III, Manufacturing Execution. As the manufacturing lead time is reduced, the gap between the time to make an item (P), and the time the customer is willing to wait (D), is also reduced. One of the axioms of forecasting is that forecasts made closer in are more accurate than those made further out. Thus, reducing manufacturing lead time leads to more accurate forecasts. As the lead time is reduced, it comes into the time interval the customer is willing to wait, and the need to forecast is eliminated. Two examples of companies that eliminated final product forecasts are Motorola and General Motors Saturn factory. Motorola reduced the manufacturing lead time of their pagers to hours, and satisfied a customer order on the day it was received. Saturn builds the final configuration of its cars in three days based on hard customer orders.

Developing Forecast Accuracy

Process

1. Form a sales and operations forecasting team consisting, at a minimum, of marketing, product managers, sales managers, materials planners, buyers, and the personnel who manage the forecasting software module.

2. Establish a sales forecasting policy. This should cover a monthly forecasting routine, including who forecasts and when; how it is input into the factory master schedule; and how forecast accuracy is measured and communicated. (Sample policy is in Appendix 5.1.)

3. Group items into families. A family may consist of items that have common product life cycles and sales characteristics. Take care not to group families based on common physical product characteristics (size, shape, form, and so on). Remember that forecasts at a group or family level are more accurate than forecasts at an individual or item level.

4. Understand the sales characteristics of the product. Does it have a stable demand, or is the demand erratic? Does it have a linear trend and seasonality? Is it consumed by individuals or by institutions?

5. Determine the stage of the product life cycle the item is in. Classify items into one of four stages: introduction, growth, maturity, or decline.

6. Classify items into A, B, and C categories based on the revenue they generate.

7. Invest in a sales forecasting software package. A PC-based program is adequate. Ensure that the forecasting packet can be bridged to communicate with the order entry software and with the master scheduling software. Most often the MRP II (Manufacturing Resources Planning) package contains a forecasting module.

8. Load up to 36 months of sales history into the module. For new groups or items, borrow the history from a similar product and scale to size. Ideally, at least 24 months of history is required to develop a good base. If there is no seasonality the historical records may be reduced to 18 months.

9. Examine, modify, or purge sales history by item. Most software has mechanisms for establishing parameters and then making a mass change. (For example, set a demand filter for A items at 1.1. This will cut off any period-to-period demand variation exceeding 10 percent.) Identify and remove all obvious irregularities.

10. Run the forecasting module. The system will usually automatically set up a forecasting model best suited to the demand history of each item. Compare the system forecast with the actual demand. Measure the forecast error. Where errors exceed the acceptable limits, fine tune the system (see the following section).

11. Obtain forecasts by family and item from the forecasting team. Emphasis should be placed on identifying special events such as planned promotions, price increases, and other such changes. Enter these into the system and after the system is run, measure the forecast error. Take appropriate action to analyze and reduce the forecast error.

12. Develop a planning bill to forecast a product that has many end items based on a combination of assemblies and/or options. Calculate the requirement of the assemblies and options through a historical usage

percentage (this will be considered in detail in the master scheduling section).

13. Generate forecasts for at least 13 months into the future.

14. Work *with* the forecasting system, not against it.

Reducing Forecast Error

Process

1. Establish acceptable forecast error by the sales characteristics of an item and its importance. Items that have a known stable demand and high volume have a lower acceptable error. Suggested acceptable errors are:

   Stable demand   A items ±5%;   B items ±10%;   C items ±20%
   Unstable demand   A items ±15%;   B items ±25%;   C items ±40%

2. Load each item's sales history to within a few months of the present. Run the forecasting module. The system will use this history to develop a forecasting model for each product or item.

3. Measure the forecast error at the family level and the item level.

$$\text{Forecast accuracy }\% = \frac{(\text{Actual} - \text{Forecast})}{\text{Forecast}}$$

Forecast accuracy can be measured for a single period or for a three-period average. The forecast used for measurement must be frozen a month or two before the current month, depending on the manufacturing lead time (as determined in the policy).

4. Determine the variability of the forecast error, as this will indicate how stable an item is. Variability is best determined by calculating the forecast error MAD.

$$\text{MAD} = \frac{\text{Absolute sum of forecast errors}}{\text{Number of forecasts}}$$

Example:

Period	Forecast	Actual	Error
1	100	110	10
2	120	118	−2
3	130	139	+9

$$\text{MAD} = \frac{(10 + 2 + 9)}{3} = \frac{21}{3} = 7$$

A high MAD is indicative of high fluctuation in an item's demand.

5. Highlight the items needing their system forecast models tuned based on errors exceeding acceptable limits.

6. Tune the system. There are many ways of tuning a statistical forecasting model. The following is an example of one such method.
   The major elements to be tuned in a forecasting model are:

   ■ Smoothing sets for the *Permanent Component* (PC), the *Trend* (T), and the *Seasonality* (S) are a set of α values that determine the reactivity of the forecast. Basically low α values (0.05 to 0.15) allow the forecast model to react slowly to the actual results and so provide a more stable forecast. High reactivity is induced by α values of 0.3 or over, and is provided for volatile products, allowing the forecast to react quickly to the actual sales.

   ■ *Demand Filters* (DF) are a means of controlling variations by providing a check that limits the amount of the variation from one period to the next. A demand filter typically is a ratio of the new demand to the average of the old demand. Thus a DF of 1.1 will eliminate variations exceeding 10 percent of the last period. Stable items should have tight DFs to maintain more control, as variations can be examined and explained. DFs also prevent input errors, as an input error will trip the demand filter and can be corrected.

   ■ *Seasonality smoothing factors* are similar to the smoothing factors except that they are specific to seasonality. The principles used in setting the general smoothing factors are equally applicable in this case.

   ■ *Tracking signal* (TS). This is the ratio of the running sum of forecast errors and the MAD.

   Example using figures above: TS at end of Period 2:

   $$TS = \frac{(10 - 2)}{6} = \frac{8}{6} = 1.3$$

   TS at end of Period 3:

   $$TS = \frac{(10 - 2 + 9)}{7} = \frac{17}{7} = 2.4$$

A value above 1.5 is an indication of bias, and a signal that an item's forecasting model needs to be tuned. Tracking signals are particularly useful for determining turning points in an item's life cycle as they indicate that the actual sales are no longer following a historical trend.

7. Run the system again and compare the system forecasts with the actual sales for the next period. (Remember that the history was not loaded to the current period: Step 2.) Tuning is an iterative process.

The system should be able to develop a reasonably reliable model of most A and B items. This will obviate the need for the forecasting team to come up with qualitative forecasts for every item. The team can then concentrate on those items with inaccurate forecasts.

Dos and Don'ts

DOs

1. Do ensure that marketing and sales are on board in the development of the forecasting process.
2. Do train all parties to understand the principles and precepts of forecasting. Ensure that users know how to use the software effectively. This is a critical requirement.
3. Do set up a multi-functional team of each major product family for obtaining the best forecasting results.
4. Do measure forecast accuracy regularly and ensure that it is analyzed and corrective action taken. Focus on the corrective action, *not* on the error.
5. Do remember that the system develops a forecasting model for each item based on its sales history. Use this model if it is accurate. Tune it if it is not.
6. Do use the forecasting team to identify special events or trends that are likely to happen.
7. Do set reasonable expectations. Expecting forecast accuracy of over 90 percent is bound to lead to frustration. Set accuracy goals based on individual product demand volume and stability.

DON'Ts

8. Don't appear to second guess the results when reviewing forecasts. Analyze and understand the reasons and take corrective action to avoid recurrence.
9. Don't have conflicting objectives between the major departments involved in forecasting. Ensure that marketing and sales and operations have joint customer service and inventory goals.
10. Don't attempt to forecast products offered with many options. Forecast these items at the subassembly level and assemble on demand.
11. Don't measure forecast error in a manner that appears to focus on blame. This is perhaps the most important caveat in implementing a successful forecasting program.
12. Don't overrule the system by dismissing the systems forecast and replacing it by an arbitrary management judgment.
13. Don't compensate for forecast error by providing large amounts of safety stock.
14. Don't forget that the best forecast is no forecast. Ask customers when they want their products. Reduce manufacturing lead time. *This is the essence of JIT.*

Appendix 5.1: Policy for Generating Sales Forecasts and Measuring Forecast Error

A. Example of Policy Letter from Head of Marketing and Sales

To: All Marketing and Sales Managers
From: Vice president, Marketing and Sales
Subject: Forecasting

You are aware that forecasts are the primary driver of the factory's schedule and have a major effect on service and inventory.

On the basis of recent discussions with our software consultant and her assessment of our progress and our internal deliberations, I would like to set forth a few guidelines for our approach to forecasting.

1. Forecasting is a marketing priority.

2. We need to review forecasts in units by item (on an exception basis) monthly.

3. Forecasts should be made strictly on our expectations of business taken, and *not* to compensate for production or inventory problems.

4. Currently we forecast three months ahead. We should attempt to extend this horizon for promotions, new product introductions, and so on, to give the factories more reaction time (even if we do not have exact figures).

5. As a corollary, our forecast for the *current* month should contain *few*, if any, changes from that forecast a month ago (changes should be made only if there is some major business change and should be authorized by me).

6. We should use the option to extend the item forecast by price and roll the result up to the product group. We will then have a more accurate means of comparing the dollar value of our item forecasts with our product group forecasts.

The software consultant has noted that Marketing and Sales have become more involved with the forecasting process, and their interaction with the system has been on a more positive basis.

This is a very encouraging trend. Please keep it up.

B. Procedure for the Forecasting System

Note: Reviews are conducted for last month's errors and forecasts made for next month and beyond sales. This month's forecasts are frozen and being worked on by the factory. Change in this month's forecasts can only be authorized by agreement of Marketing and Production V.P.s.

Monthly Cycle

1st week   Month end update of system with last month's actuals.
Item forecast reports to product managers.
Product group level report for SOP* meeting.
Error measurement report at product group level for SOP meeting.

2nd week   Item promotions, changes, etc., input and rolled up for next month and beyond. (Marketing)
SOP forecast at product group level entered for next month and beyond. (Marketing)
Results of product group forecast on items reviewed. (Marketing)
Approved item forecasts linked to factory master schedule.

3rd week   Factories review forecast for feasibility and advise marketing on items that have capacity problems. (Factory)
Problem items reviewed and finalized. (Marketing)
Final input of next month's items to master schedule made and frozen.

4th week   Work on system forecast errors. (Operations)
Work on management errors. (Marketing)

*SOP—Sales and Operations Planning. Monthly coordination meeting
Note: The system is operated by materials management (Operations). Pyramid forecasting system having product family at top level and items at lower level.

Measurement of Forecast Accuracy

Running Three period error for management and system.

$$\text{Forecast Error } \% = \frac{\text{Actual} - \text{Forecast}}{\text{Forecast}}$$

6 Production Planning

Summary of Chapter

The chapter starts with an overview and a review of basic principles governing the production planning process. Production planning uses a specific format and this is described in the next section. The chapter ends with a section on how to manage the whole production planning process to develop a single company-wide, acceptable manufacturing plan.

Contents

- Overview and basic principles
- Production planning format and technique
- Managing production planning

Relationship of Production Planning to Other Major Manufacturing Planning Processes

Production Planning receives input from the *forecasting* process, and, by establishing a factory rate by product family, provides a boundary to the *master scheduling* process. The production plan must be consistent with the company's business plan and its manufacturing strategy.

Overview and Basic Principles

Purpose and Description

The primary purpose of production planning is to establish production rates that will achieve management's objective of satisfying customer demand by maintaining, raising, or lowering inventories or backlogs, while attempting to keep the workforce relatively stable.

The APICS dictionary defines *production planning* as the function of setting the overall level of manufacturing output to satisfy the current planned level of sales, while meeting general business objectives of customer service, profitability, and productivity, as expressed in the business plan.

Executive View

Production planning is the best opportunity that top management has of ensuring that the manufacturing plans are in line with the company's strategic objectives and that there is a commitment to the strategy. If a company has a strategic objective to provide a high level of customer service, the production plan should provide sufficient inventory to ensure this. The accuracy of forecasts and the performance of the factory can both be reviewed at the sales and operations planning meeting. Special efforts to sell products should be planned and communicated. Finally, the concept and calculations required are simple and effective.

A company that does not have a production planning process, conducted through a sales and operations meeting (SOP), is missing one of the most effective means of controlling its business.

Basic Principles

1. A production plan (PP) translates a business plan into a factory run rate by product family. A production plan determines the limits to which a factory is scheduled, since the aggregate of all master scheduled items of a family should not exceed that of the production plan for that family.

   Example: Production planning may be applied to a family of economy compact cars and the SOP may determine that 800 such compact cars a day should be manufactured to meet the predicted sales. Individual compact cars—Cavalier, Baretta, Lumina—are master scheduled, and their monthly total should not exceed 800.

2. The production plan is generated at a monthly sales and operations meeting, attended by the functional heads of sales, marketing, opera-

tions, engineering, finance, and middle-level managers representing these functions.

3. The production plan is usually stated in dollars ($) of business taken or shipments planned. The preferred unit of measure should be that used by the factory's master schedule (usually units). If the plan is developed in dollars it must be converted into units using an average selling price.

4. The production planning may be applied to a make-to-stock environment, using inventories to balance sales forecasts and production, or to a make-to-order environment, using backlog of orders to match sales and production. It can use both approaches, but only one can be applied by family. Combining forecasts and customer orders is done at the master schedule level. (Note: Backlog are orders booked, but not yet due to ship.)

> Example: Consumer products purchased off the shelf is make to stock, while custom-made furniture is make to order.

5. The planned inventory or order backlog should be consistent with the business strategy of the company.

6. While determining the plan, the capacity of the factory must be considered. This is done by matching the proposed production rate against the capacity of critical resources. The capacity used should be based on past performance of the factory. (The subject is covered in detail in the Rough-Cut Capacity Planning section of Master Scheduling.)

7. The sales and operations planning (SOP) meeting is a natural vehicle for sales forecasting, as sales forecasts comprise one of the basic inputs to the production plan.

Families or Product Groups

Families are determined by grouping items that have a proportionally consistent impact on sales and manufacturing. Families should be as broad as possible, consistent with similarities in the manufacturing process of the individual members and that the product line can be handled by one sales group. Where there is inconsistency between the manufacturing process and the sales grouping, the family has to be redefined to be consistent across manufacturing and sales. The redefinition may reduce the scope and size of the family to accommodate specific requirements.

Example: A way of classifying a family of automobiles is by style—luxury, family, or economy—and within the style by size—subcompact, compact, and so on. If families are classified by size only, there may be inconsistent sales requirements across the sizes, as a mid-size can be luxury, family, or economy, each requiring different sales approaches and skills.

Similar arguments can be used for families that may have sales consistency—office furniture, but have manufacturing disconnects—steel cabinets, wooden tables, and foam chairs. In this case, the furniture group would have separate families for steel, wooden, and foam furniture.

Production Planning Format and Technique

Purpose and Description

It is necessary to develop a standard format for production planning, whether used in a make-to-stock environment or in a make-to-order environment. Such a format is described in this section.

A production plan is developed at a sales and operating planning (SOP) meeting. The purpose of a production plan is to enable management to convert a business plan into a factory production rate by product family, and determine what inventory and/or backlog the company should develop. The *production plan* is also referred to as the *sales and operations plan* or the *PSI (Production, Sales and Inventory) Plan*.

Discussion

The terms describing the process vary from *production planning* (PP) to *sales and operations planning* (SOP), to *production, sales, and inventory* (PSI) planning. In this book, the term *production planning* covers the process, while the *production plan* is the format used in the process. The *sales and operating meeting* is the forum in which the production plan is developed.

One of the most difficult issues in production planning is the unit of measure. Typically the level of attendance at the PP session and the method of forecasting (at the product group or family level) makes the use of dollars ($) the most understandable and convenient measure. In order to set production rates and check available capacity, these dollars at the family level must be converted into units. Since many items make a family and each has different prices, the only way to convert a dollar number at the family level to a unit number is by using an average price. The average price is a historical average and is subject to variation, depending on the family's item mix. Thus the method has a built-in source of inaccuracy. The alternative is to try to get family forecasts in units. This is difficult because the production plan is tied to the business plan in dollars, and most sales and marketing groups use dollars as their measurement in planning and execution.

Process for Developing and Using a Production Plan

Format and Logic for a Production Plan in a Make-to-Stock Environment:

In a make-to-stock environment, since a customer buys off-the-shelf parts, a certain quantity of finished goods inventory has to be planned for and be available. The quantity of inventory required depends on the customer service level that the company is committed to maintain, on the accuracy of the forecasts, and on how quickly the factory can respond to new orders. Thus in developing the SOP for a stocked family, many basic policy decisions have to be in place, some tied to the strategy of the company. The format of a typical production plan is shown in Table 6.1:

Table 6.1 Production Planning—Make to Stock
FAMILY: PRODUCT X

Sales	−3	−2	−1	Current	+1	+2	+3	+4	+5
Forecast	100	100	100	90	90	90	100	100	100
Actual	90	80	85						
Deviation	−10	−20	−15						
Cumulative Deviation 0	−10	−30	−45						

Production	−3	−2	−1	Current	+1	+2	+3	+4	+5
Planned	100	100	100						
Actual	110	110	101						
Deviation	+10	+10	+1						
Cumulative Deviation 0	+10	+20	+21						

Inventory	−3	−2	−1	Current	+1	+2	+3	+4	+5
Planned	46	46	46						
Actual Start 46	66	96	112						
Cumulative Deviation	+20	+50	+66						

Formula used for Sales and for Production:
Deviation = Actual − Forecast Sales or Planned Production
Cumulative Deviation = Cumulative Deviation (last period) + Actual − Forecast
Formula used for Inventory:
Actual Inventory = Inventory (last period) + Actual Production − Actual Sales
Cumulative Deviation = Cumulative Production − Cumulative Forecast

At the SOP meeting in the current month, the results of the previous months (-1, -2, and -3) are analyzed. In the case of Product X, there has been a sales decrease and the factory has overproduced, leading to an inventory increase of 66 units and an inventory balance of 112 units. Sales reduce their forecast for the next three months before bringing it back to the original forecast. They also do not want the inventory to exceed 46 units.

Planning for this scenario is as follows:

Total sales plan for next 6 months = 570 units
(3 months at 90 and 3 months at 100)
Planned Inventory at end of 6th month = 46 units
Ending Inventory = Begin Inventory + Required Production − Sales, or
Required Production = Ending Inventory + Sales − Begin Inventory, and

$$\text{Production Rate} = \frac{\text{Required Production}}{\text{Number of periods}}$$

In this case: Required Production = $46 + 570 - 112 = 504$, and

$$\text{Production Rate} = \frac{504}{6} = 84 \text{ units per period}$$

The factory production rate will be set to 84 units per month.

The reduction in factory production from 100 units to 84 units may be too steep, leading to underutilization of resources. In this case, the reduction can be done in two stages.

First three periods: Reduce the production rate to 90
Then required production = $3 \times 90 = 270$, and the expected sales is 270 (as forecast).
Ending Inventory after three periods = Inventory (last period) + Production − Sales
= $112 + 270 - 270 = 112$
Next three periods:
Required Production = End Inventory + Sales − Begin Inventory,
= $46 + 300 - 112 = 234$ and

$$\text{Production Rate} = \frac{234}{3} = 78.$$

A more reasonable reduction of the factory rate reduction from 100 to 90 over three months and then from 90 to 78 over the next three months.

Format and Logic for a Production Plan in a Make-to-Order Environment:
In a make-to-order environment, orders are started only after they are received. There is a requirement to deliver the order when the customer

wants it, and often getting an order depends on how quickly the factory can respond. Raw material or subassemblies may have to be stocked. The number of orders on the books or backlog is used to adjust the sales and production plan. In developing the production plan for a make-to-order family, many basic policy decisions have to be in place, some tied to the strategy of the company.

The format of a typical production plan is shown in Table 6.2.

Table 6.2 Product Planning—Made to Order

Sales	−3	−2	−1	Current	+1	+2	+3	+4	+5
Orders	100	100	100	90	90	90	100	100	100
Actual	90	80	85						
Deviation	−10	−20	−15						
Cumulative Deviation 0	−10	−30	−45						

Production	−3	−2	−1	Current	+1	+2	+3	+4	+5
Planned	100	100	100						
Actual	110	110	101						
Deviation	+10	+10	+1						
Cumulative Deviation 0	+10	+20	+21						

Backlog	−3	−2	−1	Current	+1	+2	+3	+4	+5
Planned	76	76	76						
Actual 76	56	26	10						
Cumulative Deviation	−20	−50	−66						

Formula used for Sales and Production:
  Deviation = Actual − Sales Orders (Planned Production)
  Cumulative Deviation = Cumulative Deviation (last period) + Actual − Orders (Planned)
Formula used for Backlog:
  Actual Backlog = Backlog (last period) + Actual Orders − Actual Production
  Cumulative Deviation = Cumulative Orders Deviation − Cumulative Production Deviation

Note: In calculating the deviation in a make-to-order environment, the production numbers are reduced from the sum of backlog and orders taken. As orders are taken, backlog is increased, while production reduces the backlog. In a make-to-stock product, as orders are filled inventory is decreased, while production increases inventory.

If the SOP forecasts orders taken in the next 6 months to be 570 units, and if the backlog at the end of the sixth month is set at 28, an *increase* of 18 units from 10 in month one, then orders taken must exceed production by 18—that is, production must be $(570 - 18) = 552$ units.

Planning for this scenario is as follows:

Total Sales Plan for next 6 months = 570 units

Planned Backlog at end of 6th month = 28 units

Ending Backlog = Begin Backlog + Sales − Required Production, or

Required Production = Begin Backlog + Sales − End Backlog, and

$$\text{Production Rate} = \frac{\text{Required Production}}{\text{Number of periods}}$$

In this case, Required Production = $10 + 570 - 28 = 552$, and

$$\text{Production Rate} = \frac{552}{6} = 92 \text{ units per period}$$

The factory production rate will be set to 92 units per month.

Note: Notice the switch in production to being subtracted from beginning backlog, as compared to make to stock, where production is added to inventory.

Dos and Don'ts

DOs

1. Do ensure that *all* attendees of the sales and operations meeting understand the calculations on which the production plan is prepared.

2. Do see that the accepted orders, production, and inventory or backlog numbers are decided and agreed upon at the meeting.

DON'Ts

3. Don't revise the sales and production numbers more than once each quarter (unless there is a crisis).

4. Don't make large changes in the month-to-month production rate. Try to work to a level or constant rate.

Managing Production Planning

Purpose and Description

The production plan provides an opportunity for the company's executives to lead the company and execute the business strategy.

The production planning process must be formalized and institutionalized to be an effective means of managing a business. The production plan represents a unified game plan for the entire company. Through a sales and operations meeting, a company is able to translate its business plan into a production plan, and develop a production rate by product family for the factory to follow.

Discussion

Many companies have the sales and operation planning (SOP) meeting led by the chief executive and have the functional directors attend, along with the appropriate departmental managers. This demonstrates the company's commitment to the planning process and to the business plan. The production planning process develops a single set of numbers every month that is called the *production plan*. This plan is reviewed, bought into, and followed by the entire company. Good production plans establish a company's backlog and inventory. Companies that have a formal MRP II (Manufacturing Resource Planning) system in place will need to incorporate the production planning process as an integral part of the whole system. There should be a formal production planning company policy approved by the chief executive. (A sample of a production planning policy may be seen in Appendix 6-1.)

Process for Production Planning

1. Determine who will attend the monthly sales and operating meeting. As noted above, to be truly effective, the chief executive and his or her immediate staff should attend along with all appropriate managers. Product and sales managers should attend, as should the factory managers, materials managers, and the master scheduler.

2. Provide adequate training to *all* the participants. Training should include an overview of MRP II and a detailed understanding of the mechanics of the production planning process. An overview of master scheduling (MS) and material requirements planning (MRP) is also very useful.

3. Define product families and group all items into a family. Items in a family must have common manufacturing characteristics and should affect costs and revenues in a consistently proportional basis. Every attempt should be made to express the production plan in the same unit of measure as the factory item schedule. This is not always easy to do, since dollars are the thinking medium of the top executives.

However, agreement should be obtained on a conversion ratio (a selling price for converting sales dollars to units).

4. Decide upon a format and ensure that all participants are familiar and comfortable with the use of the format. This should be part of the training (see the section on production planning format).

5. Decide on the production planning strategy that will be used.

    ■ Level strategy: The production rate is kept constant. This helps factory planning and production, but may lead to excess inventory.

    ■ Chase strategy: The inventory level is kept constant and the production rate is based on what is withdrawn from inventory. This can only be adopted with a flexible manufacturing process.

    ■ Compromise strategy: This combines level and chase. It helps a company to revise its planning based on actual results. Provided the production rate is not changed too frequently (once in three months), this approach is the best.

6. Develop a production planning policy and a sales and operations meeting agenda. This should include:

    ■ Objective of the production planning process (see discussion).

    ■ Personnel expected to attend (see above).

    ■ The schedule of the monthly meeting (should be on the same day and time every month).

    ■ Format to be used to develop the plan and the mechanics for calculating the inventory or backlog and the deviations (see last section).

    ■ The planning horizon in months (should be at least equal to the longest cumulative lead time for purchasing and manufacturing).

    ■ How the production rate will be verified to ensure that capacity is available and that the capacity used is the demonstrated capacity, that is, that the factory has actually produced and sustained this level. Include what authorizations are required to make additional capacity available.

    ■ A guide on what deviations from plan for which families are acceptable, and what deviations must be reviewed.

- A guide on what changes are allowed and when (time zones). Include what authorization is required to supersede these rules and make changes.

- A guide on how the production plan for a family will be reconciled with the individual items of that family. Include an understanding of how these items are planned for the master schedule.

7. Conduct a pilot production plan on a few well-known and stable families. Demonstrate the effectiveness of the process.

8. Implement full production planning. This should include:

   - Obtaining sales forecasts by family for the production planning horizon.
   - Deciding the period ending inventory or backlog.
   - Establishing capacity limitations by family.
   - Developing production rates by family.
   - Identifying unusual and new sales trends.
   - Identifying new product implementation.

9. Establish a set of measurements to be reviewed and published at each meeting. These should include, by product family:

   - Actual sales versus forecast sales.
   - Actual production versus planned production.
   - Actual inventory versus planned inventory.
   - Actual backlog versus planned backlog.
   - Changes made within a specified "no change" period (typically this is the current month).
   - Actual capacity used, including special capacity developed.

10. Issue a detailed fact sheet prior to the SOP meeting. This should be compiled by the departmental managers meeting prior to the SOP. At this pre-SOP meeting, problems should be identified and solutions formulated.

11. Conduct an effective meeting. As referred to before, this meeting can become a confrontational and fault-finding session. It is essential that no attempt is made to blame individuals or departments. Facts should be established and reasons for explaining the facts should be determined with a view to improving future planning. The focus should be on future months.

12. Follow up the meeting by publishing detailed minutes that include action items agreed upon.

A sample of a production plan policy may be seen in Appendix 6.1.

Dos and Don'ts

DOs

1. Do ensure that the executive management team of a company participates in the production planning process in a sales and operations meeting, and develops and commits to the production plan for the company. This is an imperative.
2. Do formulate and issue a production planning policy covering roles, responsibilities, rules and authorizations. (See Appendix 6.1).
3. Do try and get numbers in units. This will avoid complicated conversions from dollars to units for production numbers.
4. Do remember to check the production rates set against critical capacity constraints. Ensure that the capacity considered is the demonstrated capacity.
5. Do reconcile the master schedule with the relevant production plan family.
6. Do try to keep a level load on the factory, even if it means developing production rates by quarter.
7. Do ensure the plan is visibly supported by all functions once it is accepted.

8. Do make sure the previous month's results are reviewed. Reasons for sales and manufacturing not meeting target should be understood *in a constructive manner.*
9. Do insist that parties come to the meeting to suggest solutions to problems uncovered in a review conducted before the SOP meeting.

DON'Ts

10. Don't attempt to start the production planning process without first training *all* parties in understanding the interrelationships between the business plan, the production plan and the master schedule. Also ensure that all participants are comfortable with the format used.
11. Don't allow the meeting to develop into a confrontational session, with each party blaming the other. One of the best ways to avoid this is to make functional groups jointly responsible for common targets.

Appendix 6.1: Sample Production Plan Policy

1. The objective of the production planning process is to develop a company sales and operations strategy by product family to optimize production and capacity.

2. The production plan will be developed at a monthly sales and operations meeting, held on the second fiscal Tuesday of every month. Attendees should include:

   - General Manager—ensures that the plan is congruent with the business plan.
   - Vice President, Marketing and Sales—ensures that strategic issues are covered.
   - Vice President, Operations—checks for capacity utilization and level scheduling.
   - Vice President, Finance—watches cash flow and inventory investment.
   - Vice President, Engineering—oversees the new-product introduction.
   - Marketing Managers for all product families—provide market intelligence to forecasts.
   - Regional Sales Managers—provide forecasts by product family.
   - Factory Manager—provides schedule and capacity information.
   - Materials and Purchasing Managers—provide inventory information.
   - Forecasting Manager—coordinates and runs the forecasting process.

3. The production plan will be recorded by product family, in monthly intervals, using a standard format. For make-to-stock products, inventory balances will be used to track production; in make-to-order products, backlog will be used for the same purpose.
   The production plan will be projected by month for 12 months, and by quarter for 2 quarters.

4. The sales and operations group will establish production rates for each family, within ±10% of the demonstrated capacity of the factory. Assumptions and reasons for major increases or decreases will be recorded. The aggregate of all the production rates should be consistent with the business plan projections.

5. Increase or decrease of long-term capacity requirements (space, labor, and equipment) will be identified and decided upon at the meeting.

6. The prior months, and quarters, performance will be reviewed and major variations accounted for and recorded. The forecast accuracy (fore-

cast versus actual) and the production performance (plan versus actual) will be measured.

7. The production plan will be used as input to:
   - Project monthly cash flow and inventories.
   - Establish revenue and capital budgets.
   - Establish the aggregates, by family, for the master schedule.

7

Master Scheduling

Summary of Chapter

This section begins with an overview that provides a general understanding of the master scheduling process. Managing demand, a key input to the master schedule, is described next and is followed by rough-cut capacity planning, a means of determining if the master schedule can be implemented. The format and logic of master scheduling is detailed and a special form of master scheduling, the final assembly schedule is described. The chapter ends with a section on managing the master schedule.

Contents

- Overview and basic concepts
- Managing demand
- Rough-cut capacity planning
- Master schedule: format and logic
- Final assembly scheduling
- Other uses for planning bills
- Managing the master schedule

Relationship of Master Schedule to Other Major Manufacturing Processes

The master schedule obtains its input (demand) from the forecasting and customer-ordering processes. It determines how this demand should be satisfied through a master production schedule. The combination of all master schedules for a product family should be equal to the production plan for that family. This schedule is fed to the rough-cut capacity requirements module, where it is checked for feasibility. The final "doable" version of the master production schedule drives the material requirements planning module to determine what components should be manufactured and purchased. Underlying the above planning approach is just-in-time's concept of making a product after it is ordered by a customer.

Overview and Basic Concepts

Purpose and Description

The master schedule is one of management's principal means of planning and controlling a manufacturing business. The challenge of master scheduling is to meet the customer demand while optimally using the company's resources. A master schedule provides management with an ability to:

- Anticipate, accumulate and satisfy customer demand.
- Activate resources required to meet the demand—manpower, machines, inventory, and cash.
- Have a common plan to coordinate the activities of sales, marketing, engineering, manufacturing, and finance.

A master schedule (MS) is the manufacturing plan, by item, of a factory. It obtains requirements, usually referred to as *demand*, from forecasts, customer orders and other warehouse or service orders. It compares these requirements with what is available in stock, what is being made or purchased, and determines what is still required. Items available and items on order are usually referred to as *supply*. A master schedule thus balances demand with supply. The master production schedule (MPS) is a line in the master schedule format which indicates how many items the factory is committed to make in a given period of time. The distinction between a master schedule and a master production schedule is worth repeating. The master schedule is the entire plan and includes all the demand, supply and ordering calculations; the *master production schedule* is the line in the master schedule that lists what the factory is expected to make. This master production schedule's quantity and timetable drives the material requirements planning (MRP) module to calculate the requirements of raw material, components, and assemblies.

In the 1990s there has been a thrust to manufacture to demand, or master schedule to hard customer orders. This approach is the essence of just-in-time (JIT). A customer order signals the shop or the supplier to make and deliver the required product.

Discussion

Executive View: Master scheduling (MS) is an old label, but a comparatively new planning technique. Even today many companies on MRP II do not use master scheduling; instead the material requirements planning (MRP) module is used to satisfy the planning needs of the factory. Master scheduling offers some distinctive advantages that complement and strengthen MRP. A master schedule can distinguish between and control different types of demand—customer orders, forecasts, service orders, and so on; MRP can distinguish between different types of demand, but cannot prevent them from being combined into one gross requirement. Master scheduling has many useful techniques for controlling demand not found on MRP, such as time fences and consumption logic between forecasts and customer orders (covered in the managing demand section). The fact that the master schedule can separate and control demand enables it to be decoupled from MRP and the shop, and allows it to be assessed before deciding to release it to operations. Contrast this to MRP, where all demand is lumped into gross requirements and then is exploded into orders for shop manufacture or purchasing. Finally, master scheduling enables a company to give customers a delivery commitment by comparing what is being made to what has been booked.

Master scheduling is a negotiation process, between the principal players of a company—marketing, sales, engineering, finance, and operations. Sales and marketing must realize their forecasts with customer orders; engineering must ensure that designs are ready for manufacture as advertised; operations must manufacture to meet a customer's schedule, and finance must ensure that sufficient cash is available to make the plans work. The master production schedule (MPS) is the factory schedule, the negotiated result of master scheduling process, calculated by the master schedule's logic. The MPS should reflect not only what supply orders are required to meet the demand placed on the factory, but also that the factory has the capacity to meet the demand.

Since the master schedule develops its plans based on inventory available and order due dates, it is critical that this information is accurate and up to date. Similarly, forecasts must be accurate to enable the master schedule to satisfy customer demands effectively.

Further, it is essential that the aggregate of master scheduled items in a product family be consistent with the production plan for that family (see production planning chapter).

> Example: If the production plan of a GM car factory calls for 1000 compact cars, the master schedules for specific compact cars should not have more than 1000 compact cars—500 Novas, 200 Cavaliers and 300 Chevettes.

Unfortunately, in many industries, this check on aggregate capacity is not conducted and the result is that many factories run an MPS that is *overloaded*, that is, there is insufficient capacity to produce what is scheduled. Overloading a master production schedule leads to delivery failure and backorders. This problem will be treated in detail in the managing the master schedule section of this chapter.

The MPS drives the MRP which in turn plans all the orders on the factory and on purchasing. The MPS is also checked against availability of critical resources (rough-cut capacity planning).

Understanding a company's manufacturing environment is necessary to ensure that master scheduling is performed at the appropriate level.

Manufacturing Environments

Manufacturing environments are classified by the way a company decides to meet its customer demand and are generally divided into five categories:

- Make to Stock (MTS). These are items made to a forecast and shipped off the shelf. End items are manufactured and stocked. Consumer products are examples of this category. Master scheduling is done at the end item level. The schedule is heavily dependent on forecasts to determine what to make.

- Assemble to Order (ATO). Here a customer exercises a choice from several options, and these options are combined to form the end product. In this environment, a limited number of assemblies can be combined into many end products. Automobiles are examples of this category, where a menu of options allows the ordering of a large number of end configurations. The major subassemblies are master scheduled and, in addition to the master schedule, a *final assembly schedule* is generated. The final assembly schedule consists of the customized assemblies, configured from various options specified by customers.

- Make to Order (MTO). These items are made to a specific customer requirement. Stocking of parts is made at the raw material level and master scheduling is done using planning bills on product families. For example, in a custom-made suit, fabric is scheduled and numerous tailored suits are made from the fabric, or a family of suits is planned.

- Engineer to Order (ETO). This is a form of MTO with the item being designed, before manufacturing, based on very special needs of a customer. A bridge, or a refinery may be an example of such a product.

- Make to Demand (MTD). Items are master scheduled and made after receiving a customer order or commitment. It is the predominant concept behind the just-in-time system of manufacturing. Both make to order and assemble to order are versions of a MTD environment.

Companies typically have products belonging to more than one category. Figure 7.1 shows the environments of a matrix that compares the typical manufacturing lead time with the product variability or variety.

Item Configurations

Products can also be classified by the structure of their bills of material (BOM) and there are generally three types. Their product configuration can be pictorially represented in Figure 7.2. Companies typically have products that belong to more than one type of configuration. The type of

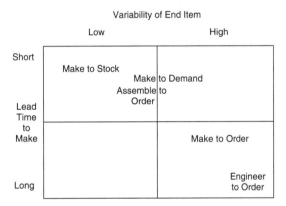

Figure 7.1. Manufacturing Environments

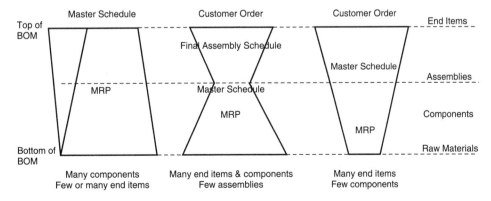

Figure 7.2. Item Configurations

configuration influences the level at which the product is master scheduled and the level at which the MRP is run.

Single-Level and Multi-Level Master Schedules

Master scheduling can be done at a single level, or a multi-level. At a single level, independent demand items are scheduled (these are usually products to be sold). This is the most commonly used level of the master schedule, and is the level that will be dealt with in this chapter. Examples include most consumer items. The multi-level is mostly used in assemble to order products, where several options exist and the customer order is configured to suit a customer's specific requirements. Multi-level master scheduling is done at the subassembly level and the end product is built on a final assembly schedule. Examples include autos. Multi-level master scheduling is also used with planning bills to avoid forecasting at an individual item level, and allows forecasting at a family level. (Two-level or multi-level master scheduling will be dealt with in a separate section.)

The format, logic and techniques used in both schedules are similar.

Process for Setting Up and Running a Master Schedule

1. Use master scheduling and not MRP to develop the factory schedule.

2. Determine the level to schedule to finished item, subassembly, or raw material.

3. Mark the items to be master scheduled. This is usually through an order policy code in the item master.

4. Minimize the number of items scheduled.

   End items for make to stock, subassemblies for assemble to order, and raw material or assemblies for make to order should be master scheduled. In addition, items that have a critical impact—such as high value or long lead time—should be master scheduled.

   The use of planning bills in two-level master scheduling will reduce a large number of individually master scheduled items.

5. Capture various types of demand.

   Typically these are forecasts, customer orders, independent demand (service orders), and warehouse orders. Ensure that marketing and sales are actively involved in this activity. Ensure that the demands are accurate.

6. Make sure that the inventory records and the ordering records are accurate and up to date. Conduct cycle count on the former and audits on the latter.

7. Determine who is authorized to make changes to the master schedule.
   Typically the master scheduler will work the mechanics of the schedule and handle routine decisions. There are times when the master schedule routines have to be superseded. For example, when giving a later order priority, or overloading the master schedule, appropriate authorization must be first obtained (usually given by the factory manager).

8. Set policies by item or family of items.
   These policies cover safety stock, order quantity (lot size), time fences, forecast consumption, and order reschedules (all these concepts will be covered in later sections). Here again there must be a schedule of authorization.

9. Run the master schedule against a rough-cut capacity plan. Provide additional capacity or reduce the master production schedule, if there is insufficient capacity.
   Use overtime, more people, more resources, subcontracting, or rescheduling to temporarily increase capacity.

10. Reschedule "overdue" or past due orders (whose delivery dates have expired). Understand that overdue orders represent a scheduling failure and need to be investigated for root cause. Even though they are rescheduled, these orders should be especially tracked as they probably represent an unhappy customer.

11. Make several iterations to check the validity of the data used and the feasibility of the master production schedule (rough-cut capacity planning is covered in the next section). Only when the master scheduler is comfortable that the master production is doable should it be released to MRP.

12. Review the performance of the master schedule weekly and monthly.
    This review should be conducted with appropriate manufacturing and marketing personnel and should analyze the performance of the factory and the accuracy of the demand forecasts.

Dos and Don'ts

DOs

1. Do provide an overview to *all* the management of the factory on how the master schedule works. Include marketing, sales, and finance in this education. Make sure all understand the relationship between the MPS and capacity.

2. Do ensure that forecast accuracy is measured and relayed back to marketing and sales. Investigate large or consistent forecast errors.

3. Do conduct cycle count of finished goods inventory to ensure that it is accurate. Also conduct audits of shop orders and purchase orders to confirm that they are correctly stated and their due dates are reliable.

4. Do ensure that the MPS is realistic—that is, it can be executed. This means tracking an MPS order's actual completion with its scheduled completion. Investigate reasons for repeated MPS failures, and take corrective action.

5. Do convince top management—the general manager or VP of operations—

not to move up the priority of an order at the request of customers or sales without the advice and consent of the master scheduler.

6. Do provide accountability for the execution of the master schedule.

DON'Ts

7. Don't allow changes to be made within the immediate future (no change should be made within one or two weeks for most short lead-time products).

8. Don't allow the master schedule to become overloaded and unrealistic. If this happens, the informal, expediting system will take over, and the master production schedule will not be followed.

9. Don't release the master schedule to the factory without first checking it against critical capacity for feasibility.

10. Don't forget that MRP II and JIT are complementary processes, with MRP having planning strengths and JIT having execution strengths.

Managing Demand

Purpose and Description

Demand drives a master schedule. In most businesses demand is uncertain and subject to variability in product, quantity and frequency. Demand needs to be managed to help optimize the performance of the master schedule.

The overview defined *demand* as a combination of forecasts, customer orders, orders for service parts and warehouse orders. Demand is what is required of the factory, and the words "demand" and "requirements" are used synonymously. Demand can be *independent*, and has to be forecast, or demand can be *dependent*, where it is related to an assembly, as determined by a bill of material, and has to be calculated. An automobile represents an independent demand, while tires used on the auto are dependent demand. Tires ordered as spares are examples of independent demand (an item can have both independent and dependent demand).

Discussion

Managing demand has two main thrusts:

- Reducing uncertainty
- Scheduling demand to the available capacity

Both these activities are needed to deliver the product in time to satisfy a customer's order.

Reducing Uncertainty

If a factory knew what and how much to make and was given enough time to make it, and *if* there were no problems with the quantity and quality of purchased or manufactured parts supply, then there would be few problems with uncertainty. Unfortunately uncertainty exists at all the above levels.

Examples of demand uncertainty include inaccurate forecasts, customers who order unpredictably in quantity and timing, unsure results of promotions and sales, and products that become popular or unpopular.

General techniques used to eliminate or reduce uncertainty are:

- Improve forecast accuracy.
- Work closely with customers to get their requirements, thereby minimizing the need for forecasts.
- Gain consensus between sales, marketing, and manufacturing on inventory or backlog goals.
- Generate orders by appropriate sales and marketing techniques to bridge the gap between forecasts and sales.
- Actively manage customer orders to meet the factory's available capacity through the use of demand filters, negotiations, etc.

The master scheduling body of knowledge has some specific tools that help to reduce uncertainty. These are:

A. Time fences
B. Lead time and lot size relationship

C. ABC stratification
D. Safety stock/safety capacity
E. Flexible capacity
F. Firm planned orders
G. Demand filters

Setting up and using each of these techniques is described below.

A. Time Fences: A master schedule time fence is a preset period of time within which only specific schedule changes are allowed to take place. Typically, there are two time fences used in the master schedule. These are:

Demand time fence (DTF): The DTF is defined as that period within the master schedule where changes are seldom permitted. Change can relate to order delivery or to product configuration. Usually a change to orders within the DTF requires the authorization of the materials or factory manager. Demand time fences are usually set to cover the assembly time of a product. Once an assembly is started any change will lead to delay or rejection or costly rework. Most master schedules do not consider forecast demand within this period and plan only on existing customer orders (unless the factory makes products to stock). The demand time fence period will depend on the product being made and on the cost/schedule impact of an order change or a product change. Demand time fences provide schedule stability as the shop orders do not usually change within the DTF.

Planning time fence (PTF): The PTF is defined as that period within which changes can be made only by the master scheduler and not automatically by the computer. Planning time fences cover the time it takes to fabricate a product's subassemblies, that is, the time from when the raw material and components are issued to make the parts of a product, until the time for the final assembly. There is usually a choice in the master schedule software to allow the demand quantity to be set equal to the greater of forecasts *or* customer orders.

> Example: An automobile usually has a DTF of two weeks, as during this time final assembly is made to a specific configuration of options, and a PTF of four weeks, during which time all major assemblies are fabricated.

	Demand time fence	**Planning time fence**	
Today	2 to 3 weeks	8 to 12 weeks	
Stage of Build	Assembly time	Fabrication time	Material Procurement time
Changes Made by	Emergency Change Factory Manager	Planned Change Master Scheduler	Open Change Computer Logic
Demand Considered	Hard Customer Orders	Customer Orders or Forecasts	All Planned Demand

Figure 7.3. Master Schedule Time Fences and their Qualification

Beyond the planning time fence: This is an unrestricted time where the master schedule software logic plans orders to meet the demand. It should cover the material procurement time. The master schedule planning horizon should extend beyond the time it takes to procure the longest lead time material.

By limiting change within the DTF and PTFs, reschedules of orders are reduced, and the master schedule is stabilized. Further working within the DTF usually ensures that only customer orders are made, and inventory carried can be minimized.

Process: Setting time fences
1. Determine the *demand time fence (DTF)* setting.

   This is the period when no change or only emergency change should be allowed. For make-to-stock items and assemble-to-order items the time fence is usually equal to the final assembly lead time, while for make to order it is usually set to cover the entire manufacturing lead time. The important consideration is the demand time fence should correspond to the period when a change to a product order would impact cost and schedule very adversely.

2. Determine the *planning time fence (PTF)* setting.

   This is the period when some change is allowed, provided it is controlled by the master scheduler. The period is usually equal to the total manufacturing lead time. The consideration is the impact of the change on the factory schedule.

3. Set the time fences by similar product groups or families.

   The software will have a screen that will prompt the user for entering a DTF and PTF interval.

4. Determine the demand policy or forecast consumption technique within the time fences.

The time fence setting affects the calculation of the master production schedule order. Usually demand within the demand time fence is limited to customer orders only, while demand within the planning time fence has options to consider the greater of forecasts or customer orders, or a combination of the two.

B. Lead Time and Lot Size Relationship: One of the most effective ways to manage the uncertainty of demand is to reduce the *lead time* required to manufacture a product. If a company can respond quickly to changes in demand, then the need to forecast is reduced and more demands of the master schedule can be based on hard customer orders. Although many factors influence the manufacturing lead time, the factor with the most influence is the *lot size*. In setting the master schedule there is a choice of options for setting the order quantity or lot size. The most common lot size choices are:

- *Economic lot size (EOQ).* This is the quantity that balances the ordering cost with the inventory carrying cost.

- *Fixed order quantity or pan size (FOQ).* This is a multiple of a machine or fixture setup quantity, as the product can best be made in these multiples.

- *Lot for lot (L4L).* The order size matches the exact demand by customer order and/or forecast.

- *Period days demand (POQ).* The order quantity is based on a preset number of days of demand.

All of these ordering techniques are covered in the materials requirement planning (MRP) chapter, but the technique that results in the smallest order quantities is the lot for lot.

Reduction of lead times by reducing lot sizes and manufacturing to customer orders are two of the principal practices of just-in-time manufacturing.

Process: Setting lot size: *The cardinal rule is to order the smallest natural lot size.* A natural lot size is one that does not cause major inefficiencies in manufacturing, and is usually equated to a day or a machine's economic setting whichever is smaller.

C. ABC Stratification: This is another extremely effective technique for managing demand. A items are the significant few (about 20 percent), that account for most of the value of the factory's production (about 80 percent), or critically affect the company's customers. It is essential that the demand on these items be carefully managed, so that there is no failure of schedule or misuse of costly product. A and B items should be tagged in the master schedule and they should have specific parameters on their DTF, PTF, optimum lot sizes, lead times, and safety stocks. Since A and B items have significant costs or are of a critical nature, they are usually manufactured on priority.

Process: The ABC concept is covered in detail in Chapter 10.

D. Safety Stock and/or Safety Capacity: *Safety stock* acts as a *buffer* from which unpredictable and variable customer demand can be satisfied. The level of safety stock maintained depends on the required level of customer service, the accuracy of forecasts and the lead time. Another technique of coping with unpredictable demand is to provide safety capacity. This implies that the master schedule is not loaded to 100 percent of the factory's capacity, but that some capacity is kept for reacting quickly to uncertain demands.

Safety stock helps stabilize the master schedule.

Process: The whole subject of safety stocks is covered in Chapter 10, and the provision of safety capacity is covered in the section MRP format and logic.

E. Flexible Capacity: *Flexible capacity* is the ability to operate manufacturing equipment and factory workers on different operations and at different rates. Flexible capacity is essential in responding to uncertainty and meeting changes in schedules. Flexibility should be extended to machines, shop operators and alternative processes. This subject is covered in the Execution section.

F. Firm Planned Orders: A *firm planned order (FPO)* is an item build or buy schedule that is fixed in both quantity and due date. The master scheduler creates a FPO or converts (firms) a computer-planned order. The master scheduler firms a planned order when it is determined that materials and capacity are available to make the items. The computer logic cannot automatically change FPOs, and, when the FPO is inconsistent with the MS or MRP logic message will be generated. A master scheduler may also respond to an action message to change the quantity or due date of a FPO, after first being satisfied that there is a real need for the change. FPOs help stabilize a master schedule, as they eliminate some of the nervousness associated with constant computerized rescheduling.

Process: Firming planned orders
1. Check planned orders which are approaching the PTF (coming within the manufacturing lead time).
2. Review action messages on the item. Ensure that the demand is still valid.
3. Determine availability of materials and capacity.
4. Firm a planned order, by using the appropriate computer screen. Usually the quantity and due date have to be confirmed.

G. Demand Filters: As the name implies, a *demand filter* is a quantity limit setting. When an order exceeds the filter setting, the filter is tripped, the order is aborted and has to be dealt with separately. If this check is not in place, a very large order can absorb most or all of the factory's available capacity leaving all other customers' needs unsatisfied, or an order can exceed the factory's total capacity and result in overdue orders. A demand filter also catches data entry errors. Having demand filters in place helps to remove some of the quantity variability in the demand and provides for a more stable master schedule. Orders that are aborted by the demand filter are negotiated with the customer to develop an even weekly or monthly schedule.

Process: Setting demand filters
1. Select the items that will be subjected to a demand filter. Items that have long lead times or are in heavy demand should be tagged.

2. Set the demand filter at around one week's production of an item.

3. Assign responsibility to a material planner to review the demand filter defaults every day, and ensure that the planner determines how much of an order can be accepted immediately and when the balance can be completed.

4. Ensure that the planned order schedule is discussed with the customer and agreement obtained.

5. Track all demand filter defaults separately.

6. Examine the possibility of increasing capacity for items that are constantly tripping their demand filter.

Dos and Don'ts (Reducing Uncertainty)

DOs

1. Do take special measures to reduce demand uncertainty.

2. Do involve customers in managing their orders.

3. Do determine inventory and backlog objectives for the company.

4. Do set demand filters for all A, B, and C items.

5. Do set the master schedule to plan small lot sizes.

6. Do involve marketing and sales in forecasts and forecast accuracy.

7. Do ensure that there is flexibility so that resources can be redeployed.

DON'Ts

8. Don't second-guess the forecasts of marketing and sales.

9. Don't set time fences uniformly for all products. Set the time fences to be consistent with the products manufacturing cycle.

10. Don't allow any changes to be made within the demand time fence, unless they pertain to safety or are needed to comply with the law.

11. Don't allow engineering changes to be made inside the planning time fence. Understand that is not always possible.

Scheduling Within Capacity

A master schedule that has enough capacity is a schedule that will be more stable, since fewer orders will require rescheduling. Checking the master schedule through rough-cut capacity planning is discussed next in this chapter.

Rough-Cut Capacity Planning

Purpose and Description

The purpose of *rough-cut capacity planning (RCCP)* is to:

1. Test the feasibility of the master production schedule before planning detailed material requirements.

2. Initiate action to adjust or increase the capacity of critical resources.

Rough-cut capacity planning (RCCP) is a technique that verifies a factory's capacity to execute the master schedule. It does this by accumulating all the work to be done in a critical resource over a period of time, and checking this accumulated workload against the capacity of that resource, over the same period.

Discussion

Most factories perform RCCP informally. The capacity of critical resources is known in terms of units of product. A master scheduler will only schedule a quantity that can be met by the critical resource, during a given period. Where there are multiple critical resources and where there is a multiplicity of product, the empirical approach comes up short.

A form of RCCP is also used to validate the production plan. When used for this purpose, it is called *resource requirement planning* and deals with longer range capacity resources, such as space and capital intensive machines. In determining the capacity of the critical resource the demonstrated capacity (the actual performance of the resource) should be considered and not the available capacity (the theoretical capability of the resource). See Table 7.1.

One of the main advantages of the rough-cut capacity plan is that it can be executed quickly, allowing several iterations of the master schedule to be made, before a final master schedule is accepted and loaded on the factory. Remember, critical resources may change every time the product mix of the master schedule changes. Master schedules that have been validated against critical resources are likely to be executed on time and this will lead to a higher customer-service rating. This process also adds credibility to the whole scheduling process as there is less change and the user is more confident of the capability of the process.

RCCP is particularly useful in repetitive production with fairly stable manufacturing processes. The technique is less effective on low volume

Table 7.1 Rough-Cut Capacity Planning and Resource Requirements Planning

Type of Planning	Validates	Time Horizon	Type of Additional Capacity Provided
Rough-Cut Planning	Master Production Schedule	Weeks to a month	Overtime, subcontracting temporary help, alternate routing
Resource Planning	Production Plan	Over 3 months	Capital purchases of equipment or facilities, permanent hiring of persons

product made in highly variable processes. This is because critical resources in such an environment are difficult to assess and normally adequate resources may temporarily become critical.

Process for Rough-Cut Capacity Planning

1. Identify critical resources. Develop a resource matrix that identifies all critical resources and the reasons why they are critical. This is called the *critical resource profile.* Typical types of critical resources are—all constraints, special tooling or equipment, special skills, and highly variable processes with possibility of rejections.

Example:	*Resource*	*Units*	*Reasons for criticality*
	Splicing	Hours	Highly specialized skill and difficult process
	Vacuum dry	Hours	Long (3-day process), limited fixtures

2. Determine the use of the key resource (operation time) in all the manufacturing operations of typical products. This is usually referred to as a product load profile.

Example:	*Product*	*Resource*	*Consumption*
	Pair	Splicing	8 splices @ 1 hour per splice
	Assembly	Splicing	8 splices @ 2 hours per splice

3. Determine when (how many days before completion) in the manufacturing cycle of the product the resource is used. This is necessary only when the product has a long lead time (over a month) to time phase the use of the resource or when the same product uses the resource at different phases in its process.

Example:	*Product*	*Lead time offset from due date*
	Pair splice	4 weeks (20 days)
	Assembly splice	3 weeks (15 days)

4. Calculate the available capacity and ascertain the demonstrated capacity (through actual performance) for all key resources.
   Remember available capacity is the theoretical or maximum capacity of a resource and demonstrated capacity is what the resource has actually performed to.

Example:

There are four pair splice stations on double shift and six assembly splice stations on single shift.

Resource	Resource area	WklyAvail Capacity	WklyDemon Capacity
Splicing	Pair Splice	$320 = (4 \times 40 \times 2)$ hrs	144 hrs = (18 pairs @ 8 hrs per pair)
Splicing	Assembly Splice	$240 = (6 \times 40)$ hrs	160 hrs = (10 assemblies @ 16 hrs per assembly)

5. Develop the total resource *requirements* for each key resource by multiplying the master scheduled quantities of all products that consume the resource by the resource time consumed by the product (from the product load profile).

Example:

Pair Weekly Ship Schedule	May	16 Pairs for 4 wks
	June	20 Pairs for 5 wks

Required Capacity   May=16 ships $\times$ 4 wks $\times$ 8 hrs/Pair
=512 hrs

June=20 ships $\times$ 5 wks $\times$ 8 hrs/Pair
=800 hrs

6. Compare required capacity with available capacity and demonstrated capacity.

Work Center Pair Splicing   Units Hours

Work Center	Units	Month	Required Capacity	Available Capacity	Demonstrated Capacity
Pair Splice	Hours	May	$16 \times 4 \times 8 = 512$	$320 \times 4 = 1280$	$144 \times 4 = 576$
		June	$20 \times 5 \times 8 = 800$	$320 \times 5 = 1600$	$144 \times 5 = 720$

Note: The process has been presumed to be uniform. Usually there will be varying capacity requirements for a product as there will be variations within the process. This may or may not be captured by the product load profile.

7. Identify potential overloads and underloads by critical resource by time period and develop plans to correct the imbalances.
(Imbalance = Required Capacity − Demonstrated Capacity).

In the example there is a shortfall of capacity in June based on the demonstrated capacity. There is sufficient available capacity. The effort should be directed at improving the demonstrated capacity. If this cannot be done the master schedule must be reduced.

8. Set up empirical limits on how much overload can be met by temporary increases of capacity and when additional resources are required. Generally up to 10 percent overload can be met with temporary capacity, while over a 20 percent overload should require providing additional resources. Action on the in-between percentages depends on how long the overload is likely to remain.

9. Implement a plan of action to increase or decrease capacity and measure results. Repeat the whole process until there is no major overload.

10. Ensure that the RCCP is run every time the master schedule mix changes, as critical resources will change with changing schedule mix.

Actions to Improve Demonstrated Capacity

1. Improve utilization or efficiency of the resource
   Improve productivity by reducing the cycle time (efficiency)
   Ensure that the resource is used for the full time available (utilization)

2. Adjust the load
   Reschedule—move orders in or out
   Split lots—reduce the lot size

3. Increase short-term capacity
   Use flexible operators from other areas
   Use overtime
   Provide additional or reduced shifts
   Engage in subcontracting
   Hire temporary employees

4. Increase long-term capacity
   Add another resource station
   Add more people

Dos and Don'ts

DOs

1. Do perform rough-cut capacity planning. Do keep it simple.

2. Do remember that product mix affects critical resources.

3. Do talk to the shop floor personnel to help determine critical resources.

4. Do concentrate on a few critical work centers.

5. Do analyze large differences between planned capacity and demonstrated capacity.

DON'Ts

6. Don't spend too much time in being too precise at any stage. This is rough cut or approximate planning.

7. Don't neglect to respond to areas that are underloaded, by training, transferring or reducing staffing.

Master Schedule: Format and Logic

Purpose and Description

The master schedule is presented in many formats, and the fact that a unique format serves a particular production environment, makes it an acceptable master schedule format. There is, however, need for a basic standard format that embodies the fundamental requirements of all master schedules.

The master schedule was earlier defined as a factory's manufacturing plan, compiled by matching supply to demand in product, quantity and delivery. It must therefore contain information on demand and supply. The master schedule format must also help the material planner to plan and finalize supply orders on the factory or purchasing, by providing information on what supply orders are in process and what orders need to be added, changed or deleted. Finally the master schedule should help determine when new customer orders can be satisfied, by providing information on open orders that have not been promised to customers.

Discussion

A common basic format was developed and is used by APICS in its certification examination on master scheduling. This format covers the following information:

■ Product identification
■ Scheduling guidelines
■ Time-phased variables

It is important to note that the term *master schedule* refers to the entire chart or format, whereas the term *master production schedule* is a single line on the format. The master production schedule line is the planned factory schedule or supply line. The terms master schedule and master production schedule are often used synonymously, but their difference is useful and is followed in this book.

Product Identification: This information is found in the header of each item. Every master scheduled item has a unique part number identifying it, together with a short description. Other information can include the revision level of the part number, the unit of measure (UOM), the buyer/planner (BP) responsible for the item, the ABC code, and a product or group code.

Scheduling Guidelines: This information is also contained in the header of each item. The master schedule is usually displayed in time buckets, usually weeks, the time interval being an option in most systems. Order due dates are usually a shop calendar date. The report also identifies the run date. It is useful to show weeks in the near time and months and quarters out in the planning horizon. Other data that governs the operation of the master schedule includes: current on hand inventory quantity and its location, safety stock, ordering policy or lot size rule used, and scrap or yield loss.

The setting of the demand time fence and the planning time fence, both from period one, is captured in the scheduling information, along with the item's lead time. Lead time can consist of various parts, such as assembly lead time, fixed lead time, unit lead time and dock to stock lead time. Finally this section contains the planning horizon and the reschedule sensitivity. All of these elements have been described in Appendix 7.1.

Time-Phased Variables: *Time phasing* is the technique of expressing future supply and demand by time period. The master schedule is a time phased planning chart, with the time periods at the top of each column.

The information rows making up a basic master schedule are: forecasts, customer orders, and the factory supply or master production schedule. In addition, there are two lines that are calculated by the master schedule program and these are the *Projected Available Balance (PAB)* and the *Available to Promise (ATP)*. Both concepts will be described in the next section.

Master Schedule Format

A sample master schedule is shown below:

Sample Master Schedule

Item: 12345   Date Run: 1/1/94
Description: Sample Part

Lead Time 2 periods **Order Quantity** 20 **Demand Time Fence** 2
On Hand 12 periods **Safety Stock** 0 **Planning Time Fence** 4

PERIOD	1	2	3	4	5	6
FORECAST	12	14	8	10	12	14
CUSTOMER ORDERS	10	9	7	6	3	0
Projected Available Balance —On Hand = 12	2	13	5	15 / −5	3	9 / −11
Available to Promise	2	4	0	11		20
Master Production Schedule (MPS)		20		20		20

Figure 7.4. Sample Master Schedule

Master Scheduling Logic

The numbers in the sample master schedule (Figure 7.4) are used to illustrate how the master schedule logic works. When the master schedule is actually updated, the calculation is done on the computer using the software's logic algorithm.

As mentioned above, the two lines calculated by the master schedule are the Projected Available Balance (PAB), and the Available to Promise (ATP).

Projected Available Balance (PAB): This calculated line is used to determine when a replenishment order is to be placed and for how many. It also projects the expected availability of inventory in the future. The master scheduler uses this line to plan receipt of new orders and/or reschedule or cancel existing orders. New orders are planned to be received (MPS line), in the period when the PAB is projected to fall below the safety stock level, or below zero where there is no safety stock. Existing orders should be rescheduled first, before placing new orders. Orders should be rescheduled out when the PAB is above the lot size and there are no immediate demands.

PAB is simply a variation of a checkbook equation. The PAB is the ending balance for a period and is:

Ending Balance = Opening Balance
+ MPS scheduled receipts − Demand

The PAB is cumulative.

Demand Policies (or forecast consumption techniques) determine how demand is to be calculated. Usually within the demand time fence (DTF) only customer orders are considered in a make to demand or made to order environment and the greater of forecasts or customer orders in a make-to-stock business. In the planning time fence (PTF) there are various options, one of the more common options being to consider the greater of customer orders or forecasts.

For the master schedule shown in Figure 7.4, given that customer orders only are to be considered inside the DTF, and the greater of customer orders or forecasts are considered inside the PTF:

Periods 1 & 2. PAB = On Hand Inventory + Scheduled Receipts (MPS) − Customer Orders (since the periods are within the DTF).

Period 1. PAB = 12 + 0 − 10 = 2

Period 2. PAB = 2 + 20 − 9 = 13

From Period 3. PAB = Beginning Inventory + Scheduled Receipts (MPS) − Greater of Forecasts or Customer Orders (periods within the PLT).

Period 3. PAB = 13 + 0 − 8 = 5

Period 4. PAB = 5 + 0 − 10 = −5. A new order receipt is planned. This will be released in period 2, and, as the lead time is 2 weeks, the order will be scheduled for receipt in period 4. After receipt:

Period 4. PAB = 5 + 20 − 10 = 15

Period 5. PAB = 15 + 0 − 12 = 3 and so on

Available to Promise (ATP): The ATP is a calculated inventory balance and is used to make delivery promises to customers. It is important that sales and marketing understand the logic of ATP and work with operations to use the ATP in selling or booking customer orders. It represents the uncommitted portion of the master schedule's inventory or projected inventory. Only hard orders such as customer orders, field orders and warehouse orders (not forecasts) are used for demand in the calculation of the ATP. The ATP is calculated over a time span that starts with the receipt of a master production schedule (MPS) order and extends up to the period before the receipt of the next MPS order. It is the MPS order quantity less the sum of the customer orders booked before next MPS order receipt. For the first period the on hand inventory is included. ATP can be cumulative or noncumulative, with the noncumulative setting preventing the possibility of overbooking the product.

The ATP is calculated as follows:

ATP = Scheduled Receipts (MPS)
        − Sum of Customer Orders before the next MPS receipt

For the sample master schedule ATP is noncumulative (Figure 7.4).

Period 1. ATP = 12 + 0 − 10 = 2 (For the first period the on hand inventory is included).

Period 2. ATP = 20 − (9 + 7) = 4 (9 customer ordered quantity in period 2 and 7 customer ordered quantity in period 3)

Period 3. ATP = 0

Period 4. ATP = 20 − (6 + 3) = 11 and so on.

Note: If a cumulative ATP was used each period would include the balance of the last ATP period. Thus

Period 3. ATP = 4

Period 4. ATP = 4 + 20 − (6 + 3) = 15

An interesting use of ATP is to reserve capacity for customers who have a partnership agreement. A fake customer order is entered on the understanding that the customer partner will place a hard order within a specified time of the date the order is due, or they will lose their reserved capacity. Receipt of the hard order cancels the fake order.

Dos and Don'ts

DOs

1. Do train all company personnel—operations, marketing, sales, and particularly customer service—how to read the master schedule. They should understand the ATP (available to promise) line.

2. Do make sure that the order entry group responsible for booking customer orders uses the ATP for determining delivery dates for these orders.

3. Do make sure that the abbreviations used in the master schedule format are understood.

4. The master schedule settings must be acceptable to all operations staff who have a stake in the master schedule (planning horizon, lead times, lot sizes, safety stock, etc.).

5. Do get consensus on the setting of the DTF and PTF. Marketing and sales must understand and agree to the duration's set. They should also determine the demand policy used, as this determines the manner in which the master schedule aggregates demand.

6. Do understand the significance of safety stock and the scrap factor, as both will increase inventory.

7. Do set a lot size at a minimum, consistent with the process of the item being master scheduled.

8. Do verify that the lead times set are being achieved. If not, modify them. Lead times drive the start date of the MPS and incorrect lead times result in orders being completed too late or too soon.

DON'Ts

9. Don't set up and work the master schedule as part of production control only. Ensure that the factory supervisors and the marketing and sales personnel are involved in the process.

10. Don't promise orders without an ATP quantity being positive.

See Appendix 7.1 for an example of an actual master schedule.

Final Assembly Scheduling (Assemble to Order)

Purpose and Definition

Final assembly scheduling minimizes the inventory of finished products and reduces customer delivery lead times, as only subassemblies are stocked and assembled.

Final assembly scheduling is a technique where the final product is assembled to a customer's specific combination of features and options. In preparation for this, product family requirements are forecast and subassemblies planned in a process called two-level master scheduling. Final assembly scheduling is performed in an assemble to order (ATO) or make to order environment.

Discussion

Final assembly scheduling is applied to products that can be configured in many ways, using features and options. Automobiles are the most common example of these types of products. A car may offer a customer many choices, such as:

- 4 cylinder or 6 cylinder engines (2 options)
- with or without air conditioning (2 options)
- several types of tires—normal, radial, or steel belted (3 options)
- variations of the audio system—AM/FM; AM/FM stereo, AM/FM stereo cassette (3 options)
- different types of upholstery—vinyl, tapestry, or leather (3 options)
- transmissions—manual or automatic (2 options)
- brakes—power or ABS (2 options)

Even for the few options considered there will be $2 \times 2 \times 3 \times 3 \times 3 \times 2 \times 2$ = 432 configurations of the end product. Considering that there are many more options (several colors, different trims, cruise control, power controls, etc.), the number of configurations for any particular car can run into the thousands. The variety of combinations can only be partially reduced by grouping the product into families—cars are classified by size into sub-compacts, compacts, etc. All these combinations cannot be stocked, nor can the components of the car be ordered and built after receiving the customer order unless the customer is willing to wait for many weeks.

How then will an auto company know what to build, and build it in time to satisfy a customer? This problem is resolved by:

- using a planning (or modular) bill of material to link the end product with its features and options,
- developing a two-level master schedule to build subassemblies, and
- using a final assembly schedule to make the customer specified product.

Companies move from a make to order environment to assemble to order because of increasing similarity between the products and because of a need to reduce the customer delivery lead time. Companies move from a make to stock environment to assemble to order when customers want many features and options. To stock all these combinations would lead to a large finished goods inventory.

Planning Bill of Material

A planning or modular bill of material (BOM) has a nonbuildable product at the top level, and buildable features and options at the next level. The planning bill can have more than two levels, to suit different product configurations. Critical to formulating a planning BOM is the expected consumption of the second level assemblies and components. This estimate of consumption is either forecast or based on historical records.

The following example will help clarify this. On the top level the planning BOM has a generic nonbuildable product—compact car, while at the second level the bill has features and options such as engines, transmission, etc. The key feature of a planning bill is the expected consumption of the features and options—that is the percentage mix of the options that will be sold in these cars. The engines are forecast at 70 percent for 4 cylinder and 40 percent for 6 cylinder, while the transmissions are estimated to sell at 35 percent manual and 70 percent automatic and so on. (Note the percentage can exceed 100 percent. This is a way of hedging or covering the likelihood of having wrong forecast percentages for the master schedule items). See Figure 7.5.

Two-Level Master Scheduling

The top level of the two-level master schedule corresponds to the top level of the planning bill, while the second level of the schedule contains the features and options planned to be built in the percentages specified in the planning bill.

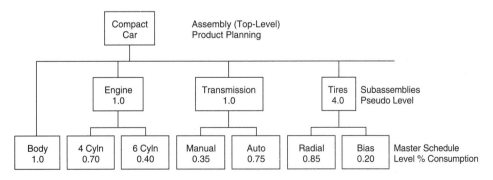

Figure 7.5. Planning Bill of Material

The quantity planned for the top level usually comes from a production planning process and is a forecast of the expected sales of the family. Thus a 100 compacts forecast and scheduled for a given week in the top level, generates a second-level *Master Production Schedule* (MPS) of subassemblies, based on their planning bill option percentages (see Figure 7.5).

Top Level	Compacts	100
MPS Level	Body	100
	4 Cylinder	70
	6 Cylinder	40
	Manual Trans	35
	Auto Trans	75
	Radial Tires	340
	Bias Tires	80

Isolating one assembly, the 4 cylinder engine and forecasting a production plan demand for 4 periods, the two-level master schedule will appear below:

PHASE I: Before Receiving Customer Orders

LEVEL 1: Compact Car/Period	1	2	3	4
Production Plan/Forecasts	100	100	50	50
Customer Orders				
Available to Promise (ATP)	100	100	50	50
MPS	100	100	50	50

Master production schedule orders are set up to meet the production plan demand. (In Period 1 an MPS order of 100 is firmed to balance the production plan demand of 100, and so on.)

LEVEL 2: 4 Cylinder Option (70%)	1	2	3	4
Production Forecast	70	70	35	35
Customer Orders				
Projected Available Balance *On Hand=0*	0	0	0	0
Available to Promise (ATP)	70	70	35	35
MPS	70	70	35	35

The top level ATP is exploded by the option percent, to the second level as the production forecast.
ATP (first level) × percent option = production forecast (second level).
The MPS is planned to meet that demand.
(In Period 1 MPS order = 70 to meet production forecast of 70, and so on.)

PHASE II: On Receiving Customer Orders:

LEVEL 1: *Compact/Period*	1	2	3	4
Production Plan/Forecasts	100	100	50	50
Customer Orders	60	40	50	40
Available to Promise (ATP)	40	60	0	10
MPS	100	100	50	50

The customer orders consume the MPS. The balance or ATP is exploded to the second level by the original mix percentage specified in the planning bill.
(Period 1 Customer Order of 60 leaves a balance of 40 compacts to be translated to 4 cylinder production forecast requirements at 70% or 40 × 0.7 = 28; and so on.)

LEVEL 2: *4 Cylinder Option (70%)*	1	2	3	4
Production Forecast	28	42	0	7
Customer Orders	30	20	20	20
Projected Available Balance *On Hand=0*	12	20	35	43
Available to Promise (ATP)	40	50	15	15
MPS	70	70	35	35

The level 2 of the master schedule is executed like a single level master schedule. Customer orders are usually added to the production forecast in calculating the PAB.
(Period 1. PAB = 0 + 70 − (28 + 30) = 12, and Period 2 PAB = 12 + 70 − (42 + 20) = 20 and so on) and (Period 1 ATP = 0 + 70 − 30 = 40, and Period 2 ATP = 70 − 20 = 50 and so on).

In the above example the 4 cylinder engine option which was estimated at 70 percent is selling below the planned percent and the PAB is excessive. The master scheduler has clear visibility of this and can push out the MPS orders in periods 3 and 4. Thus the master scheduler can adjust the MPS to meet the changing consumption of options based on *actual* customer order configurations.

Summarizing the functioning of the two-level master schedule:

There are two distinct levels—the top or generic level for an assembly, and a second level at which options are scheduled and built based on an option's percent consumption of the top level. The top level forecast is usually referred to as the production plan as it comes from the production plan's planning bill. Using an available to promise logic similar to that used in the single-level master schedule (ATP = MPS − Actual Demand up to the next MPS), the top level ATP is multiplied by an items option's percent to form the requirement of that item at the master schedule's second level. The calculated quantities are referred to as the production forecast. The second-level calculations of the PAB and the ATP are the same as the single-level master schedule. The production forecast is usually added to the customer orders when calculating the PAB. The MPS line of the second-level schedule represents assemblies and components that will be manufactured by the factory. A master schedule is prepared for each feature and option.

Final Assembly Schedule

In the above example, when the customer order for a configuration is booked, a sales order is created. The configuration is assigned a temporary part number (usually the sales order number), that has as its BOM the combination of features and options ordered. The configured BOM is used to pick parts to satisfy the order. The part number and BOM are used in the final assembly schedule with the required delivery date of the customer order. The process of creating the temporary part number links both the sales order and the product family. In the above example:

Sales Order	Quantity	Assembly	Component	QPA	Quantity
SO-1 (Reqd 3/01)	10	SO1 (Part No)	Body	1.0	10
			4 Cylinder	1.0	10
			Auto Trans	1.0	10
			Radial Tires	4.0	40

The Final Assembly Schedule uses the assembly lead time to determine a start date of the assembly to meet the customer's need date (required date). The quantity of each assembly and component entered against a customer order, consumes the master production schedule order, and leads to a revising and recalculation of the balance ATP.

Process for Two-Level Master Scheduling

1. Establish product families that can benefit from two-level master scheduling. Two criteria should be used: a) cumulative lead time of the assembly exceeds the time the customer is willing to wait and b) the product's final sold configuration cannot be forecast.

2. Set up the planning BOM, in consultation with sales. Decide which items are common and which items are options.

3. Determine the percentage consumption of all items on the planning BOM, using history or sales forecasts.

4. Obtain a forecast from sales or marketing for the top level generic product.

5. Based on the ATP explosion of the top level, use the second-level master schedule to make the option subassemblies or parts.

6. Develop an order entry process that prompts a customer to work through the options available and select those he would like to order. The order entry process should establish the sales order and tie it to a temporary BOM which consists of the configuration of the sales order.

7. Release a final assembly schedule which links the sales order specifying the assemblies and components that must be assembled to make the customer's order.

Dos and Don'ts

DOs

1. Do solicit and obtain sales and marketing involvement from the inception of the process.

2. Do set up the sales and ordering function to offer the customer the options reflected in the planning bill.

3. Do have the ordering logic set up the sales order number and the bill of material simultaneously, since both reflect the unique customer order configuration.

4. Do overplan any assembly where the demand is most likely but the mix is uncertain (that is, make the sum of the percentages for that option exceed 1.00).

5. Do use the ATP logic to promise orders. It adjusts to inaccurate product mix forecasts.

6. Do expect the implementation of the final assembly process to take time (around a year).

DON'Ts

7. Don't expect to implement the process as part of the day-to-day operations. Provide dedicated resources to make it happen.

8. Don't use planning bills for products that have lumpy demand or small sales volumes.

9. Don't provide safety stock for items whose demand has been overplanned (percent over 1.0).

Other Uses for Planning Bills

The section above described the use of planning bills in final assembly scheduling. Planning bills are also used when there is difficulty forecasting the master schedule demand for individual items in a product family. This is particularly relevant when there are a large number of end products within a family. In this environment forecasts are made for a product family only and the individual item demand is determined by multiplying the historical demand percent of the item by the family forecast. A simple example will illustrate this:

> Product family: Neckties   Forecast requirements: 10,000 per week.
> Planning Bill item percentages: Red 20%, Navy Blue 35%, Yellow 5%, Black 40%
> Master schedule weekly forecast:
> Red 2000, Navy Blue 3500, Yellow 500, Black 4000

Managing the Master Schedule

Purpose and Description

The master schedule must be managed to satisfy a customer's order while using a factory's resources productively. A master schedule must be managed effectively to increase a company's responsiveness to the market and increase its long-term profitability.

As seen in the section on format and logic, the master schedule is a computerized process that attempts to balance supply with demand and then make recommendations on action to be taken. A master scheduler needs to look at the recommendations made and exercise judgment in deciding what should be done. If a master schedule has demand that is repetitive or whose workload can be balanced, the master schedule is called a uniform schedule and is used in just-in-time manufacturing.

Discussion

The master schedule is one of the most important sets of information that is under management control. It is the conversion of actual and/or forecast demand into a factory schedule. It is also the means of consuming a factory's inventory and utilizing its capacity, both people and machines. It should provide a real handle on a company's business. In view of its importance a master scheduling policy must be set. This document should reflect management's understanding of how the business should be run. The policy should include:

- Management's views on stability of employment—stable employment will entail level loading the shop based on existing capacity, even if this may not satisfy some customer orders, as they have to be pushed out to level the load.

- The strategic approach of the company—is the company trying to be a low cost producer or a high service provider, or some mix. This strategy has to be translated by the master scheduler into an appropriate master schedule. A low cost producer may plan for minimum inventory and this may adversely affect customer service. The strategy is also needed to resolve differences between marketing, finance, and operations—all having different goals—on levels of inventory and appropriateness of response.

■ The authorizations for changing the master schedule—Who can change orders within the demand time fence? Who can decide to overload the master schedule and who should review overdue customer orders? Will the master schedule plan for forecasts in addition to customer orders or for the greater of the two? What value of forecast orders can be committed or how far out in the future can the factory build orders?

■ Numerous routine procedures—How often should the master schedule be run? Who should review it? Who should sign off the schedule? How should the performance of the master schedule be measured? Should there be a monthly review of the results of the master schedule—by whom?

Unless these and other considerations are clearly thought through and spelled out, the master schedule will not succeed in giving the company a competitive edge.

An example of a Master Schedule Policy is attached at the end of this chapter (see Appendix 7.2).

Process for Managing the Master Schedule

Managing a master schedule should cover the following considerations:

1. Determine which items will be master scheduled.

   Depending on the product environment an item will be scheduled as an end item sold from stock, or a subassembly which will be assembled to order, or a component/raw material that will be used in a make-to-order product.

2. Determine which dependent items have independent demand as service parts and decide whether these should be master scheduled.

3. Ensure that all master schedule's item parameters are correctly set. These should include:

   ■ The duration of the demand time fence (DTF) and the planning time fence (PTF)

- The demand policy to be followed—how the different types of demands are to be treated in the DTF and PTF (also called forecast consumption technique)

- The setting of the ATP to cumulative or noncumulative. Noncumulative is the more conservative rule as it ensures that inventory cannot be over promised.

- The lot size rule to be followed—lot for lot, days supply, EOQ, and so on

- The level of safety stock

- The planning horizon to be used

These settings can be done by ABC code within a product family rather than by individual item.

4. Remember the production plan is upper management's agreement on the execution of the business. Check that the sum of all the master production schedules (factory orders), for a product family, is equal to the production plan for that family. This equation should be verified before finalizing the schedule.

5. Check that the master production schedule does not violate any critical capacity resource by validating it against the rough-cut capacity plan.

   If there are periods of inadequate capacity, the ability to increase capacity must be confirmed from management or the master production schedule must be adjusted. Iterate the process until balance is achieved.

6. Check that past due orders will be completed in the current period. If not, reschedule the orders.

7. Change the order status to reflect the decision taken as a result of the master scheduling process.

   Generally computer-planned orders are in the *planned order* status. To firm these orders implies that the master scheduler has decided that the factory is to execute these orders. Such firmed orders are referred to as *firmed planned orders,* and the scheduler has to make a transaction that converts them from planned status to firm status. Once an order is firmed only the master scheduler can reschedule it. This is in contrast to the planned order which can be rescheduled by the computer. In many master schedules a separate transaction is required to initiate action by the shop to execute the order, and this transaction is referred to as "releasing an order." In these cases an order is first firmed and then released. Orders in process in a factory are usually referred to as open factory orders (see Figure 7.6).

Type of Factory Order	Created and Scheduled By	Time Frame	Purpose
Planned Order	Computer	Outside Planning Time Fence (PTF)	Balance demand with supply
Firm Planned Order	Master Scheduler	Inside PTF	Set up for execution
Released Order	Master Scheduler	Inside PTF & DTF	Factory plans execution
Open Factory Order	Shop Supervisor	Inside PTF & DTF	Factory starts to work on

Figure 7.6. Master Production Schedule Order Types

8. Set up a demand filter (see Managing Demand).

   The filter should prevent any large orders from consuming all available stock. Orders aborted by the demand filter *must* be dealt with on priority, or there will be the possibility that the best customers will be penalized.

9. On a weekly basis review the performance of the factory by generating a master schedule performance report (see Table 7.2).

   The report should measure the factory's completions against the schedule. The review should also address the trend of forecast consumption and forecast accuracy. The weekly meeting should be attended by members of marketing, finance, and operations. Decisions should be made to ensure that orders scheduled are completed per schedule and that forecasts made are being realized.

Table 7.2 Master Production Schedule Performance

Month week	Total Orders Scheduled	Total Orders Completed	% Complete	Reasons for Shortfall
Jan Wk1				Checked by specific order
And so on				

The weekly review has a primary function of resolving resource contentions that the master scheduler could not resolve within the policy guidelines. The master scheduler should present solutions that help to resolve the problem.

Dos and Don'ts

DOs

1. Do educate all affected parties of the company—operations, sales, marketing, and finance—on how the master schedule works and what trade-offs are involved.

2. Do check that the aggregate of all the master production schedules for a product group equals the production plan for that group.

3. Do ensure that a master schedule review is conducted with all affected functions—marketing, sales, finance, materials, and operations.

4. Do understand that trade-offs have to be made—between finished goods level and customer service; between lot size and lead time; between subcontracting and overtime.

5. Do ensure that sales and order service use the ATP (available to promise) properly when making customer promises.

6. Do ensure that marketing/sales and customer service are informed, as early as possible, when customer order dates cannot be met.

7. Do ensure that data input into the master schedule is accurate (on hand and on order).

8. Do work with the Order Action report generated by the master schedule. Take action to order, schedule out and in and cancel orders according to the report.

9. Do measure the master schedule performance and analyze reasons for poor performance.

DON'Ts

10. Don't ignore exception messages.

11. Don't work with an overloaded master production schedule. Only schedule orders up to the demonstrated capacity of the factory.

12. Don't work with "past due" orders. Reschedule past dues within the availability capacity, or reschedule existing orders to accommodate past dues.

13. Don't work without a defined "demand time fence" in make to order or make to demand environments in which only customer orders are considered. Make sure that this period corresponds to a phase in the manufacturing cycle when only emergency changes should be made.

14. Don't work the master schedule as part of production control only. Ensure that all affected parties—operations, sales, marketing, and finance are involved in the policies and performance of the master schedule.

15. Don't accept top management adding to a fully loaded schedule without rescheduling some other orders.

Common Problems in Master Scheduling and Possible Reasons

Overloaded: The master production schedule far exceeds the capacity of the factory on a long term basis. Some reasons are:

- The production plan is overloaded
- The schedule is not processed through rough-cut capacity planning
- The master schedules messages are not honored
- No clear policy on past due orders and lack of capacity
- Lack of management review

Front End Loaded: Too many past due orders are scheduled in the first few periods of the schedule. Some reasons are:

- No demand filter to trap large orders
- Rough-cut capacity not checked
- No demand time fence, and urgent orders continuously included
- No policy for prioritizing orders
- Factory not making the schedule so past due orders accumulating
- No policy on past due orders

Unstable Schedule: Master production schedule order quantities are not uniform, but have very erratic highs and lows. Some reasons are:

- No demand filter to trap large orders
- No demand time fence, and urgent orders continuously included
- Forecasts very inaccurate and no forecast accuracy review

Symptoms of Poor Master Schedules

- Unreliable delivery promises
- Persistent past due orders
- Excess inventory
- Expediting
- Excessive schedule changes
- Use of "hot lists," "red priority sticker," and other manual priority artifacts
- Upper management intervention
- Excessive overtime or idle time
- End of the month shipping surge
- Lack of accountability, finger pointing
- No regular management review and planned action for improvement

Appendix 7.1: An Example of an Actual Master Schedule

Master Schedule Format

An actual Master Schedule is shown below. It is busy but does contain *all* the variations that a practitioner is likely to find.

		MASTER SCHEDULE				Run Date 1/01
Item No. A1234	**Dem.Pol** 1	**FLTime** 10 days	**DTF** 4 weeks	**Safety** 10	**QOH** 30	
Desc. 10uF Capac.	**SC**	**ULTime** 2 hours	**PTF** 10 weeks	**Lot Size** 50	**InLoc** B1	
Grp. Cd. Cap	**PlnHoriz.** 18 mnths	**DTS** 4 days	**AccATP** N	**Shrinkage** 0%		

		DTF							PTF			
Period Ending	**Past Due**	1/08	1/15	1/22	1/29	2/05	2/12	2/19	2/26	3/05	3/12	3/17
Forecasts		20	20	30	30	30	30	30	30	30	30	40
Customer Orders		30	20	25	20	10	5		5	3		
Projected Available		50	30	55	35	55	25					
Available to Promise		30		5		35						
MPS		50		50		50		50				

Scheduled in   Rescheduled Out

DEMAND						SUPPLY					
Order No.	Demand	Ty	Qnty	Reqd Date	Ref	M	Order No.	Status	Where Reqd	Qnty	Due Date	Need
SO12	CusOrd		20	1/05	Will		On hand					
SO13	CusOrd		10	1/06	Hyman		On hand					
SO14	CusOrd		20	1/12	DEC	S	121	WIP	SO14	50	1/04	1/12
SO15	CusOrd		25	1/20	DEC				SO15			1/20
SO16	CusOrd		20	1/27	Boer	I	122	OFO	SO16	50	1/29	1/27
SO17	CusOrd		10	2/05	Cat				SO17			2/05
F01	Forecast		30	2/05	Mkt	S	123	FPO	F01	50	2/05	2/05
SO18	CusOrd		5	2/10	TRW				SO18			2/10
F02	Forecast		30	2/12	Mkt				F02			2/15
SO19	CusOrd		5	2/10	IH				SO19			2/20
F03	Forecast		30	2/19	Mkt	O	124	PLN	F03	50	2/12	2/19

Message Key
S   On schedule
I   Schedule In
O   Schedule Out

Even though there are many forms of master schedules a basic format should contain most of the following information and have a format shown above.

Explanation of abbreviations used in a typical Master Schedule

Run Date: The date the report was run. Corresponds with the first time period.

Item No.: Item Number. This is the unique identification of the item.

Desc.: Brief description of item.

Grp.Cd.: Group Code. Used where items belong to a product group or family.

SC: Source Code. Used where more than one factory is scheduled.

Dem.Pol.: This is a letter or number that represents how demand is to be treated. Example: Code 1 may mean that only customer orders are to be considered within the DTF, and outside the DTF the greater of customer orders or forecasts will be considered.

PlnHoriz.: Planning Horizon. This indicates the length of time the master schedule covers.

FLTime: Fixed Lead Time. This is the lead time of an item that is independent of the quantity made. Set up time is included in this time.

ULTime: Unit Lead Time. This is the time to make one piece.

DTStock: Dock to Stock. This is the time taken from receipt, through inspection, to delivery in the storeroom.

DTF: Demand Time Fence. This is a period of time normally set to correspond to the item's final assembly time and is thus a period where few changes are made. Only customer orders are considered within this period, except in a make-to-stock business where forecasts are considered.

PTF: Planning Time Fence. This is a period of time usually set to correspond to the item's manufacturing time. All changes have to be made by the master scheduler and not the computer.

AccATP: Accumulate Available to Promise. The code is Y (yes), or N (no).

Safety: Safety Stock. The master schedule is replenished when the balance goes below the safety stock.

Lot Size: Indicates the lot size that is used in reordering. Can also be a code that indicates the method to use.

Shrinkage: This is a built-in yield factor that the master schedule uses to increase the quantity of the supply (MPS) order. If the demand is for 100, an MPS order of 110 will be calculated to provide for a scrap factor of 10% ($100/0.9 = 110$). Also referred to as a yield factor.

QOH: Quantity on Hand. This is the number of pieces of the product in the factory.

InLoc.: In Location. The locations where the quantity on hand is located.

Calculations

Projected Available Balance (PAB)

The PAB was defined as:

$$PAB = \text{On-Hand Inventory} + \text{Scheduled Receipts (MPS)} - \text{Demand}$$

For the master schedule exhibited:

PAB for Period ending $1/08 = 30 + 50 - 30 = 50$
PAB for Period ending $1/15 = 50 + 0 - 20 = 30$
PAB for Period ending $1/22 = 30 + 0 - 25 = 5$

Note: An MPS order has to be received in period $1/22$, because the balance is going down to 5, which is below the safety stock level of 10. An MPS order #122 has to be scheduled in from period $1/29$ to be received in period $1/22$. The master schedule will provide such a message

Revised PAB Period ending $1/22 = 30 + 50 - 25 = 55$
PAB for Period ending $1/29 = 55 + 0 - 20 = 35$

Note: Since period $2/05$ is within the planning time fence, as per the Demand Policy, the demand is determined by considering the greater of forecasts or customer orders.

PAB Period ending $2/05 = 35 + 50 - 30 = 55$

Calculate the PABs and MPS quantities required to be received for the remaining periods on the sample master schedule. The completed sample master schedule is shown on page 177.

Available to Promise (ATP)

The ATP is a calculated balance used to make delivery promises to customers. It represents the uncommitted portion of the master schedule's inventory or projected inventory. Only customer orders are used in the calculation of the ATP. The ATP is calculated over a time span from the receipt of a master production schedule (MPS) order to the period before the receipt of the next MPS order. Each period in which a master scheduled order is received, generates an ATP.

The ATP is calculated as follows:

ATP = Receipts (MPS)
       − Sum of Customer Orders before the next MPS receipt
(For the first period On-Hand Inventory is added)

For the sample MS (with the second rescheduled MPS order #122 due in period 1/15) the ATP calculations are:

ATP for Period ending 1/08 = 30 + 50 − (30 + 20) = 30
ATP for Period ending 1/15 = 0 (it is part of the above period)

The ATP is shown only in the first period of the time span up to the next MPS order. It is used to promise customer orders in any of the periods up to the next MPS order. Thus the ATP for period 1/08 is 30, and customer orders up to 30 can still be promised in the period 1/08 or 1/15.

ATP for Period ending 1/22 = 0 + 50 − (25 + 20) = 5
Note: The ATP is set (see header) to be noncumulative—AccATP = No, and so the starting available inventory for each new period is 0.

If the ATP was set to be cumulative—AccATP = Yes, then:

ATP for Period ending 1/22 = 30 + 50 − (25 + 20) = 35

that is the available to promise quantity of 30 from the first time span is carried over (accumulated) to the next time span. The merits of each of the accumulated versus the nonaccumulated was discussed in the section on "Managing the master schedule."

ATP for Period ending 1/29 = 0 (within the 1/22 MPS time span)
ATP for Period ending 1/29 = 0 + 50 − (10 + 5) = 35

Calculate the ATPs for the remaining periods of the master schedule. See page 177 for the completed Master Schedule.

Pegging

Pegging is a very useful feature in a master schedule. It links the master schedule demand numbers to specific forecasts or customer orders, and MPS supply numbers to specific shop orders. It can be thought of as "where used" order information. Looking at the sample master schedule, the pegging information is below the master schedule and is divided into two parts: demand and supply. See Master Schedule Format.

Under the heading titled Demand, the columns are:

Order No.: The order number can be coded to represent type of order.

Demand Type: Indicate type—forecasts or customer or warehouse or service orders.

Quantity and Reqd Date: How much and when required.

Ref.: Reference. The actual customer or forecaster can be coded here.

Under the heading titled Supply, the columns are:

M: This is a message to take action. Examples are: S means on schedule, I schedule in, O schedule out, C cancel, P place order, etc. Different systems will have different codes and cover different messages.

Order No.: This is the Shop Order number.

Status: Indicates whether the order is in process (WIP), or has been released to the shop (OFO), or is still in the planning stage (FPO or Planned Order). Again, different companies have different order status and codes.

Where Used: This is the demand the shop order seeks to satisfy.

Quantity, Due Date and Need Date: This shows the quantity, when the order is due or scheduled to be completed and the date when the items are needed. This last date coincides with the demand required date.

Exception messages

Exception messages are usually displayed in the pegging section against the supply orders or under the pegging information. These messages indicate an out of balance condition of the master schedule. Typical messages *in order of priority* (or order of resolution) are:

Negative Available: This means the PAB is below zero and the master schedule is out of balance. An existing MPS order(s) must be scheduled in to meet the shortfall.

Negative Order: Here the ATP is negative and customer orders have been overbooked. Existing MPS order(s) must be scheduled in to meet the shortfall. In extreme cases a customer may be rescheduled.

Past Due: A customer's order due date is earlier than the report date, i.e., an overdue order. Expedite an existing MPS order to satisfy the overdue order.

Cancel: Order not required. Cancel the order.

Reschedule In or Out: Move order in or out.

It may be seen that pegging contains a lot of very useful information and can provide visibility to the numbers of the master schedule.

MASTER SCHEDULE   Run Date 1/01

Item No. A1234   **Dem.Pol** 1   **FLTime** 10 days   **DTF** 4 weeks   **Safety** 10   **QOH** 30

Desc. 10uF Capac.   **SC**   **ULTime** 2 hours   **PTF** 10 weeks   **Lot Size** 50   **InLoc** B1

Grp. Cd. Cap **PlnHoriz.** 18 mnths   **DTS** 4 days   **AccATP** N   **Shrinkage** 0%

DTF .... PTF

Period Ending	Past Due	1/08	1/15	1/22	1/29	2/05	2/12	2/19	2/26	3/05	3/12	3/17
Forecasts		20	20	30	30	30	30	30	30	30	30	40
Customer Orders		30	20	25	20	10	5		5	3		
Projected Available		50	30	55	35	55	25	45	15	35	55	15
Available to Promise		30		5		35		45		47	50	
MPS		50		50		50		50		50	50	

Scheduled in

DEMAND						SUPPLY						
Order No.	Demand	Ty	Qnty	Reqd Date	Ref	M	Order No.	Status	Where Reqd	Qnty	Due Date	Need
SO12	CusOrd		20	1/05	Will		On hand					
SO13	CusOrd		10	1/06	Hyman		On hand					
SO14	CusOrd		20	1/12	DEC	S	121	WIP	SO14	50	1/04	1/12
SO15	CusOrd		25	1/20	DEC				SO15			1/20
SO16	CusOrd		20	1/27	Boer	I	122	OFO	SO16	50	1/29	1/27
SO17	CusOrd		10	2/05	Cat				SO17			2/05
F01	Forecast		30	2/05	Mkt	S	123	FPO	F01	50	2/05	2/05
SO18	CusOrd		5	2/10	TRW				SO18			2/10
F02	Forecast		30	2/12	Mkt				F02			2/15
SO19	CusOrd		5	2/10	IH				SO19			2/20
F03	Forecast		30	2/19	Mkt	O	124	PLN	F03	50	2/12	2/19

Message Key
S   On schedule
I   Schedule In
O   Schedule Out

Appendix 7.2: Sample: Master Schedule Policy

1. The master production schedule by family must be equal to the production plan for each period.

   A deviation of $\pm 10\%$ cumulatively may be allowed for any one month. Any deviations greater than this must be approved by the factory manager.

2. The master production schedule must be achievable. It must be checked for material and capacity availability.

3. A demand time fence has been set at one week. This represents the assembly time of the order and, as such, changes to orders in this time frame should be made only in an emergency.

4. Sales and order service must use the ATP (available to promise) numbers and make customer promises only in those periods where there is projected or actual product availability.

5. A planning time fence has been set at 4 weeks. During this time frame all orders are firmed and released by material planner. Orders should be changed by the master scheduler, only if there is good cause.

6. Planned orders outside the planning time fence are scheduled by the computer to meet existing demand, unless firmed to stabilize the schedule.

7. Lot sizes should be lot for lot or period days demand. In the case of days demand, A items should have 15 days, B items 30 days and the lot size for C items may be set at 90 days of demand. These are maximum lot sizes and should be reduced if the setup allows making smaller lot sizes efficiently.

8. Measurements should be made and reported by the master scheduler. These should include:
   - Master production schedule performance which measures the percent of orders completed per schedule to the number of orders scheduled.
   - Forecast accuracy which compares the sales forecast with the actual orders taken, by item, for the month.
   - Number of changes made within the demand time fence with reasons for the change.

- The master production schedule summary which compares the schedule load by product family with the production plan.

9. The weekly performance of the master production schedule should be reviewed by the factory manager, the operations manager and the materials manager and coordinated by the master scheduler.

10. A monthly review should be held between operations, sales, marketing, and materials to review the factory performance and the forecast accuracy. This meeting should also be coordinated by the master scheduler.

8 *Materials Requirements Planning (MRP)*

Summary of Chapter

The chapter starts with an overview of the basic concepts of materials requirements planning (MRP). A standard format and the MRP algorithm is described next, followed by a description of the types of lot sizes used and their application. A note on capacity requirements planning is provided, and the chapter ends with a guide on managing the MRP process.

Contents

- Overview and basic concepts
- MRP format and logic
- Determining lot size
- Capacity requirements planning (CRP)
- Managing MRP
- Executing MRP

Relationship of MRP to Other Major Manufacturing Planning Processes

Materials requirements planning (MRP) obtains its requirements from the *master schedule* and plans the *shop floor* and *purchasing* supply in order to make or buy parts.

Overview and Basic Concepts

Purpose and Description

MRP should improve customer service, reduce inventory investment, improve cash flow, and provide accurate information to help make sound planning decisions.

The APICS dictionary defines *materials requirements planning* as "a set of techniques that uses bills of material, inventory data and master production schedules to calculate requirements for materials. It makes recommendations to release replenishment orders for material. Further, since it is time phased, it makes recommendations to reschedule open orders when due dates and need dates are not in phase. Time-phased MRP begins with the items listed on the MPS and determines (1) the quantity of all components and materials required to fabricate those items and (2) the date that the components and material are required. Time-phased MRP is accomplished by exploding the bill of materials, adjusting for inventory quantities on hand or on order, and offsetting the net requirements by the appropriate lead times."

Executive View

The master schedule is a factory's plan to satisfy customer orders. MRP is a technique that translates this game plan into specific actions that must be taken by the factory and purchasing to produce the materials needed by the master schedule.

The master schedule addresses: What do we have to make? How much? and when? by each saleable end item.

MRP addresses (see Figure 8.1):

- What makes up the master schedule item through a bill of materials.
- How many of the parts are in house through inventory records.
- How many of the parts are on order and when we will get them through the on-order status for both manufacturing and purchasing.
- When we need more through the projected available balance and planned-order receipts.
- When we should order more through the planned-order release date, which backs off the item's lead time from the required receipt date.

Prior to MRP, the order point technique was used to order materials. Order point, as the name implies, allowed the inventory of an item to fall to a predetermined quantity before placing an order for replenishment. The inventory at the order point was calculated to be enough to cover demands of the item until the new order was received. MRP requires a

Table 8.1 Order Point Compared to MRP

Characteristic	Order Point	MRP
Scope of items covered	Individual parts	Assemblies and their parts
Ordering basis	Historical demand	Current and future demand
Method of change	Manually	Dynamically by computer
Order quantity	Predetermined	Can be based on dynamic rules
Inventory order time	Always at order point	Ordered only when required
Determination of order delivery	Lead time determines supply	Order due date determined from order need date backing off lead time

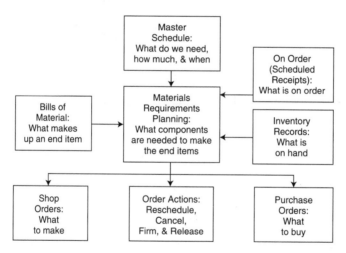

Figure 8.1. Schematic of MRP Relationships

lot of data and computing capability. Prior to the computer becoming an available resource, this technique could not be used effectively. See Table 8.1 for a comparison of MRP and order point.

The main advantage of MRP is that it develops a unified plan to build or buy all the parts of an assembly in time to satisfy the requirement of the master schedule. It is able to plan well into the future and can easily update the plan to reflect changes in orders and/or changes in factory and supplier performance. MRP is usually part of a manufacturing system that will include master scheduling, capacity planning, purchasing, and shop floor control. The whole system is usually referred to as *manufacturing resource planning (MRP II)*—an unfortunate duplication of an acronym.

Just-in-time (JIT)

In the last decade a manufacturing approach developed in Japan has gained popularity. This approach focuses on shop floor execution, using controlled work in process, minimum lot sizes, and visual signals to communicate. Its principal objective is to minimize lead time while making a high-quality product. It involves the labor force in continuous improvement. MRP does not contradict the JIT approach. In fact, MRP can be complementary to JIT when using MRP to provide the material planning and JIT to provide the execution. (JIT is described in detail in Unit III, along with how MRP and JIT, combined in a hybrid system, are used to control the manufacturing process.)

Dos and Don'ts

DOs

1. Do remember that materials requirement planning (MRP) is not a system. It is a technique that uses computer software logic to convert a master production schedule into the subassembly and component orders required to make the ordered items. As discussed above, MRP helps to plan the future supply of product dynamically to meet expected demand.

2. Do realize that the logic and functioning of MRP is not understood. Do not neglect training, and make sure that top management is also educated and that their education is visible to the organization. The computer only does the arithmetic. People have to manage the outputs of the computer program.

3. Do ensure that data are valid and accurate. Data include master schedules or demand for items, bills of material on how items are structured, inventory records of on-hand material, and ordering status information. Most problems with

MRP arise out of data being incorrect. The only way to ensure that data are accurate is by constant and detailed monitoring of inventory and ordering information. There is no shortcut.

4. Do check the master schedule and MRP for available capacity before issuing orders to the factory. If this check is neglected, the factory can be overloaded and the schedule will be adversely affected, with the consequent impact on customer service.

5. Do select a software system carefully. Large MRP systems cost from $500,000 (for a 32-user factory) to $1 million and up, and take about 18 months or more to implement. Much has been claimed for MRP and the benefits it confers on those using it. History has not supported the claims made on its behalf, and it is estimated that many of the installations fail to realize any advantage. There are horror stories about the cost and time taken to implement MRP II. Consider

(*continues*)

Dos and Don'ts

DOs (*continued*)

PC-based MRP packages as an inexpensive alternative.

6. Do select an MRP system that is applicable to the product and process characteristics of the factory with only a minimum of alteration.

7. Do be aware that consultation, software modification, and integration costs often exceed the cost of the software. This makes the need for selecting an appropriate software package even more critical.

8. Do ensure that MRP is implemented correctly. A poor implementation can devastate a business with the confusion it creates by having inaccurate and unreliable records.

9. Do involve the software supplier in the implementation as part of the purchase contract. The best help is an experi-

enced project manager of the software company.

10. Do review the safety-stock levels for validity. Wrong quantities of safety stock can lead to stock outs and missed deliveries.

DON'Ts

11. Don't neglect the system. Surveys indicate that the most common reason given for not implementing MRP effectively is lack of upper-management support.

12. Don't expect quick implementation, reliable information, and quick results. Do expect that inventory will be reduced, space will be saved, and customer order delivery will be improved (all by 50 to 100 percent over two years).

MRP Format and Logic

Purpose and Description

MRP must have an easily understood and universally accepted planning format to enable all users to relate to its output effectively.

MRP systems determine what materials should be ordered and when, by first getting input in the form of end-item demand from a master schedule, and then exploding this demand through a bill of material to determine the gross requirements of assemblies and components. After reducing the gross requirements by quantities on hand and on order, MRP makes recommendations for making or buying additional assemblies or components to satisfy the total requirements.

Example: If an assembly A consists of two pieces of B and one piece of C, exploding a demand for 100 As will give gross requirements of 200 Bs and 100 Cs. This gross quantity is reduced by materials in the factory and on order to get a net required quantity in a process known as *netting*. So, if there are 50 Bs on hand and 50 Cs on order, the net requirements will be 150 Bs and 50 Cs. MRP then uses the assembly and component lead times to determine when to start to manufacture or purchase the parts in a process known as *lead-time offsetting*. If A takes one week to build, B takes two weeks to buy, and C takes a week to make, an order should be placed to buy B three weeks before A is needed and an order to make C one week before A is needed (see Figure 8.2).

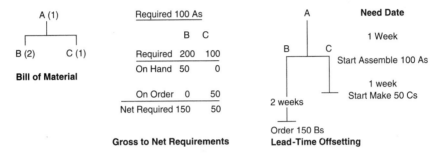

Figure 8.2. Logic of MRP Calculations

Independent and Dependent Demand

A demand represents the need for an item. An item has an independent demand if it is unrelated to the requirements for other items. Independent demand must be forecast or booked. Finished goods and service parts are examples of independent demand.

A dependent demand is a supply order, pegged to a parent as determined by a bill of material, and generated to satisfy the demand of that parent. Dependent demand is calculated, and the preferred means for this is MRP.

Components and raw material of any finished product are examples of dependent demand. In the example of Figure 8.2, A has independent demand, while B and C have dependent demand.

An item can be independent and dependent at the same time, as a component can be sold as a spare part and form part of an assembly.

Example: Automobile tires are both independent (ordered for replacements), and dependent (4 per auto).

Discussion

At the end-item level, MRP gets its demand from the master schedule. Where there is no master schedule, it accepts demand from many sources—such as customer orders, forecasts, warehouses replenishments, and so on—and combines them into *gross requirements* by period. MRP explodes these requirements through the bill of material and generates requirements for subassemblies and components. It reduces these requirements by the material on hand and on order, time phases the net requirements, and recommends releasing or rescheduling shop work orders. This is the supply. MRP can plan well into the future and can update the plan whenever there is any change in any of the inputs to the plan. MRP constantly balances the demand with the supply at every level of the bill of material.

The *exciting value* of MRP is that it is able to develop and dynamically adjust dependent demand at all levels of a bill of material, balancing every level by the quantities per assembly contained in the BOM. Before MRP, the order-point techniques could deal only with a single item at a time with its independent demand; that is, each item had to have a forecast for only its expected demand and the parent-child (assembly-component) relationships had to be manually calculated to support production schedules.

MRP issues action messages to:

- Schedule and release planned orders, recommending the quantity and date of order placement.

- Reschedule existing orders—pulling them in if the required date is earlier than the supply date, or pushing them out if the required date is after the supply date.

- Recommend canceling orders if they are not needed. This is another opportunity to reduce inventory investment that is neglected.

Care must be exercised to keep from being inundated with action messages or the real needs will not be addressed. To this end, sensitivity filters must be set to respond only when the plan deviates by a meaningful amount of time—usually more than a week. Attempts must also be made to minimize change to the MRP, and techniques to achieve this will be discussed.

Time Phasing and Lead-Time Offsetting

Time phasing is the key concept of MRP. It is the technique of expressing future projected demand, supply, and inventory by time period (days or weeks). Time phasing is usually displayed in a horizontal (bucketed) format, which is the format used in this section. It can also be displayed in a vertical (bucketless) format using date/quantity information (see Appendix 8.1 for samples of both formats).

Lead-time offsetting is scheduling the start of a shop order or issuing a purchase order by providing enough time (the lead time) from the required or receipt date to allow the order to be made or bought.

Supply and Demand: Netting

MRP compares the demand (gross requirements) with the supply (on-hand and planned-order receipts) for every level of the bill of material of the required end item and determines the net requirements, plus or minus. It does this for every item on the bill of material of a product. If net requirements are negative, MRP will plan an order to bring the balance back to zero or slightly positive.

Format

There are many valid formats. The *horizontal format* used below is probably the simplest and it follows the logic of MRP. It is the traditional method of representing and calculating MRP. It must be stressed that the examples given below are meant to illustrate how MRP works. In reality, all the calculations are done by the software and the practitioner is guided by *action messages* to *order, reschedule,* or *cancel*.

Example:　　Part No 1234　On Hand=50　OrderQuan=20
SafetyStock=0　LeadTime=2 periods

Table 8.2 Format of a Material Requirements Plan

Period	1	2	3	4	5	6
Gross Requirements	20	30	20	20	30	10
Scheduled Receipts		20				
Projected Available On Hand 50	30	20	0	0	10	0
Planned-Order Receipt				20	40	
Planned-Order Release		20	40			

Explanations of the MRP Format:

- Gross Requirements: Requirements from all sources—independent and dependent.
  Note: Gross requirements must be met either by scheduled receipts (open order) or by a planned-order receipt.
- Scheduled Receipts: Existing open orders due to be received.
- Projected Available Balance (PAB) = Begin On Hand
  + Scheduled Receipts − Gross Requirements
  + Planned-Order Receipts.
  (The beginning on hand may be the last period's PAB.)

  Note: PAB quantities represent status *after* planned-order receipt. Projected on hand should never go negative or below safety stock.

- Planned-Order Receipt: The required receipt of the plan order to en-sure that the PAB remains positive or above the safety stock.
- Planned-Order Release: Planned-order receipt offset by lead time.

Calculations:

Projected Available = Begin On Hand (or PAB Last Period) + Sched-uled Receipts − Gross Requirements + Planned Order Receipts

Period 1: Proj.Avl = 50 + 0 − 20 = 30
Period 2: Proj.Avl = 30 +20 − 30 = 20
Period 3: Proj.Avl = 20 + 0 − 20 = 0
Initial Period 4: Proj.Avl = 0 + 0 − 20 = −20
Plan Order Receipt = 20, Plan Order Release Period 2
Final Period 4: Proj.Avl = 0 + 20 − 20 = 0
Period 5: Proj.Avl = 0 + 0 − 30 = −20
Plan Order Receipt = 40, Plan Order Release Period 3
Final Period 5: Proj.Avl = 0 + 40 − 30 = 10
Period 6: Proj.Avl = 10 + 0 − 10 = 0

Showing the Effect of Safety Stock:

Part No 1234   On Hand=50   Order Quan=20   Safety Stock=5
Lead Time=2 periods

Table 8.3 Format of a Material Requirements Plan Showing Effect of Safety Stock

Period	1	2	3	4	5	6
Gross Requirements	20	30	20	20	30	10
Scheduled Receipts		20				
Projected Available On Hand 50	30	20	20	20	10	20
Planned-Order Receipt			20	20	20	20
Planned-Order Release	20	20	20	20		

Note: When the projected available balance falls below the safety stock level, an order must be placed. This results in ordering more material for the same gross requirements. Many MRP formats subtract the safety stock from the projected available balance. This is misleading, as the planner often forgets that the projected balance does not include the safety stock and will expedite receipt of a regular order when it is not required. By including the safety stock as part of the projected available balance, the actual balance or projected balance is always visible.

Vertical Format

Today MRP's software is moving toward a bucketless vertical format. In this format the supply demand balance is calculated on a continuous time stream, with each demand separately represented. The logic and calculations are the same as the horizontal format, but there are no time periods. An example of a vertical format MRP is attached in Appendix 8.1.

Determining Lot Size

Purpose and Description

The lot sizing policy selected will have a major bearing on the amount of inventory carried and the level of customer service maintained.

The MRP algorithm has a setting, usually called a *lot size policy*, which determines how the order size is calculated. Lot sizing is one of the key elements in manufacturing execution (this is also covered in Unit III).

Discussion

Much of the discussion on how much to order (lot size) has always centered around the trade-off between the cost of setup or ordering and the cost of carrying inventory. The economic order quantity (EOQ) model captures this approach. (See Appendix 8.2 for a detailed discussion of this model.) Unfortunately, the assumptions made in the EOQ model on the inventory and ordering costs are not valid. Furthermore, the EOQ does not take into consideration other important factors such as shelf life, obsolescence, and process and packaging considerations.

None of the lot sizing algorithms evaluate the effect of building items early, thereby misusing capacity that could be more effectively applied. The safest approach is using a lot size that minimizes the inventory carried and does not require committing an order to cover requirements too far out. Lot sizing techniques should provide stability to MRP and should not be susceptible to constant quantity changes (an argument against some of the complex techniques and dynamic order quantities).

However, there is an equally important consideration in determining the lot size: the effect of lot size on lead time. The larger the lot size, the longer the lead time, either from the supplier or from the factory. Lead time directly affects meeting customer delivery dates and must be considered in the lot size decision. A key thrust of JIT manufacturing is reduction of lot sizes so that lead time is minimized.

The commonly used lot size techniques are described below.

Lot for Lot (L4L)

Process: APICS defines *lot for lot* as "A lot sizing technique that generates planned orders in quantities equal to the net requirements for each period. Synonym: Discrete order quantity." The equation for determining L4L is:

Net requirement or PAB = Gross requirements
$$- \text{(On hand + Scheduled receipts)}$$

and corresponds to the projected available balance on the MRP format, with no remainder.

Table 8.4 Lot for Lot (L4L)

Period	1	2	3	4	5	6	Total
Net Requirements or PAB	25	10		20	5	10	70
Order Quantity	25	10		20	5	10	70

Lot-for-lot ordering is used for costly (A-class) items, as the technique minimizes purchasing excess material. It is also used where demand is erratic and unpredictable. This technique eliminates the numerous assumptions made in the EOQ lot size, and is thus more reliable. Finally, lot for lot minimizes the purchasing and ordering lot size and ensures that the lead time is also minimized.

When dealing with purchased items, there may be a fear of placing an excessive number of orders. This should be dealt with by placing a blanket order and issuing a release to the supplier whenever material is required. When dealing with manufactured items, the setup time required is a limitation in applying the technique. Where setup time is considerably larger than run time, the size of the manufacturing lot becomes a consideration, and lot for lot ordering may be inappropriate.

Fixed-Period Quantity (FPQ) or Number of Periods Demand

Process: This technique determines the lot size by equating it to a prescribed number of periods of future net requirements. An example for two period demand lot size follows:

Table 8.5 Fixed-Period Quantity (FPQ)

Period	1	2	3	4	5	6	Total
Net Requirements	25	10		20	5	10	70
Order Quantity	35		20		15		70

This technique is useful for ordering costly items as it can be set to minimize inventory costs. For example, costly or A items are ordered to few periods, typically 2 to 3 periods, while inexpensive or C items can have a lot size set for 20 periods or more. This technique also helps to smooth out the lumpy or irregular demand that may result from lot-for-lot orders. However, the number of periods selected is based on judgment and may be arbitrary or biased.

Period-Order Quantity (POQ)

Process: The period-order quantity is similar to the FPQ except that the ordering interval is computed from the EOQ logic. For the example being used, assume the EOQ has been calculated as 50 with an annual demand of 150. Then

$$\text{Orders per year} = \frac{\text{Annual Demand}}{\text{EOQ}} = \frac{150}{50} = 3$$

$$\text{POQ} = \frac{\text{Periods per year}}{\text{Orders per year}} = \frac{12}{3} = 4$$

and the order quantity (OQ) = Requirements for 4 periods.
For Period 1. OQ = 25 + 10 + 0 + 20 = 55 (4 periods)

Table 8.6 Period-Order Quantity (POQ)

Period	1	2	3	4	5	6	Total
Net Requirements	25	10		20	5	10	70
Order Quantity	55				15		70

Like the FPQ, POQ minimizes inventory ordered as excess is not ordered. Since consideration is given to setup costs (in the EOQ formula), this technique is useful when dealing with items with significant setup costs.

Fixed-Order Quantity (FOQ)

Process: This lot sizing technique is usually used where a fixed quantity is ordered by considerations other than MRP. Usually the FOQ is dictated by a fixed-supply lot, a customer fixed-order quantity, limited shelf life, or processing or packaging requirements.

Table 8.7 Fixed-Order Quantity (FOQ)

Period	1	2	3	4	5	6	Total
Net Requirements	25	10		20	5	10	70
Order Quantity	40			40			80

Table 8.8 Summary of Lot-Sizing Techniques

Technique	Lot for Lot	Fixed-Period Qnty	Period-Order Qnty	Fixed-Order Qnty
Calculation	Order = Demand	Order = Demand for fixed number of periods	Order = $\dfrac{12}{\text{orders per year}}$	Order = Fixed Qnty
Use	For expensive items with irregular demand	For irregular demand and control	For items with large setup costs	Machine or customer wants fixed quantity
Effect	Minimum order size	Control by type of item	Balances setup to size	Particular quantity set

Lot Size Adjustments

1. Maximum and Minimum: Maximum lot sizes are usually imposed by management, such as "not to exceed a 6-month supply." Minimums are usually set by a supplier not willing to supply less than a certain quantity, or when the requirements are too small to be ordered practically, such as ordering 16 ordinary screws.

2. Pan Size or Multiples: These lot sizes are dictated by processing considerations, such as the number of pieces held in a machining fixture, or the packaging lot. The lot size is calculated using one of the regular rules and rounded off to a multiple.

3. Yield Allowance: A yield or scrap allowance is a quantity added to a lot size to make good expected loss in the process. This will ensure that

the required quantity of acceptable material will be made. Yield allowance is usually expressed as a percentage and should be decided by item at the lowest appropriate level of the BOM. Yield allowance is particularly important in lot-for-lot sizing as there are no additional pieces planned.

Dos and Don'ts (Order Quantities or Lot Sizes)

DOs

1. Do eliminate or minimize the cost of ordering or setup. For example, if a blanket order is placed on a supplier, then there is only a one-time ordering cost and no additional orders need be placed. Supply is made by FAX, EDI, or a telephone call. Setup time should also be reduced so that it is a minor cost, thereby allowing lot sizes to be minimized.

2. Do order what is required to meet customer requirements, using a lot-for-lot or smallest-possible lot size ordering rule, especially on A items.

3. Do be aware of the adverse impact of large lot sizes on shop order manufacturing lead time.

4. Do develop and apply ordering parameters based on an ABC classification of material, where demand is erratic or small. Typically order-size parameters are set at days of future requirements: A items at 15 to 30 days, B items at 30 to 45 days, and C items at 90 to 180 days. Whenever possible a large order (annual) should be placed with a supplier partner and the delivery regulated as required by customer orders.

5. Do make sure the quantity of material ordered is factored by a yield percent.

Yield factors indicate the percentage of items that will be lost in the manufacturing process.

6. Do take into consideration specific marketplace realities, such as minimum purchase size and purchasing multiples—pallet loads, drums, dozens, and so on.

DON'Ts

7. Don't order per the calculated EOQ without checking the quantity with individuals who are aware of the specific item and its future use. Instead, determine the equivalent periods and use the POQ.

8. Don't try to determine an exact or very accurate order quantity, other than making sure the customer order is covered. The timing of the order is more important. A general ABC approach is usually adequate based on sample EOQ studies.

9. Don't use a complex lot-sizing algoorithm unless there is a specific proven reason for doing so. Many algorithms require a lot of calculations and lead to schedule nervousness.

10. Don't expect any lot-sizing technique to work satisfactorily if the demand is variable.

Capacity Requirements Planning (CRP)

Description and Purpose

Where it can be used effectively, CRP will ensure that the materials requirements plan can be executed by the factory in time to meet a master scheduled order, which in turn will satisfy a customer's order.

Capacity requirements planning (CRP) is the process of determining, in detail, how much labor and how many machine resources are required to execute MRP, and comparing this capacity requirement with capacity that is available. Open shop orders and planned orders in the MRP system are input to CRP, which "translates" these orders into hours of work by work center by time period. CRP tests MRP's schedule of planned and released orders against the actual capacity available, and validates the feasibility of the MRP schedule.

Discussion

The above stated description and purpose is the conventional MRP II approach. In actual practice, for most discrete products there is too much variability on the shop floor to allow an accurate evaluation of available capacity to be made. Available capacity is a dynamic, unpredictable variable and cannot be captured by a static, calculated work-center capacity number. The result is that CRP does not correctly reflect a work center's capacity to process MRP's requirements. As empirical confirmation of this fact, today few factories use CRP to evaluate their future capacity requirements.

Instead, factories attempt to manage shop floor variability by:

- Identifying process constraints and providing material buffers to absorb the variability
- Using a "pull" or Kanban approach of responding to process and product variability
- Providing excess capacity for absorbing variability
- Providing worker and/or machine flexibility to permit developing additional capacity when required

All of the above practices, including a detailed discussion on the subject of variability, are covered in Unit III—Manufacturing Execution.

Suggested Approach to Capacity Planning

There is need for capacity planning, particularly for evaluating and providing for long-term resources. Earlier, in the chapter on master scheduling, the technique of rough-cut capacity planning (RCCP) was discussed and mention was made to resource requirements planning (RRP). These are both very effective techniques. RRP is useful for long-term resources of capital, equipment, buildings, and the like. RCCP checks the master schedule against the factory's critical resources, to determine that there is sufficient capacity at a gross or average level. It is recommended that this form of rough-cut capacity planning be conducted against every major change of the master schedule.

Managing MRP

Purpose and Description

A well-managed MRP program will plan the components and parts required to make assemblies in time to satisfy master scheduled orders.

MRP is a calculation technique whose success depends heavily on the accuracy of its inputs and on the response made to the action messages it generates. Inputs include the bill of materials, the inventory records, and the ordering records. Further MRP is loaded with master schedule demand, and this demand must be within the capacity of the factory. All of these activities confirm the need to manage the MRP process. To run a successful MRP program requires meticulous and sustained attention to detail. MRP in itself is a software algorithm. Its inputs must be managed and its outputs must be responded to.

Discussion

For successful execution of an MRP program the following prerequisites must be met:

- A valid master schedule containing what is required, how much, and when. The schedule must be checked to ensure that it is doable. An overloaded master schedule will lead to a materials requirement plan that cannot be implemented.

- Accurate bills of material having the correct quantities per assembly and structured to reflect the way the product is assembled.

- Accurate inventory records of all material in the factory.

- Accurate information of parts on order, both manufactured and purchased, with order quantities and due dates.

- Dependable lead times to procure or manufacture parts. If lead times are unreliable, the time phasing of MRP will be incorrect and schedules may not be met.

- Controlled master schedule changes and limited product engineering changes. Too many changes will lead to a large number of action messages and this may overwhelm the planner.

All of the prerequisites require basic disciplined monitoring and maintenance. *This is the crux of a successful MRP program.*

Data Requirements for MRP

1. Every part must have an *M* (manufactured) or *P* (purchased) *identification.*

2. Every part must have a *lead time.* Purchased parts' lead time should include planner and buyer preparation time, order placement time, suppliers' manufacturing time, and dock to stock time. Manufactured parts' lead time should include order preparation, parts picking, move, queue, setup, and run times. Run times should be per piece. There should be a default field where a manual composite lead time can be input.

3. Every part must have an accurate quantity of the *stock on hand,* and a means of maintaining it through inventory transactions covering receipt, issue, and shipment.

4. Decisions must be made on which parts and locations should *not be considered available* for MRP calculations (such as rejected parts, or floor stock).

5. A decision should be made on which parts should have *safety stock* and how much. (A detailed discussion on safety stock may be found in Chapter 10.)

6. Parts that are likely to have losses during production should have a yield factor or percentage loss.

7. The lot size policy should be determined by part.

8. Every part should have a cost attached to it, for simulating projected material costs.

9. Every manufactured item must have an accurate and valid bill of material. Bills must have the correct quantity per assembly for the components used and the appropriate unit of measure. The bill of material is used to determine the amount of material required.

Tools for Managing the MRP Process

Dynamic Order Quantities: Dynamic demand and supply is beneficial when the demand reflects the changing order environment and the supply can be adjusted to meet what is actually needed. However, there has been a lot of innovation in determining how to set the supply quantity or lot size. Perhaps some of this enthusiasm has been misplaced. Elaborate lot-sizing techniques can lead to an excessive number of reschedules in/out and the order action report becomes so busy that the important actions may become obscured.

A lot-for-lot (L4L) order size technique should be used wherever possible, with lots being combined into the weekly time buckets of MRP. This means that only the quantity required in the period will be planned in that period. This method minimizes inventory and improves material flow. At the lowest or purchased level of the bill of materials, a *fixed period quantity* *(FPQ)* or number of periods of demand technique may be favored. Here the order size is equal to the demand placed on the item over the period specified. Thus for expensive A items, the FPQ may be fixed at 2 periods (2 weeks), whereas for inexpensive C items, the FPQ may be fixed at 28 periods.

ABC Stratification: This is another extremely effective technique for managing demand. A items are the significant few (about 20 percent), that account for most of the value of the factory's production (about 80 percent). It is essential that the demand on these items be carefully managed so that there is no misuse of costly product. A and B items should be tagged and they should have specific parameters on their optimum lot sizes, lead times, and safety stocks. Since A and B items have significant costs or are of a critical nature, they are usually manufactured on priority. (The ABC concept is covered in detail in Chapter 10.)

Dynamic Lead Time: There has been much discussion on the value of dynamic lead time as part of MRP's enhancements. Typically MRP has a lead-time field for every item and this can be a manually input number (the default), or it can be calculated using a fixed setup and queue time and a time per unit. The system calculates the lead time based on the lot size determined for the item. The lead time is:

Lot Lead Time = Fixed Lead Time + (Unit Run Time
                        × Number Pieces in Lot)

The decision to use the dynamic lead time should depend on the product environment, with make-to-stock orders having large lot sizes favoring the dynamic lead-time setting.

Safety Stock: The use of safety stock has been discussed in detail in Chapter 10 on inventory management. Where the supply of the purchased parts or the manufacturing process is unpredictable, safety stock must be provided to ensure that order due dates are met.

In a make-to-order environment, safety stock should be provided at the lowest level only, or else there is danger of a cascading effect down the bill of material, resulting in excess stock being built. In a make-to-stock environment, safety stock is usually provided at the finished goods level through the master schedule.

Understand that safety stock will distort the MRP calculations, so it should be provided judiciously. In MRP a planned-order receipt is made whenever the projected on hand falls below the safety stock.

Safety Capacity: Safety capacity is planned to provide additional capacity needed to meet supply/demand variability (referred to above). The advantage of having safety capacity is that material does not have to be committed in advance in the form of safety stocks. Furthermore, safety capacity does not distort MRP quantities and dates.

Safety capacity may be assessed empirically as a percentage of available capacity. Typically 10 to 20 percent of available capacity is reserved as safety capacity. The use of safety capacity can be monitored using an input/output report (covered in Unit III).

Sensitivity dampers: The material analysis report and the exception messages it generates is a key part of MRP. If the volume of messages generated is too great, it will obscure those where it is critical to take immediate action. To help reduce the number of unimportant messages, sensitivity logic should be applied.

Example: Schedule messages are generated only when the supply due date differs from the demand required date by more than one week.

The logic can also be applied as a percentage of lead time. If the gap between supply date and demand date exceeds 15 percent of the parts' lead time, then an exception message is generated.

Pegging: *Pegging* was discussed in the master scheduling chapter as a pointer to determine which shop order satisfied which customer order(s). In MRP, pegging provides the capability of tracing an item's gross requirements to its parent. MRP explodes through an items product structure from top to bottom, calculating gross and net requirements at every level for the appropriate time period. Thus pegging is used to determine which parent items are affected by shortages or problems at a component level. Specific actions can then be taken as: rescheduling the parent order, making priority allocations, changing the lot size, or subcontracting the item in trouble. Pegging can be single level, where one level at a time can be traced, or multi-level, where several levels are traced in one inquiry. Generally single level is adequate for most purposes. Pegging is usually contained in the MRP report or the material analysis report (see below).

Order Types and Status

Planned and Firm-Planned Orders: Planned orders are generated by the computer, and have a quantity and schedule that balance the demand for the item. These orders are recalculated and reset whenever there is change in the item's demand or supply. As the orders approach the present, there is need to stabilize the schedule. *Firm-planned orders* are a means of preventing the computer from rescheduling the order. By firming an order, a planner takes it off auto pilot and puts it into manual control. By firming an order, a planner affirms the recommendation of MRP, and thus signs off or approves MRP. Firming an order is a very useful and important technique and should be used to advantage. The computer will not plan any order inside a firm-planned order. When MRP detects a potential shortage condition inside the firm-planned order, it will issue an action message to schedule in or recommend placing another order. Firm-planned orders may be added to, changed, or deleted. Firm-planned orders may also be changed to planned or released status.

Released Orders and Allocations: When an order is released—that is, it is decided to make or buy the part—the components attached to the parent are reserved or allocated. For purchased components and at parent levels, the quantities shown as order start and order release are deleted, and the quantity is transferred to the scheduled receipts line. The components required by these parents have to be reserved until they are disbursed or picked. To ensure this, the requirements generated on components are allocated. When the quantities are physically issued, they are reduced from the projected available and the allocated fields. The balance between the levels of the product is thus maintained. This process is done by the computer program, and is transparent to the user.

Released orders are also known as *open factory orders* (*OFO*) and can be added to, changed, or closed. In most systems, an order cannot be deleted in the OFO status. If deletion is allowed, it can be done only if there has been no activity for some time. Some systems allow an open order to be reversed to firm if no material has been issued against the order. Once an order is returned to firm status it will not be part of MRP's on-order balance.

A released order is assigned a work order number, either through the system or by the user. It also generates a pick list for storeroom parts, and can develop a material availability check on all parts and components that go to make up the assembly required based on the bill of material configuration that exists at the time of order release. A shortage list may be issued based on this check. Orders can be released in batches by time or individually.

Table 8.9 Types of MRP Orders and Their Properties

Order Type	Explodes to lower levels	Resched- ules auto- matically	Generates action messages	User control Quantity/Date
Planned	Yes	Yes	No	No
Firm Planned	Yes	No	Yes	Yes
Released or Scheduled Receipt	No re-explosion	No	Yes	Yes, with restrictions

Executing MRP

Process

A material planner is usually responsible for maintaining a materials requirements plan. This planner must ensure that purchased material is available at the point of use for the factory to start production.

1. Check that the appropriate MRP parameters have been set by item or product family.

   See data requirements for MRP above. All these settings affect the calculations of MRP and can result in significant variations in the planning results.

   Ensure the planning horizon is set for at least as long as the longest cumulative lead time (preparation time + purchasing time + manufacturing time + assembly time), of the parts being planned.

2. Run the MRP.

   With today's computing power, a daily MRP update is recommended. The frequency of running action and exception reports depend on the planner's capacity to deal with the messages. At a minimum, MRP's recommendations should be reviewed weekly.

3. Examine and respond to the Material Exception report.

   Example:
   Report ID: MRP200   Materials Exception Analysis Report
   07.30am 7 July '94   P1

Part Number	Description	M/P	QTY	Reqd Dt.	Type	Due Dt.	Reference	Action
824326193	Cone machine	P	104	5/22/94	PO	7/15/94	SL200854	Resch In
847053683	Chassis	P	120	6/15/94	PO	8/15/94	SL200768	Resch In
844243402	Clamp	M	400	9/21/94	WO	7/10/94	6457	Resch Out
840304562	Screw=112	P	1200	9/10/94	PL	7/6/94	7/07/94	Cancel
824322820	Ty Wrap	P	569	6/13/94	PL	6/28/94	7/07/94	Past Due

Symbols used: M/P represents manufacture or purchase.
The Reqd Dt. is the required or MRP calculated date.
The Due Dt. is the due or promised date from the supplier or shop.
Type is the type of order with PO being purchase orders, WO work orders, and PL planned orders.

4. Take action on the report. "Past Dues" and Reschedule In" are the most important as they may result in a stock-out or line stoppage. "Reschedule Out" and "Cancel" are important too, as these lead to reduction in inventory and investment, and improvement in cash flow.

5. After taking action on the "Past Due" and "Reschedule In" parts, a "Parts Status" screen should be checked to determine the actual stock of these parts. Parts that show as stocked out or close to stock-out

should be cycle counted. If there is no stock, the parts should be expedited. MRP advocates often imply that MRP eliminates expediting. The reality is that due to variability there will invariably be some orders that are late. These have to be expedited to prevent the customer order from having a late delivery.

6. Run a Purchase Order Action report to determine what orders to firm. There is a similar report that indicates what orders to release.

   Example:
   Report ID: Pur300   Purchase Order Firmed Action Report
   7.30am 7 July '94   p1

Part Number	Description	UM	QTY	Firm Dt.	Delq. days	Reqd Dt.
846781458	Cover	EA	250	5-23-94	52	8-25-94
106988223	Fiber, Green	M	360	7-15-94		9-05-94

Symbols used: UM is unit of measure.
Firm Dt. is the date when the order should be placed.
Reqd Dt. is the required or MRP calculated date.

For A and B items, before placing the order, it is advisable to check the demand to ensure that the order is valid and is still required per the MRP date. Very often the "Schedule Out" messages are not acted on by entering new dates in the system, and as a result the purchase order action is premature. The material planner should de-expedite and cancel orders as recommended.

7. Examine a report showing purchase order coming due. Ascertain the criticality of the parts based on-hand availability and their effect on the order completion.

8. Measure the effectiveness of MRP on a weekly basis in terms of stockouts and inventory investment.

9. If possible, set up inventory targets and an inventory budget. The scheduled receipts line of the MRP represents work in process. The projected available line is the on-hand inventory for each period, and the planned-order receipt and planned-order release lines represent future work in process.

Dos and Don'ts

DOs

1. Do emphasize and re-emphasize that MRP requires detailed, meticulous attention.
2. Do understand that the validity of MRP depends on the accuracy of the inventory and ordering records. It is essential that an active cycle counting program and regular verification of the order status is in place.
3. Do watch out for units of measure. A wire assembly measured in pieces having a unit of measure of *each* may have 50 *inches* (its length) entered in the quantity required, thereby planning 50 pieces instead of one.
4. Do ensure that error messages indicating that the bill of materials is incomplete or not available are corrected.
5. Do determine frequency for updating (regenerating) MRP. This will depend on the variability of the product-sales environment with more change in demand and supply requiring more frequent updates.
6. Do respond to action messages. If the action messages appear to be too numerous, go back and change the sensitivity factors.
7. Do monitor MRP with an ABC approach. A items should be set up for minimum lot size and safety stock. They should also be monitored by the most experienced planner or the materials supervisor.
8. Do clean up invalid demand. Shipped or canceled orders all too often remain and continue to drive requirements. Similarly parents can be issued or canceled without the components being in tune, and thus the components demand remains open. Demand should be reviewed periodically from the top down to ensure that it is valid and accurate.
9. Do have a regular program of validating shop inventory. All manner of input errors are made on the shop floor. These result in the work-in-process inventory being misstated and consequently an incorrect MRP being generated.
10. Do understand how MRP treats inventory balances. Usually MRP has a separate inventory file that contains only those inventories that MRP uses to make its calculations. For example, rejected material or material under review should not be included in this file. The users can decide what inventory they want MRP to use in its planning.
11. Do try to use the power of the MRP program to calculate an inventory budget and a projected cash flow.

DON'Ts

12. Don't neglect to train personnel engaged in working with MRP. These should include not only the materials management group, but also the engineers who develop and monitor the processes.
13. Don't fail to ensure that lead times are correct. If lead times are overstated, more demand is created up front and the program tends to have too much work in process. Additional work in process increases lead times, and the vicious cycle begins. On the other hand, if lead times are under-

(*continues*)

Dos and Don'ts

DONT's (*continued*)

    stated,orders will be late and customer service will suffer. Lead times must be continuously monitored. Here again the ABC approach is recommended.

14. Don't neglect to respond to messages to move out or cancel orders.

15. Don't allow demand to be overstated. The MPS should be validated and the MRP should be run through a critical

constraint check before it is loaded on the shop.

16. Don't let the informal system of expediting through "hot lists" to become prevalent. This will happen if MRP does not call the right priorities.

17. Don't forget that the single biggest reason for ineffective MRP is inaccurate records, particularly inventory balances. Continuously check stock on hand.

Appendix 8.1A: MRP Output: Horizontal Report

Materials Requirements by Part

Report No: MRP 100   Date 01/03/94   Run: 1/3/94 8.00A

Part No. A220   Descrp: Coupler 10.5   UoM: Each   M/B: B   C.Code: E   ABC: A

QOH: 50   InLoc: Y31   Alloc: 0   Open Ord: 20   GrossReq: 170

FLT: ULT: DStk: Vendor: 2 wks   BC/PC: BRW   Safety: 0   OrderPol: Lot 20

S u p p l y	Order No.	Q	Value	Vendor	DueDt.	MrpNeed	STA
	Corn 194	20	$20000	Corning	01/08/94	01/14/94	P.O.

	Past Due	01/10/94	01/17/94	01/24/94	01/31/94	02/07/94	02/14/94	02/21/94
Gross Requirements		20	30	20	20	30	10	20
Sched. Receipts			20					
Projected Available		30	20	0	0	10	0	0
Order Due					20	40		20
Order Start			20	40		20	20	

	02/28/94	03/07/94	03/14/94	03/21/94	03/28/94	04/04/94	04/11/94	04/18/94
Gross Requirements	20							
Sched. Receipts								
Projected Available	0							
Order Due	20							
Order Start								

	04/25/94	05/02/94	05/09/94	05/16/94	05/23/94	05/30/94	06/07/94	06/14/94
Gross Requirements								
Sched. Receipts								
Projected Available								
Order Due								
Order Start								

D	Where Required	DemandTyp	Quan	Date
e	AMP2	WO	20	01/08/94
m	AMP#	WO	30	01/15/94
a	RD 1	CO	10	01/14/94
n				
d				

Abbreviations used

Part No.: Part number of item

Descrp.: Description of part

UoM: Unit of measure

M/B.: Make or Buy

CCode: Commodity code: dividing parts by commodity

ABC: Classification by value or importance of part

BC/PC: Buyer or Planner code

Order Pol.: Determines what lot size technique is to be used

QOH: Quantity on hand: This is the number of pieces in the factory

InLoc.: Location where on-hand quantity is located

Alloc.: Quantity on allocation, that is reserved for an order

OpenOrd.: Supply Orders open (Scheduled Receipt)

Gross Req.: Total Gross Requirements against the part

FLTime: Fixed lead time. This is the time independent of the quantity made

ULTime: Unit lead time. This is the time to make one piece
DTStk.: Dock to stock. This is the time taken from receipt, through inspection to storage
Vendor: Vendor lead time
Safety: Safety Stock

Pegging Information

Supply
Order No.: Work order or Purchase order number
Quan: Quantity on order
Due date: When the order has been promised by shop or vendor
MRPNeed: Date when MRP needs the part
STA: Status. Order Planned, Firmed, Released as Work Order (WO) or Purchase (PO)

Demand
Where required: Order number of the demand
Typ: Type of order for which the item is being made—forecast, customer order, warehouse order, etc.
Quan: Quantity on order
Date: Date required

Appendix 8.1B: MRP Output: Vertical Report

Courtesy: Interactive Software, Burlington, MA

Part No. 840305098   Descrp: Screw, BHSC, 138-32x.500   UoM: EA   Qty on Hand: 5125

Order Pol: L Time Bucket   M/P: P   Qty in MRB: 0

Lead: 140   Safety Stock: 900   Qty in RII: 0

Begin Bal: 4225

Due Date	Type	Reference	Pegged Ref	Supply	Demand	Balance
09/23/93	RQ	(W)2135	N		20	4205
12/09/94	RQ	(P)275436	Y		20	4185
05/17/95	PO	SL201275		14000		18185

05/22/95	RQ	(P)765612			48	18137
05/31/95	RQ	(P)765618			20	18117
06/01/95	RQ	(P)766637			120	17997
06/08/95	RQ	(P)766853			40	17957
06/15/95	RQ	(P)766854			40	17917
06/22/95	RQ	(P)767424			20	17897
06/29/95	RQ	(P)767425			40	17857

Special Keys
F5–Pegging Refs   F6–Requirement detail   F7–Print   F8–Another part

Example: Using F6 to get RQ 275436 (Requirement of 12/09/94 above)

Order Type	PL(Planned)
Order No.	275436
Parent Part	A-838490-ACBB
	AMP PAIR UNIT ASSY, ACBB
Order quantity	1
Warehouse	LS 200

Abbreviations used

Part No.: Part Number   **Descr:** Description of part
UoM: Unit of Measure
OrderPol.: Order Policy: L Time Bucket = Set at number of days (FPQ)
M/P: Make or Purchase
Lead: Lead time in days
Qty on Hand: Quantity on hand
Quantity in MRB: Quantity in material review board
Quantity in RII: Quantity in receiving inspection
Begin Bal.: Beginning balance
Due Date: Date required by MRP
Type: Type of order: **RQ**—system-generated demand; **PO**—purchase
      order; **PL**—system-planned purchase
Reference: Order number of demand or supply: **W**—Work Order; **P**—
      Planned Order
Pegged Ref.: **Y** yes there is a pegged reference to parent part number
Supply: Receipt of parts form work order or purchase order
Demand: Requirement of part
Balance: Level of planned stock (excluding safety stock)

Appendix 8.2: Economic Order Quantity

Process

The *economic order quantity* (EOQ) is a fixed-order quantity (lot size) rule that specifies the number of units to be ordered each time an order is placed. The EOQ is the quantity that balances the cost of ordering with the cost of storage. The calculation assumes that the ordering costs and the carrying costs are known and remain constant and vary uniformly with lot sizes. The calculation also assumes that the rate of demand for an item is known and is constant.

It is obvious that there are very few, if any, situations in which the elements of cost and demand are known with any degree of certainty, and they are seldom constant. The EOQ calculation should be used as a guide and to understand the trade-off involved. As the order quantity increases, more average inventory is stored, with the consequent increase in storage costs; but the number of orders placed is reduced, with the consequent decrease in ordering costs.

A graph on the costs makes the trade-off clear:

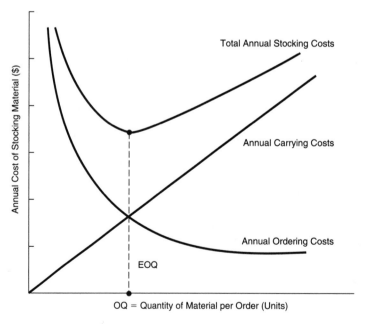

Figure A8.2-1. EOQ: Ordering Costs versus Storage Costs

Calculation of the EOQ is as follows:

D = Annual demand, in units
U = Unit cost
S = Ordering or setup cost, in dollars
I = Inventory carrying cost, in %, cost of money
OQ = Order quantity

$$\text{Number of orders (Setups)} = \frac{D}{OQ}$$

$$\text{Cost of ordering} = \left(\frac{D}{OQ}\right) \times S$$

$$\text{Average inventory} = \frac{OQ}{2}$$

$$\text{Cost of storage} = \left(\frac{OQ}{2}\right) \times U \times I$$

Then the optimum order quantity (EOQ) is the quantity at which the ordering cost is equal to the carrying cost, or:

$$\frac{D}{OQ} \times S = \frac{OQ}{2} \times U \times I$$

$$OQ^2 = \frac{2D \times S}{U \times I}$$

$$OQ = \frac{\sqrt{2D \times S}}{U \times I}$$

It is worth repeating that the EOQ should not be used to determine lot size as the assumptions behind the formula are unrealistic. The EOQ equation is useful as it illustrates the trade-off between inventory carried and number of orders or setups. It can be converted to the more useful period order quantity (POQ).

The approach to the EOQ should be to make the cost of the setup or ordering insignificant, so that the size of a lot is determined by the size of customer order. This would minimize the carrying costs and ensure that there will be no excess inventory.

III

Manufacturing Execution Overview

There is a lot of literature on manufacturing planning but comparatively little has been written on manufacturing execution. Publications on manufacturing execution have been in the context of MRP II and include shop floor control, input output control, and order prioritization. These approaches do not deal with *how to* execute. This book deviates from the traditional approach by attempting to describe how to implement and use specific practices and techniques needed to manufacture effectively.

This book primarily addresses the discrete product industry. Within this industry all manufacturing approaches will be covered—make to stock, assemble to order, make to order, and engineer to order. It does not specifically address the process industry, although there are many similarities between the process and the discrete product industry.

There are two main chapters in the unit on manufacturing execution. The first chapter is Lead-Time Management. One of the key objectives of an effective manufacturing operation is to reduce lead time while maintaining or improving the required throughput. There is a view of manufacturing that relates *all* effective manufacturing activity to reducing lead time, and determines the value of an activity by its effect on lead time. This view is legitimate if it is qualified with making a quality and profitable product. This chapter is based on this view and the execution practices chosen in the chapter all directly affect manufacturing lead time.

The second chapter is Inventory Management. Inventory is the medium in which manufacturing operates, and its management largely determines the profitability and return on investment of the operation. It also directly affects lead time. Inventory management is thus critical to manufacturing effectiveness.

The chapter Lead-Time Management first reviews *just-in-time* (JIT) as a complete approach to manufacturing. The overview of *process relationships* examines the relationships between queues, arrival and service

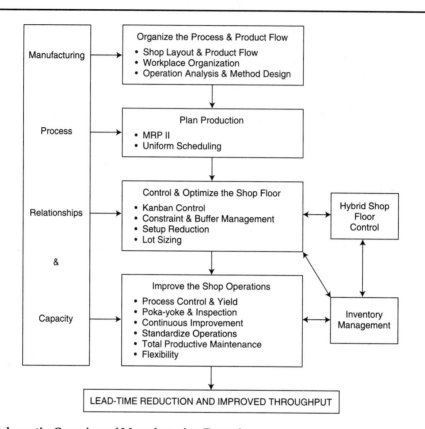

Schematic: Overview of Manufacturing Execution

variability, and throughput and capacity. These relationships are common to all manufacturing activities.

Practices that improve the flow of product and the process are first described. *Shop layout* and *product flow* are the primary means of minimizing the time spent on transportation and product movement. *Workplace organization* and *work method design* ensure that operations are set up to be efficiently executed.

Shop scheduling is covered with *MRP II shop floor control*'s use of dispatch lists and input-output control, and with *uniform demand scheduling*'s synchronization of cycle times.

Activities that optimize operations on the shop floor are described next. *Kanban control*, a form of scheduling, is covered as a means of work-in-process control, and is followed by *constraint and buffer management*, a priority-setting technique. Several lead-time reduction activities are covered including *lot sizes and transfer lots* and *setup reduction.*

A *hybrid shop floor control system* that combines the planning strength of MRP II, and the execution strength of Kanban and constraint management is described as an optimum system to operate a manufacturing factory.

Activities for improving the shop floor execution include two means of reducing process variability: *statistical process control and yield* and *poka-yoke*, and some general practices such as *work standardization, continuous improvement*, and *total productive maintenance* all leading to greater *flexibility*. The chapter is reviewed and all the techniques described tied together with a section on *manufacturing lead-time reduction.*

The chapter on inventory management provides the practitioner with specific methods for controlling inventory in the factory or warehouse. The chapter starts with an overview of inventory, followed by the *ABC technique* that is one of the most effective ways of prioritizing approaches to controlling inventory. The section on *material ordering* covers general ordering principles and explains the need for and the calculation of safety stock. The process for obtaining *record accuracy* is described, followed by specific techniques to *reduce inventory*, including obsolescence and a section on *storeroom management.*

Two general sections on material cost concepts and performance measurement criteria are covered. A section on an inventory management integrates the whole inventory control program. A brief treatment of purchasing in the form of supplier partnerships and best purchasing practices closes this chapter.

As in Unit II, the emphasis of Unit III—Manufacturing Execution Overview—is on *how to* execute manufacturing practices. As mentioned before, there is a lack of information on how manufacturing is executed. The competitiveness of Japanese products in the 1980s and the focus of their system (just-in-time) on shop floor execution has provided new emphasis on manufacturing at the factory level. This is a healthy development. In the United States there may have been too much emphasis on planning and replanning manufacturing using computerized systems. Most of the published literature on manufacturing execution has been on Japanese practices or applying Japanese techniques to U.S. companies. This is not to imply that such techniques are not being practiced in the United States.

There is need for better shop floor management techniques, particularly in environments subject to frequent demand changes, with small production runs and with product and process variations. Currently there is no satisfactory means of manufacturing effectively under these conditions in the discrete product industry. Perhaps there will be a factory floor management system in the future that will anticipate or recognize demand, product, and process variabilities. Such a system would also provide the factory floor with alternative approaches to making optimum decisions on what jobs to process and where to process them. In the meantime, it is essential that best manufacturing practices are instituted and propagated.

The purpose of this unit is to describe *how to* execute some effective manufacturing and inventory practices.

9

Lead-Time Management

Summary of Chapter

The essence of manufacturing execution is lead-time reduction. This chapter describes the practices and techniques used to manage and reduce manufacturing lead time. Sections that cover the organization of the product and process flow are described first, followed by sections on schedule planning. Techniques to control and optimize the shop floor are described next, along with a hybrid system for controlling the whole operation. The chapter ends with sections describing practices that improve shop floor operations.

Contents

- Executive overview of just-in-time
- Manufacturing relationships
- Shop layout and product flow
- Workplace organization
- Operations analysis and work methods design
- MRP II shop floor control
- Uniform scheduling and cycle time
- Applying a kanban control system
- Using different kanban systems
- Constraints, bottlenecks, and buffer management
- Lot sizes and lead time
- Transfer lots and lead time
- Setup reduction
- Hybrid shop floor control systems
- Statistical process control (SPC) and yield
- Poka-yoke and other forms of inspection
- Continuous improvement and problem solving
- Standardizing operations

- Total productive maintenance (TPM)
- Flexibility
- Reduction of manufacturing lead time: summary
- Manufacturing execution elements and lead-time reduction

Relationship of Lead-Time Management to Other Chapters in the Book

Unit II on manufacturing planning developed the factory schedule and ensured that there were material and capacity available. Chapter 9 implements this schedule, after first ensuring that the shop floor and workplace are properly organized. As a result of the execution of the master schedule, inventory is built and has to be managed. Lead-time reduction is also critical to customer service, as it ensures on-time delivery of customer orders.

Executive Overview of Just-in-Time

Just-in-time (JIT) was first formally practiced by Toyota, who called their manufacturing control system the "Just in time and respect for people system." This title was too long, so the "respect for people" part was dropped! There is also a story that after World War II there were two major Japanese auto plants. Nissan, the more affluent, purchased specialized machinery from the West, while Toyota had to make do with old general-purpose machinery and was forced to develop a system that had to make many different models on the same machine. This condition led to manufacturing with limited inventory, small lots, and quick setup changeover—three key characteristics of JIT.

JIT has philosophical underpinnings and a practical set of best practices. The philosophical approach is related to reducing "waste," while the execution portion relates to reduction of lead time and using a minimum of inventory.

The most common definition of just-in-time is that JIT is an operating philosophy with a basic objective of elimination of waste. *Waste* is defined as " . . . anything other than the minimum amount of equipment, materials, parts, space, and workers' time, which are *absolutely essential*." This is the definition of Taiichi Ohno, the founder of Toyota and the individual most responsible for the development of the Toyota production system. Waste includes any operation that does not add value to the product and includes such commonplace activities as incoming receiving, waiting, transportation, queuing, inspection, storage, and making transactions! It is estimated that most processes have only 10 to 20 percent of value-added time in their total time.

Perhaps JIT is best considered as a statement of manufacturing objectives and effective manufacturing techniques. The "just-in-time" comes from one of the objectives, which is to make material available to the production line *only* when needed. This underscores a focus JIT has of working with a minimum of inventory. Shigeo Shingo considers that the most significant difference between Japanese and American production systems is that the American system condones stock. Shingo says, "In the Japanese production system, stock is an *absolute* evil [his italics]. The basic Japanese strategy is to eliminate all factors that necessitate stock" (Shingo 1987).

Inventory reduction is a key objective of JIT. The other key objective of JIT is the reduction of lead time in all phases of manufacturing. To achieve these two objectives there are several effective manufacturing techniques, including *Kanban* or *pull* for shop floor scheduling; product and process *layout;* synchronized *supplier* deliveries; *small lot* production and *setup* reduction.

A feature of JIT is that it provides visible signals to control operations and move material—cards, lights, flags, banners, and methods of stacking material are all used as signals. Labels identifying material and indicating its disposition are also used extensively.

JIT's objectives are attained in an infrastructure that demands:

- Uncompromising quality with control maintained by the operator
- Respect for people by driving decision making to the person executing the job
- Continuous improvement in all activities

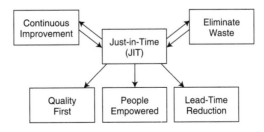

Figure 9.1. Overview of Just-in-Time

Change

JIT embodies several major organizational cultural changes that have to be institutionalized. Among these are:

- Top management support. This must be visible, vocal, persistent, and continuous. Merely endorsing the JIT philosophy verbally or in a commitment statement is not enough. Effective top management support is difficult to sustain, but is the most influential factor in achieving the JIT objectives.

- Middle management's understanding and commitment. This is almost as difficult to secure and sustain as the support of the top management. Here again the commitment has to be visual and continuous. In spite of all the gains in team empowerment and delegation, the local managers do retain a strong influence on the direction a company takes.

- A change from managers directing and approving to managers advising and coaching. Personnel must be allowed to make their own work decisions, preferably in a team.

- A companywide culture of encouraging open, honest communication and relationships within the company and with suppliers and customers.

- The willingness to learn new techniques and practices and work with them until they are proven and productive.

- The willingness to develop and accept a new set of financial and performance measurements.

Summary of JIT

JIT is not a panacea to all the problems of manufacturing. It has limited multi-level material planning or look-ahead capability, and has to react to change. JIT is most effective when operating with standardized, uniformly loaded, repetitive products, and cannot respond too well to changing demand. In a dynamic environment with variable demand and variable lead times, JIT may require more inventory to meet customer need dates. Without formal pegging capability JIT cannot easily trace components to product, or product to customers and vice versa. This makes it difficult to control complex, multi-level products with JIT.

The use of hybrid systems, combining the planning capability of MRP II and the shop floor responsiveness and control of JIT, are becoming increasingly popular and are discussed in a later section. Finally, the techniques used in JIT are not Japanese but are universally representative of effective manufacturing.

Manufacturing Relationships (Variability, Queues, Capacity, Lead Times, and Throughput)

An Overview of Manufacturing

The purpose of manufacturing is to deliver a quality product, when needed by a customer, at a profit.

- Manufacturing planning, the second part of the book, addressed planning and ordering the required quantity of the right products.

- The third part of this book addresses execution, with a primary focus on reducing the lead time to manufacture—*lead time* being defined as the time required to perform a process—from start to finish.

In most companies, over 80 percent of the time a product spends on the factory floor is idle time—that is, it is not being worked upon. Winning in today's hotly competitive market demands maximizing responsiveness to our customers. The emphasis is on reducing the total order fulfillment time, from order entry to shipment. This, in turn, demands that we minimize the manufacturing time in the cycle. This not only reduces the cost of the product, but also helps improve customer satisfaction by reducing promised delivery times and by increasing the probability of on-time delivery.

An ideal manufacturing process is one where raw materials and components are input at one end, flow through their conversion process, and are delivered as finished goods at the other end. A good analogy is the processing of product through a series of pipes in an oil refinery. Two critical factors govern the rate of flow of product through the pipes:

- First is the *rated capacity* of the pipes and will include such factors as size, working pressures, and so on. This is comparable to the equipment and plant of a factory.

■ Second is the *variability* of the process, or, using the pipe analogy, impediments in the pipe or the product that prevent flow as planned. This is comparable to the day-to-day operations.

Improving either or both of these factors will usually lead to a reduction in lead time. In manufacturing the focus is to control the variability of the process.

Another critical goal of manufacturing is improving throughput. *Throughput* is the rate of product through a facility.

Given sufficient product in a line, reducing lead time will increase throughput. For most practical purposes, reducing lead time and increasing throughput may be treated synonymously.

Inter-Arrival and Service Times: Utilization Ratio

Some basic terms need to be defined. The term *arrival rate* describes the amount of work that arrives at a given workstation per unit of time. The related term *inter-arrival time* describes the time between arrival of work at a given workstation.

Example: Auto bodies reach paint at an arrival rate of 30 per hour. The time between arrival of each auto body, or inter-arrival time, is 2 minutes.

The term *service rate* describes the amount of work that is completed at a given workstation per unit of time. The related term *service time* (i.e., *operation time*), describes the time it takes to process a unit of work at that workstation.

Example: Auto bodies are painted at a service rate of 60 per hour. The service time per auto body is 1 minute.

For manufacturing to be effective over the long haul, the service or operation time must be *less* than the inter-arrival time.

Example: If it takes 1 minute to paint a car and cars arrive every 2 minutes, there should be no delay. If, on the other hand, the painting time increases to 2 minutes or more, the painting of cars will start to be delayed.

There is a special term to describe the relationship between service time and inter-arrival time:

$$\text{Utilization Ratio} = \frac{\text{Service Time}}{\text{Inter-Arrival Time}}$$

The *utilization ratio* may be thought of as an indicator of capacity available. As long as the utilization ratio is less than or equal to 1 (i.e., the service time is less than the inter-arrival time), manufacturing can process work without delay. If, however, the utilization ratio is greater than 1, manufacturing does not have the capacity to process the work and delays occur.

Variability and Queues

When variations in the service time and inter-arrival time occur, the picture becomes complicated.

To understand the effect of variability on an operation, the phenomena of *queues* must be understood. In any operation, as long as the service (operation) time is less than the time between arrivals, *and there is no variation*, all product will be processed as soon as it arrives, there will be no waiting time, and the lead time will be equal to service time. At lower utilization ratios there is sufficient capacity to absorb variations in the service time and/or inter-arrival time. As the utilization ratio approaches 1.0, any increase in service time (operation takes longer), or decrease in inter-arrival time (persons arriving more quickly), results in processing delays. Delay results in queues being formed—that is, in people or work waiting—and in lead time being increased.

> Example: To illustrate this concept, consider a hamburger stand. If the time to make a burger is 2 minutes and a person arrives every 3 minutes, there will be no waiting to be served. The stand has sufficient capacity to meet the current demand. If, however, a person arrives every 2 minutes, then the additional capacity to serve is reduced. Now any increase in service time (burgers not being on hand, grill malfunctioning), or reduction in inter-arrival time (such as a busload of people arriving), will lead to people waiting and an increase in the total time to serve a person (lead time). Figure 9.2 illustrates this.

In manufacturing, the main sources of process variability are variations in inter-arrival times of jobs reaching work centers, and variable service or operation times. The arrival of work at a given workstation may fluctuate for many reasons, such as uneven receipt of customer orders,

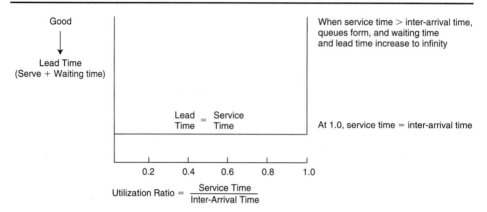

Figure 9.2. Relationship between Lead Time, Service Time and Inter-Arrival Time

bunched release of orders from other work centers, defective product, late delivery by suppliers, and so on. The service time (operation time) may fluctuate for many reasons also, such as a large quantity of parts in an order, equipment problems, variable setup times, lack of operator skill, and quality problems.

Variability is the enemy of all manufacturing, as it reduces capacity, leads to the formation of queues, and increases lead time. It is the main culprit of delayed customer order shipment. In order to reduce lead time it is critical that the source of variability be determined and corrected.

Figure 9.3 illustrates the effects of variability on lead time. The vertical axis measures lead time in any units, such as days or minutes (waiting time + service time), and the horizontal axis is the ratio of service time to time between arrivals—the utilization ratio. The curves drawn are for increasing levels of variability, with 0.1 being low variability and 2.5 being high variability. Observe that lead time increases abruptly for utilization ratios above 0.8, suggesting that, at an 80 percent utilization, a process with almost any variability leads to a substantial increase in lead time. Table 9.1 summarizes the effect of variability shown in Figure 9.3.

Arising out of the relationship between variability, utilization, and lead time two general rules may be stated:

■ *As variability increases, lead time will increase and/or capacity (utilization) will be lost.*

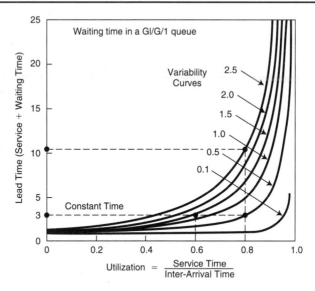

Figure 9.3. Relationship between Variability, Utilization, and Lead Time

**Table 9.1 Relationship between
Variability, Utilization, and Lead Time**

Variability	Utilization	Lead Time
0.5	0.8	3
2.5	0.8	10.5
0.5	0.8	3
1.5	0.6	3

■ *Small changes in service time and inter-arrival time will lead to large changes
  in lead time.*

Process for Dealing with Variability

1. Observe the work centers where queues form consistently. These
   work centers are likely to be constraints in a process.

2. Determine the type of variability the constraining work centers are
   subjected to.

- Record the times when new work orders arrive and keep a log of the work orders in queue at the constraining work center. This will determine arrival variability.

- Record how the work center performs and note times and reasons for the work center not being productive. This will determine service variability.

3. Control and reduce arrival and service variability. Among the principal reasons for variability are:

Table 9.2 Arrival and Service Variations and Possible Corrective Actions

Arrival Variability	*Action to Be Taken*
Erratic or unpredictable order arrivals	Improve scheduling by working with customers
Expediting "hot" orders	Generate and follow a schedule
Nervous or unstable schedule	Improve master scheduling techniques
Large batch quantities	Reduce lot sizes

Service Variability	*Action to Be Taken*
Machine breakdown or malfunction	Perform preventative maintenance
Operator unproductive or erratic	Train and motivate operator
Defective process	Apply statistical process control
Unstable or variable process	Provide buffers and safety stocks
Manufactured material not available	Improve MRP process. Improve constraint management
Purchased material not available	Improve supplier performance
Operators learning	Balance training with stable operators

(Specifics of action to be taken are detailed in subsequent sections of this chapter. See Appendix 9.10 for the calculation of variability.)

4. If variability cannot be reduced below a certain point, understand the trade-off between the lead time and the utilization of capacity. Higher levels of utilization cannot afford variability.

5. Determine the expected manufacturing lead time based on the existing variability and level of available capacity utilized. Use this for lead time planning.

6. Measure the lead time of the process. Analyze increases in lead time and take corrective action.

Dos and Don'ts

DOs

1. Do observe where queues consistently form. These are constraining work centers.
2. Do remember that the output of the entire process will be determined by the throughput of this constraining work center.
3. Do understand the trade-off between variability, capacity, and lead time.
4. Do understand the effects of variability. An otherwise capable process may become a constraint because of arrival fluctuations.
5. Do start to improve constraining work centers first. Remember, a small improvement in service time will lead to a large reduction in lead time.

DON'Ts

6. Don't expect major improvement without reducing variability.
7. Don't expect a process to have no variability.
8. Don't operate on a utilization ratio of over 80 percent. Allow a 20 percent capacity reserve for variability.
9. Don't neglect to set a lead time goal for a process and measure the lead time continuously.

Shop Layout and Product Flow

Purpose and Description

The purpose of an effective shop layout is to minimize the handling and transportation of product. This in turn will reduce manufacturing lead time. A secondary benefit is to improve the space utilization.

In most factories, material-handling equipment moves material between different operations and processes. Material also moves to testing stations, inspection points, and storage areas. Very often all these work areas are not laid out for an overall efficient product flow. As a consequence material may travel long distances, in many haphazard directions, before the product is manufactured. This movement consumes unnecessary time and labor and costs money.

Discussion

The origins of shop layout and product flow go back to the 1880s and Frederick Taylor, who instituted *time and motion study* with his work on shoveling coke at the Bethlehem Steel works. The approach grew into a function called *industrial engineering* and was, in part, responsible for the phenomenal productivity of the United States. One of the cornerstones of motion study was *method improvement*, a technique that included *work flow* as a basic requirement. One has to read a standard work such as *Motion and Time Study* by Ralph M. Barnes, first published in 1937, to appreciate the precise methods established on laying out a workplace and ensuring that the most productive work method was used. Sadly, *Time and Motion Study* fell out of favor in the 1970s and only in the late 1980s has there been renewed interest (the NUMMI plant in California has helped to reestablish the approach). The Japanese, on the contrary, place great emphasis on shop layout and product flow, and their focus has helped to revive this important technique. Shigeo Shingo, in *Non-Stock Production* (1987), states that the West treats processes and operations as being on the same axis (type of work). He contends that instead they are on different axes. Shingo defines *processes* as "being the flow of products from one worker to another, that is, the stages through which raw materials gradually move to become finished products." He further defines *operations* as a "discrete stage at which a worker may work on different products." He includes *processing, inspection, transport,* and *delay* in processes. Finally, Shingo says quite categorically, *"In improving production, process phenomena should be given top priority"* [italics his].

The two principal methods used to record a product or process layout are *flow diagrams* and *process charts.*

Flow Diagrams

In its simplest form, this technique traces the path a product traveled, with a pencil line or a string, on a scaled drawing of a factory floor (with all facilities in place). Such a graphical illustration of product flow clearly demonstrates movement inefficiencies. The method also measures the distance traveled and the time taken. Facilities are rearranged, operation sequence is changed, and movement is reduced, effecting considerable economy in time and money. An example of a flow diagram is:

AFTER

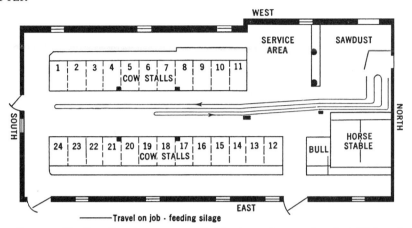

Flow diagram of feeding silage to cows on small dairy farm—improved method. Distance traveled, 199 feet.

BEFORE

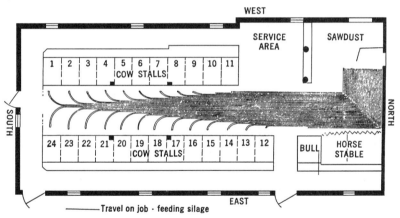

Flow diagram of feeding silage to cows on small dairy farm—old method. Distance traveled, 2070 feet.

Figure 9.4.   Example of a Flow Chart Applied Before and After Layout Improvement Reprinted with permission John Wiley & Sons, Inc. *Motion and Time Study* Ralph M. Barnes, 1990.

Process Charts

A process chart graphically records all the tasks performed in a process. All tasks are classified in one of the following categories: operation, transportation, inspection, delay, and storage. Each task is represented by a

The activities of flow charting have been expressed in symbol form and are widely accepted throughout industry. They are as follows:

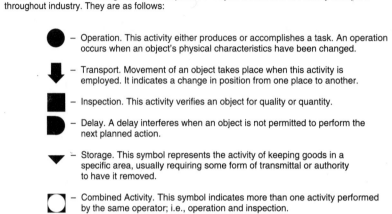

   – Operation. This activity either produces or accomplishes a task. An operation occurs when an object's physical characteristics have been changed.

   – Transport. Movement of an object takes place when this activity is employed. It indicates a change in position from one place to another.

   – Inspection. This activity verifies an object for quality or quantity.

   – Delay. A delay interferes when an object is not permitted to perform the next planned action.

   – Storage. This symbol represents the activity of keeping goods in a specific area, usually requiring some form of transmittal or authority to have it removed.

   – Combined Activity. This symbol indicates more than one activity performed by the same operator; i.e., operation and inspection.

These symbols have been standardized by the American Society of Mechanical Engineers.

Figure 9.5.   Process Chart Symbols

separate symbol: Process charts are either *man type* or *material type,* depending on whether the task traces the movement of a person or a product. The charts can be applied to whole factory layouts or section layouts. It can also be used to track a person or a group of people. The distance moved and the time taken at each task should always be recorded. A typical process chart used by General Electric in the 1970s is shown in Figure 9.6. Observe that in addition to tracing the type of process step used, the chart has provision for questioning the step by seeking to combine, eliminate, or change the step.

Process for Preparing Process Charts

1. Determine the process to be studied. Decide whether the subject to be followed is a person or product.

2. Choose the starting and ending points to determine the scope of the chart.

3. Have a good understanding for the steps of the process and how they can be classified into operations, transportation, inspection, delay, and storage.

4. Chart the process. Make sure to include the distance traveled and the time taken at every step.

GENERAL ELECTRIC COMPANY

FLOW PROCESS CHART

NO. ___1___
PAGE ___1___ OF ___2___

Analysis					QUESTION EACH DETAIL
Why					
What	Where	When	Who	How	

SUMMARY

	PRESENT		PROPOSED		DIFF.	
	NO.	TIME	NO.	TIME	NO.	TIME
○ OPERATIONS	1	1				
▷ TRANS.	6	43				
☐ INSPECTIONS	0	-				
D DELAYS	2	9				
▽ STORAGES	5	-				
DIST. TRAVELED	1195 FT.		FT.		FT.	

JOB ___BLANK UNDERCOVER INSULATION___

☐ MAN OR ☒ MATERIAL ___KRAFT INSULATION___
CHART BEGINS ___AT BUTLER BLDG.___
CHART ENDS ___AT ASSEMBLY___
CHARTED BY ___J. SMITH___ DATE ___8/9/59___

	DETAILS OF (PRESENT) METHOD	Symbols	Dist. in Ft.	Quantity	Time	Possibilities (Eliminate / Combine / Sequence / Place / Person / Improve)	NOTES
1	At Butler Bldg.	○▷☐D▽					All Raw Stock Storage
2	To Receiving Dock	○▷☐D▽	600	20		X	2 Trips/Day
3	At Dock	○▷☐D▽			X		Roll Stock on Pallets
4	To Cover Room	○▷☐D▽	285	8			
5	At Storage Rack	○▷☐D▽				X X	Double Handling
6	To V & O Reel	○▷☐D▽	30	2			
7	Blank	○▷☐D▽		1			
8	At V & O Punch Press	○▷☐D▽		7			In-Process Storage
9	To Storage Rack	○▷☐D▽	30	6		X	In Prove Time
10	Storage	○▷☐D▽			X X		
11	To Pre-Treat Area	○▷☐D▽	200	5			
12	At Pre-Treat	○▷☐D▽					Double Handling
13	To Assembly	○▷☐D▽	50	2			Material Handler
14	At Assembly	○▷☐D▽	1000				In-Process Storage Tote Pans
15		○▷☐D▽					
16		○▷☐D▽					
17		○▷☐D▽					
18		○▷☐D▽					

Figure 9.6. Example of a Process Chart from a General Electric Internal Guide

5. Include on the chart a summary of the number of operations, moves, inspections, delays, and storage. Include the time taken against each category. Note the distance moved.

6. Use a basic approach to improve the process, reducing the distance traveled or the time taken, by:

   ■ Eliminating all unnecessary work—including handling, inspection, and so on
   ■ Combining operations or elements
   ■ Changing sequence of operations
   ■ Simplifying operations

7. Chart the improved process and compare the distance traveled and the time taken for the new method.

   A complete example of a process chart exercise may be seen in Appendix 9.1.

Product versus Process Layouts

Discussion: Traditionally factories were laid out to support a process or function, and had similar types of machines grouped together. This layout leads to a lengthy and convoluted movement of products. A process layout also causes accumulation of work in process inventory, excessive material handling, increased lead time, and defects being detected long after they are caused.

In contrast, a product-oriented layout is organized by the sequence of processes the product undergoes. This results in processes being closely linked and the product moves through less distance than it does

Table 9.3 Summary of Process Layout versus Product Layout

Characteristic	Process Layout	Product Layout
Distance traveled	Usually long and convoluted	Minimized
Inventory	Inventory accumulated at handoffs	Minimized
Lead time	Tends to be longer than product	Minimized
Material handling	Usually excessive	Minimized
Supervisory control	Minimized	May need more supervision
Operator requirement	Can have limited skills	Must be multi-skilled
Machine utilization	Utilization maximized	Tends to be underutilized

in the process layout. However, such a layout does lead to reduced supervisory control and individual equipment may not be used to full capacity. Operators also have to be flexible and trained to perform many functions. In addition to reduced material movement, this layout also reduces lead time, work-in-process inventory, and material handling. A product layout is also called a *focused factory*.

Product layout depends on operators with multiple skills and general-purpose machines or machines that are small and can be dedicated to a single line.

Process Mapping

Process mapping combines the elements of flow diagrams and process charts to form a map of the process. The physical flow of the product is traced and the operations and their times are noted, making it easy to see unproductive movements and their magnitude.

The U Layout

Many companies use a *U layout,* not only to reduce material handling, material movement, work-in-process inventory, and lead time, but also to provide a means of dealing with variations in work load by changing the number of workers in a cell. The fact that workers can communicate with each other directly is considered a major advantage of this layout. Parts come directly to the first operation and then get moved through with very little, if any, backtracking or redundant movement.

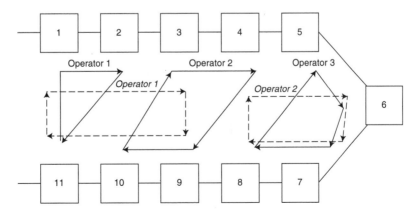

Figure 9.7. U Layout Showing Reduction from Three to Two Workers

The above process is normally operated with three operators. Operator 1 works on Operations 1, 2 and 11; Operator 2 on Operations 3, 4, 9 and 10; and Operator 3 on Operations 5, 6, 7, and 8. If there is a reduction of product demand, the process can be worked by two operators. In this case, Operator 1 will handle Operations 1, 2, 3, 9, 10, 11, while Operator 2 will handle Operations 4, 5, 6, 7 and 8.

Combining U Layouts

The logic of using combined U layouts is similar to reducing operators in a single U layout. In Figure 9.8 (left), one operator works all the operations in the four processes. If the demand for products is reduced, two operators can operate two processes each, a reduction of one or two operators (Figure 9.8 right).

Combining a U Layout with a Focused Factory

A U cell making components can be combined with a focused-factory assembling product. This layout is most useful when a product combines standard parts and subassemblies (made in a focused factory), with a variety of end products (assembled in cells).

Process for U Layouts (Cells)

1. Ensure that the layout supports the manufacture of a product or product family.
2. Purchase equipment that has capacity to meet the likely demands of a single product line (not highly specialized, high-production machinery).

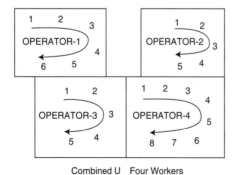

Combined U   Four Workers

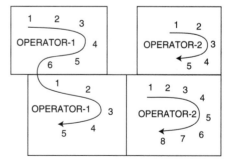

Combined U   Two or Three Workers

Figure 9.8. Combining U Layouts

3. Set up equipment so that it can be moved easily. Ensure quick connection of power, air, and data lines.
4. Avoid the use of mechanized material-handling equipment.
5. Avoid AS/RS (automated storage and retrieval systems).
6. Ensure that operators are trained and capable of performing most of the operations on most of the machines.
7. Plan the layout placing products with similar operations side by side, in parallel. This will ensure that the same operators can work more than one product line.

Dos and Don'ts

DOs

1. Do lay out a product's proposed or existing process of manufacture with a flow diagram.
2. Do use a process chart to understand the time and distance that the products will travel.
3. Do develop an improved layout based on 1 and 2 above and ensure that the processing factors of transport, inspection, and delay are minimized.
4. Do develop a U layout, where possible.
5. Do ensure that the layout allows flexibility in processing and readily permits change in the future.
6. Do ensure that operators are cross trained.
7. Do improve a product's layout first and then improve the detailed operations.
8. Do remember that a U layout empowers operators and this in turn significantly enriches their job and their productivity.

DON'Ts

9. Don't install machinery and plant in immovable positions.
10. Don't underestimate the value of a product layout on reducing lead time.
11. Don't install utilities (power or air lines) and communication cables under the floor as this makes relocation difficult. Ideally, utilities and communication lines should be part of an overhead grid that has quick connect/disconnect capability.

Workplace Organization

Purpose and Description

The purpose of organizing a workplace is to reduce cycle time and hence manufacturing lead time. Secondary benefits include improved quality, reduced waste, and better morale.

A workplace must be organized so that there is no loss of productivity. Disorganized workplaces lead to wasted efforts because of having to find materials and tools and runs the risk of contributing to defective product. Further disorderly surroundings do not motivate personnel to take pride in their work. A clean and organized workplace, where there is "a place for everything and everything in its place," may be a cliché, but when neglected it leads to wasted time and effort.

Discussion

The systematic organization of the workplace was a practice followed by Taylor and his disciples, but, like process layouts, this practice fell into disuse. The Japanese, through their insistence on meticulous workplace cleanliness, and through their demonstration of the correlation between workplace organization and worker productivity, revived the importance of this simple but effective discipline. Tools and fixtures are clearly identified and located for easy retrieval. Work-in-process inventory is labeled and stored in predetermined locations, with maximum and minimum levels clearly marked, to enable any worker to know what is to be done. Storage should allow ease of loading and retrieval. The Japanese use the term *seiton*, which means "to neatly arrange to identify things for ease of use."

The gain is threefold:

- First there is the obvious improvement in productivity by having everything at hand.
- Secondly, personnel receive a psychological boost when working in clean surroundings.
- Finally, problems become more identifiable as something out of place usually represents a possible process or product shift.

Good housekeeping rules and workplace organization are essential to effective manufacturing.

Process

1. Organize for a clean factory floor by providing receptacles for garbage and insisting that litter is properly disposed.

2. Provide for the work area to be cleaned, either by having a cleaning agency or by allowing the workers time to clean their workplace.

3. Organize the workplace with a definite storage space for tools and fixtures at the point of use. A shadow board is a very useful fixture for storing tools and gauges.

4. Locate all tools and materials close to the point of use.

5. Use color codes to associate specialized tools with their equipment. (Keep a blue tool with blue machine, for instance.)

6. Hang frequently used tools in racks or overhead with their handles foremost. Ensure that the retrieval of frequently used tools requires minimal movement. Design tools for continuous use with one hand.

7. Label each tool and fixture with its description and reference number. Also label all material storage space and provide clear marking for maximum and minimum levels of stock.

8. Ensure that calibration stickers are attached to gauges and tools to confirm that they are accurate.

9. Provide bins within arm's reach for easy picking of small items. Move parts close to the worker or introduce inclined planes or chutes to reduce walking or body movement.

10. Set up storage of material to allow the first-in, first-out (FIFO) loading and unloading.

11. Provide special purpose racks to take advantage of gravity. Parts should be fed down an inclined plane, so that no move and pick is necessary.

12. Set up tools and materials to support the operations sequence.

13. Minimize the time spent on material handling. There is a handling progression that goes from having to open up, lift, and carry parts to having parts ready for use. Toyota has an index of liveliness that measures how much material handling is required on a part. It is essential to minimize the time spent on handling parts.

14. Provide a large, colored (red) label on inventory that has been lying unused and collecting dust. Arrange to examine and eliminate all unnecessary inventory.

15. Where possible completed parts should be dropped and not lifted and placed.

16. Provide good lighting and ventilation.

Dos and Don'ts

DOs

1. Do walk around and pay attention to the cleanliness of the workplace. Make cleanliness a visible priority.
2. Do make housekeeping everybody's business.
3. Do ensure that there are definite places for all tools, fixtures, and materials.
4. Do use color coding and colors to identify tools easily.
5. Do ensure that the gauges, tools, and fixtures for the job at hand are identifiable and easily accessible.
6. Do locate materials at point of use, within arm's reach.
7. Do ensure that the material handling of parts is minimized.

8. Do pay attention to the location, storage, and identification of material. Use first-in, first-out, clearly labeled storage devices with indicators for maximum and minimum levels.

DON'Ts

9. Don't make an operator bend to pick up parts, tools, or accessories.
10. Don't treat housekeeping as just another chore. It has effects that far outweigh its apparent lack of importance.
11. Don't neglect to ask the worker which part of his job is onerous and how it can be improved.

Operations Analysis and Work Methods Design

Purpose and Description

The purpose of improving work design and work methods is to reduce the time taken on an operation and to reduce the manufacturing lead time.

Operations are performed by people and machines. For work to be performed productively:

- Operations must be improved by conducting a method study and redesigning the work method.
- Operations must incorporate the principles of motion economy.
- Proper procedures must be in place describing how the operation is to be performed.

Operation analysis and work method design is most effective on high-volume repetitive product.

Discussion

Even though much of the pioneer work in operations improvement was done in the United States (the Gilbreths), it remained neglected for many decades. Method design and motion study have been out of vogue till quite recently. This is unfortunate because the function provided a means of continuous improvements in workplace productivity. When the Japanese publicized their *kaizen* (continuous improvement) approach to manufacturing operations, many companies embraced the concept and the practice. It is ironic that the technique had to be disguised to be reintroduced in the United States!

In *Non-Stock Production* (1987), Shigeo Shingo defines an *operation* as "a discrete stage at which a worker may work on different products." He makes a strong argument that processes and operations are not similar. Processes cover the flow of material and consist of five elements, namely processing, inspection, delay, storage, and transport. Operations are the activities performed by machines or people. Operations consist of *preparation* or *setup*, *main* and *incidental* operations, and *allowances* such as fatigue, personal hygiene, operation, and workplace. All incidental operations and preparation must be eliminated and the main operation should be improved. Shingo writes, "Operation analysis is done principally by *motion study*—devised by Gilbreth, and *time study*—devised by Taylor" (1987). Again it is ironic that one of the Japanese gurus of manufacturing quotes Gilbreth and Taylor and their techniques in advocating productivity improvements, while in the United States their methods have been neglected.

Paul S. Adler (1992) describes how the GM-Toyota joint venture NUMMI has succeeded in using an innovative form of Taylor's time and motion regimentation on the factory floor, not only to create world class productivity and quality, but also to increase worker motivation and satisfaction.

Analysis of Work Methods

The best opportunity for coming up with an effective manufacturing method is when a new product is being designed or developed. (Design for manufacture has been covered in Chapter 2 of this book.) Process charts and flow diagrams, as described in the section on product layout, help to improve the process. Work method design seeks to ensure that an operation is performed in the shortest possible time with the least effort.

To do this, the fewest possible body members and body motions should be used through the shortest distance. The job should result in the least expenditure of energy and mental stress. There are specific techniques for improving operations, the most commonly used being activity charts and operation analysis charts.

Activity Charts

Activity charts plot operations against a time scale. A simple example is shown in Figure 9.9. It may be seen that wasted activity becomes apparent immediately. Another variation of the activity chart is a *worker-machine activity chart*. Here the activities of the operator and the machine are plotted against time. The chart suggests ways of eliminating idle machine time and helps balance the effort of the operator with the running of the machine.

Operation Analysis

After improving the use of a machine, a detailed review of the motions of the operator is often conducted. This is known as *motion study,* and its purpose is to record and analyze the motions performed by an operator, with a view to eliminating all unnecessary motions and rearranging necessary motions in the most efficient manner. All hand motions are classified into 17 motion elements, or *therbligs* (Gilbreth spelled backward). These include search, select, grasp, transport empty, transport loaded, hold, release load, position, pre-position, inspect, assemble, disassemble,

Description of activity	Time (minutes)	Description of activity	Time (minutes)
Select 2 books for a shelf	2.0	Select 20 books for shelves	10.0
Walk to appropriate shelf	1.0		
Put books in proper place	2.0		
Walk back to book desk	1.0	Load in book trolley	5.0
For 20 books repeat 10 times		Walk to appropriate shelf	1.0
		Put books in proper place	10.0

Old Method   Total   60 min   New Method   Total   26 min

Figure 9.9.   Example of an Activity Chart

use, unavoidable delay, avoidable delay, plan, and rest for overcoming fatigue. (Refer to any standard text on industrial engineering for a detailed description and discussion on the elements. Ralph M. Barnes's *Motion and Time Study*, 1968, is a good reference.)

Operation Charts

These usually record the elemental motions of the hand. Two symbols are commonly used in making these charts:

- A small circle represents transportation—moving the hand, reaching, carrying, and so on
- A large circle represents action—grasping, positioning, assembling, and so on

Even for the simple example shown, the power of this technique is evident in showing the motion of the hands. This process is usually adopted on assemblies involving both hands (such as a nut to a bolt). The idleness of the hands is clearly visible. There are standard checklists that can be used to analyze each of the fundamental hand motions used. (See Appendix 9.2 for an example of an operation chart.)

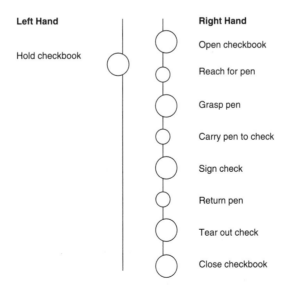

Figure 9.10. Operation Chart (Showing the Motion of Two Hands in Signing a Check)

Micro-Motion Study

A video camera records the motions of the operation. The frames are set against a timeclock, so that the time taken for each element is recorded. The film can be analyzed and unnecessary motions eliminated and necessary motions balanced effectively. This technique is particularly helpful in setup reduction.

Process for Improving Work Methods

1. Involve the union or the operators in the plan to improve the work methods. Stress the fact that the reason for the study is to improve productivity and not to eliminate jobs.

2. Provide education on the techniques that will be used.

3. Select constraining or bottleneck operations (see the section on constraint management).

4. Perform an activity analysis of the operation and determine redundant, nonvalue-added activities. Determine unbalanced operations between worker and machine and idle time being spent by either.

5. Eliminate, combine, and change activities to develop the most productive operation. Consult with the operator during this process, and preferably have a team involved in the determination of what should be changed.

6. For Steps 3 and 4, use the worker's input on how to improve the operation.

7. Conduct a motion study of the operation. The use of a video camera, with the worker's consent, is the preferred method.

8. Apply the principles of motion economy to analyze and improve the operation. The main principles that should be applied are:

   - The two hands should operate in balance and unison—start, move, and complete the operation. Movement should be smooth and continuous.

   - Motions of the wrists, hands, or arms should be made together in opposite directions.

- Hand and body motions should be performed in a sequence that allows the least expenditure of energy and results in the least fatigue. The sequence that should be followed starting with the lowest effort is:

  finger motions

  finger and wrist motions

  finger, wrist, and forearm motions

  finger, wrist, forearm, and upper-arm motions

  finger, wrist, forearm, upper-arm, and shoulder motions

  finger, wrist, forearm, upper-arm, shoulder, and trunk motions

  Use the fewest motions at the lowest possible end of the effort scale.

- Product momentum should be used where possible to help the worker. Gravity should be used to advantage.

9. Locate tools, and material to permit the best sequence of motions as indicated in Step 7 above. The height of the workplace table should also support the minimum use of motion effort.

10. Ensure that jigs and fixtures are used to relieve the hands of as much work as possible. Use leg-operated devices where possible.

11. Use as few gauges and tools as possible and avoid having to put them down and pick them up. Combine gauges and tools. Preset gauges and tools where possible.

12. Locate all operating handles, levers, wheels, switches, and so on, so that the operator can use them with a minimum of effort. Here again the sequence of least-effort motions should be applied.

13. Ensure that dimensional readings, if taken, have an ergonomically designed screen or gauge.

14. Review the entire operation with a checklist to ensure that the motions of each operational step have been thoroughly analyzed and improved or eliminated. (See Appendix 9.3 for a checklist of fundamental hand motions.)

16. Ensure that the workplace is maintained and well organized.

17. Measure reduction in the time taken for the operation, before and after the improved method is implemented.

Dos and Don'ts

DOs

1. Do conduct an activity and motion study on all operations that are associated with a bottleneck or potential bottleneck.
2. Do involve the operator or the area improvement team with the effort to improve the operation. They should be the owners of the new job design.
3. Do use the motion study as the prime driver of method design and workplace layout.
4. Do use a standard checklist (see Appendix 9.2) for improving the operation.
5. Do document the improved process as a work standard to be followed.

6. Do use a video camera recording if possible (obtain union and operator consent first).
7. Do keep the work at a level that does not require reaching or stretching.

DON'Ts

8. Don't allow an activity where the body has to be bent, unless it is unavoidable.
9. Don't conduct work method improvement before improving the process (see the section on shop floor layout and product flow).

MRP II Shop Floor Control

Purpose and Description

The purpose of MRP II shop floor control is to help the shop make a required quantity of a product in time to satisfy the demands of MRP.

In manufacturing planning, the path of a customer requirement was tracked from forecasting through to material requirements planning, where the components to make the end product were determined. The output of the planning process is a shop order or a purchase order authorizing a specific quantity of a specific part to be made or purchased by a due date. It must be stressed that this section covers the *shop floor* control part of MRP II, and not the planning modules. (The planning modules have already been described in Unit II of this book.)

Discussion

A work order for a part is usually released when MRP issues an action message to release the order (based on the required date offset by the parts manufacturing lead time). A work-order packet is issued. This contains a *stores pick list*, a *routing sheet* (or *traveler*) for the part, *job* or *labor time tickets,* and a *header sheet.* All documents cover the details of the part to be made, the quantity required to be made, and the date when the order is to be completed. The packet is sent to the store, where the parts are picked, and the parts and order are sent on to the shop floor. The work-order packet follows the material, as it gets machined or built from work center to work center. As each operation is completed, the routing sheet gets signed off by the operator and the quantity and date get recorded. Inspection is also recorded on the routing sheet, with parts rejected being reduced from the order quantity. In most MRP II systems, the parts are recorded on the system as they move along their routing. For such systems, MRP also issues a daily priority listing, called a *dispatch list,* of the work orders awaiting execution at each work center. The dispatch list advises the shop supervisor what to work on. Various options are available for determining how the order priority should be established, such as by *order due date,* by *shortest processing time,* or by a *critical ratio* (comparing the time remaining with the work remaining). By planning shop orders in anticipation of customer orders, MRP *pushes* production through a factory. This is in contrast to JIT where production is initiated or *pulled* in response to a final assembly schedule that in turn is generated in response to a customer order. Today MRP systems exist where the whole process is paperless or electronic. This is the preferred process.

MRP shop floor scheduling is based on predetermined lead times for each part's operation or assembly. Required dates are calculated by backing out the lead time from a customer required date, at each level of the bill of material, to the level of the part being considered. This is called *backward scheduling.* Unfortunately, the lead times used are at best an average, as they are subject to schedule and process variability. These variations invalidate the priorities established by the dispatch list. This is probably the biggest weakness of MRP, and it makes the shop floor control system unreliable. The lead times that MRP II uses to schedule cannot be dynamically adjusted to actual variability. This is the reason that so many companies run the shop floor with *hot lists,* as manual dispatch lists reflect the latest changes on the shop floor. The limitations described above are generic and the remarks made are not meant to be critical of the performance of practitioners.

The shop floor module of MRP II is recommended for use only in environments where lead times are stable and predictable. Since most products and processes have variability, manufacturing should be executed by techniques that make provision for such variability. These techniques (such as Kanban pull, constraint management, and the hybrid system) are considered in later sections of this chapter. The suggestion for not using the shop floor does not extend to the rest of MRP, which, as will be seen, provides planning capability not found in JIT. In fact, the hybrid shop floor control system combines the planning strengths of MRP with the execution strengths of JIT.

Many companies do use the MRP shop floor control module. The process described below will help to optimize their efforts.

Process for MRP II Shop Floor Control

1. Ensure that every manufactured part has a routing that details the sequence of operations, the work centers, and the labor requirements (persons and time) to which a part has to be manufactured.

2. Release work orders based on MRP's action message (based on the lead time required to manufacture the part). This in turn should generate a work-order packet with a routing, labor tickets, and a stores pick list. The stores pick list is copied from the current bill of material for the part.

3. *Do not release* a work order without first determining that material for the part is on hand (*part status*) and that capacity to make the part is available (*work center load*). If material or capacity are not available, action has to be taken to get the parts (expedite suppliers) and develop additional capacity (overtime or subcontract). If neither material nor capacity can be made available, the work order should be rescheduled. If material is available for only part of an urgent order, make a partial order.

4. If a factory has an order backlog, the orders should not be released to the floor until they are required to be released per the action message of MRP. Backlogs should be kept in the office.

5. Ensure that the parts required for the work order are picked by the storeroom and sent with the order packet to the shop floor.

6. Determine the priority rules to be adopted in developing the dispatch list—that is, the sequence in which orders are executed. Some of the rules are:

   ■ First-in, first-out. This is a simple rule, but it does not consider the required date of the order.

- By required date. This uses the priorities of MRP.

- Shortest processing time. Here the orders that take the least time are worked on first. This helps to clear most orders in a time period. It does not consider the required dates of the order.

- Critical ratio. This is the time remaining divided by the work remaining. The smaller the ratio, the higher the priority. A critical ratio of less than 1 technically means that the order cannot be completed in time.
  Set the priority rules in the MRP parameter setting.

7. Run MRP daily, but run exception reports only as often as action can be taken on its messages. Also run MRP off peak hours.

8. Attempt to follow the dispatch list. If there are major variations from the planned operation time or lead time, inform the planner to reschedule the order. Similarly, operation times or lead times that are constantly above or below that specified should be changed in the system.

9. Ensure that the shop floor system is updated—accurately and soon after the operation is completed.

10. Establish an *input-output* report for the constraining work centers only. Maintain the report tracking and controlling the work entering and leaving the work center. The work load maintained should be commensurate with the work center's capacity to complete the work as scheduled. (A sample of an input-output report is shown in Appendix 9.4).

Dos and Don'ts

DOs

1. Do run MRP as frequently as there are changes in planned events or the order placed. Since updating time is now minimal in most systems, MRP can be run daily.
2. Do generate a dispatch list listing the priorities of all work orders by work center.
3. Do constantly update the routing standards of operation times based on actual performance. Record actual lead times for completion of all products and update the master record if necessary.
(continues)

Dos and Don'ts

DOs (*continued*)

MRP does all its scheduling based on these inputs.

4. Do control inputs into the factory floor by maintaining a check on capacity through an input-output report.

5. Do keep work in process to the minimum. It determines lead time and throughput. Do work with the smallest lot sizes that are consistent with the set-up time.

DON'Ts

6. Don't allow a manually generated hot list to be followed. The entire MRP system will lose credibility if this is allowed. If MRP is wrong, correct the error and the reason for the error.

7. Don't allow orders to go past due. Either provide more capacity to work them (overtime, subcontract, and so on), or reschedule them.

8. Don't schedule orders a factory cannot make because of insufficient materials or capacity.

Uniform Scheduling and Cycle Time

Purpose and Description

The purpose of uniform scheduling is to reduce arrival variability and queues, and hence to reduce lead time.

The main cause of increased lead time is the formation of queues, which in turn is a function of variability. *Variability* has two main forms—*arrival*, or the rate at which product arrives to be worked upon; and *service*, or the time taken to complete an operation. Arrival variability is due to material arriving at workstations at irregular intervals, or arriving with uneven lot sizes. Uniform scheduling is an attempt to level load a factory by scheduling a consistent amount of work, commensurate with the capacity of the factory. Uniform scheduling also seeks to develop a common cycle time (operation time) across all operations of the entire work process.

Discussion

Customer demand must be managed as it does not come in evenly. It is essential that the random patterns of customer demand are organized into a uniform schedule. There are two parts to demand: the type of product and the volume. Uniform scheduling seeks to even out the irregularities of volume and combine product mix to develop a level, constantly increasing or decreasing (linear), weekly and/or daily workload.

For repetitive, make-to-stock, make-to-demand, and assemble-to-order products, the schedule is formulated around a dominant operation cycle time or *drumbeat*. All operations are synchronized to this drumbeat, which should also be equal to the shipping rate.

For nonrepetitive, make to order products, establishing a level or linear schedule is much more difficult and can only be achieved if there is sufficient backlog to enable leveling of the schedule. The mix of parts and the options may vary, as long as a level schedule or steady rate of production can be established for a segment of time (preferably two weeks to one month).

Once this drumbeat is established, all operations are *synchronized* by making them equal to, or in multiples of the drumbeat cycle time. Cycle times may differ, but by making them almost equal to or in multiples of the drumbeat, the entire process can be synchronized. Synchronization of operation cycle times can be achieved by:

- Providing additional resources, such as people and machines
- Improving the work methods
- Changing the schedule when products consume more time

Uniform scheduling does not work well in a make to order environment or with frequent demand changes, as it needs a linear demand. Make to order items of the type described should be processed using constraint management in a hybrid manufacturing system (covered in a subsequent section of this chapter).

Process for Developing a Uniform Schedule

1. Establish an expected product-delivery rate among production, marketing, and sales. The rate should be commensurate with the factory's capacity and the likely sales. Remember that a uniform schedule can be developed only from a regular linear flow of repetitive product bookings.

2. Use a production plan developed from a sales and operations meeting to establish a product delivery rate. (See Managing Production Planning, Chapter 6.) Planning bills can be used to estimate the quantity of individual products from a product family forecast.

   Example: Product family ABCD has a forecast of 28,800 units a month. The planning bill shows that the product family has four products: A, B, C, and D. The historical rate of consumption of these products (planning bill percentages), are A—50%, B—25%, C—15%, and D—10%.

3. Determine the drumbeat or dominant cycle time, based on the required delivery rate.

Example: Family ABCD with monthly sales of 28,800.

Product	%	Monthly Sales (units)
A	50	14,400
B	25	7,200
C	15	4,320
D	10	2,880
Total		28,800

At 20 days = 1,440 units per day
At 8 hours = 180 units per hour
= 3 units per minute
**OR 1 unit every 20 seconds. This is the
drumbeat or cycle time.**

Where the products A, B, C, and D have different work content, they should be converted into equivalents of the smallest time. The products can then be converted to a common base (the product with the least time), and the drumbeat calculated.

4. *Ensure that all operations in final assembly have a cycle time of 20 seconds or less, or multiples of 20 seconds.* This may require dramatic setup time reductions (see section on setup time reduction).

5. Until all operations have cycle times in multiples of the drumbeat, uniform scheduling *cannot* be fully applied. If operations have a cycle time of more than 20 seconds, reduce the cycle time by improving the operation, providing additional resources, or changing the operation.

6. Develop a product mix schedule to ensure that the products are made in the ratio they are sold.

Taking the above example: The ratio of the monthly sales is 5 As to 2.5 Bs to 1.5 Cs to 1 D. On a daily basis, since 1,440 units are produced, this works out to 720 As, 360 Bs, 216 Cs, and 144 Ds.

7. Determine a lot size that will allow the products to be made in a daily sales mix. Here again the setup time can be a deterrent to reducing the lot size, and may have to be reduced first.

Using the same example: The daily schedule of 1,440 units can be made in 20 lots of 72 each. Then a daily product mix schedule for 10 As, 5 Bs, 3 Cs, and 2 Ds can be:

A B A B A C A B A D A B A C A D A C A B

This is called *mixed model scheduling.*

8. If the lot size is too large, reduce it, maintaining the ratio of the products.

   If lot size is reduced to 18, then the product mix schedule can be the same as above, repeated four times.

9. Determine the crew size or the number of workstations, to meet the daily workload, for each operation. This is dependent on the labor content of the operation.

   Example: Assume labor content for an operation is 2 worker minutes per piece or 120 seconds per piece.
   Then number of people required

   $$= \frac{\text{labor per piece}}{\text{cycle time per piece}} = \frac{120}{20} = 6 \text{ persons}$$

   If only 1 person can work on the operation, then there must be 6 workstations to meet the load.

   The multiples of the drumbeat are accommodated by providing proportionately more resources.

10. Synchronize the feeder operation times to meet the drumbeat or final assembly cycle time.

    Example: Assume there are two (2) components per assembly
    component cycle time =

    $$\frac{\text{final assembly cycle time}}{\text{component quantity per assembly}} = \frac{20}{2} = 10 \text{ secs.}$$

    Here again it has to be ensured that the component operations capacity can meet the required cycle time. If this cannot be met, additional resources must be provided or the operation improved to meet the time.

11. Establish rules for allowing change in schedule only after negotiation with marketing. The schedule's product mix should be frozen for two weeks to a month, while the drumbeat or production rate should be changed only after two to three months' notice and planning.

12. Develop and maintain charts in each area, showing the required output and the actual output. On each area chart record the problems and issues, assigning responsibility for taking action.

13. Satisfy short-term increase in demand by working overtime or sub-contracting. Meet short-term decrease in demand by transferring workers to other lines, conducting training, or performing maintenance.

14. If there is demand for a very different product that would disrupt the drumbeat of the regular product, build the irregular product on a special line (even if it is made inefficiently).

15. Use layout and worker flexibility to help overcome variations in the work flow (see Shop Layout and Product Flow and Flexibility).

Dos and Don'ts

DOs

1. Do remember that uniform scheduling cannot be applied to all products and all processes, only to those products with repetitive and steady demand and stable processes.

2. Do remember that customer orders determine the cycle time or drumbeat of the product. If the drumbeat is likely to be changed, ensure that there is sufficient time to synchronize the line to the new rate.

3. Do ensure that some quantity of each product is made every day if the quantities are large and at least weekly if the quantities are small.

4. Do attempt to develop a mixed-model schedule. It helps to stabilize the process and reduces scheduling variability.

5. Do work with sales and marketing to develop and ensure a consistent flow of repetitive product bookings.

6. Do ensure that workers are cross-trained. Multi-functional workers are the best means of providing the flexibility needed to maintain a uniform schedule.

DON'Ts

7. Don't change the schedule in the current week or two to three weeks out (depending on the type of product). The line cannot adjust to unplanned changes without some disruption. The success of uniform scheduling depends on the ability to freeze the schedule for a minimum of two weeks, preferably a month.

8. Don't attempt to make product having problems (and needing rework) on the uniform scheduling line. Deal with such products on a separate line/area.

Applying a Kanban Control System

Purpose and Description

It was noted that MRP II shop floor control worked with fixed lead times and did not dynamically react to variability. The purpose of Kanban control is to ensure a synchronized flow of material that dynamically adjusts to demand and process variations. Kanbans also minimize lead time by limiting the work in process inventory.

Kanban loosely translated means *signal* or *card*. A kanban control system is a means of identifying, authorizing and controlling inventory, primarily work-in-process inventory, by using kanbans or signals. Kanbans also dynamically synchronize the flow of material.

A kanban can be any physical entity or electronic message used to signal or authorize the movement of material and/or the start of production. Cards are a common form of kanbans and contain information on what and how many pieces are to be moved or made. Kanbans can also be fixtures on which product is made, taped spaces on a workbench or floor, a trolley carrying material, and so on. Regardless of the type of kanban used, it must represent a unit or lot of material, and it must be the only authorization for material movement and/or production. Without the appropriate kanban, material *must not* be moved or worked on.

Kanban movement is initiated at the end of a production line, usually with a customer order consuming an assembly and releasing the assembly kanban. The released assembly kanban authorizes work on another assembly, by allowing subassemblies to be pulled. This in turn releases their kanbans to start work on more subassemblies and so on, all the way back to raw material being pulled from suppliers. The whole system is a *pull* process, with downstream or following work areas pulling from upstream or leading areas. The entire pull process should be initiated by a customer ordering or the consumption of a finished part or product. For make-to-stock products, the distribution center plays the role of the customer.

Kanban control is only one part of the entire JIT system.

Discussion

There is an anecdote that Taiichi Ohno, the creator of the Toyota Production System, developed the kanban system from a model of the U.S. supermarket. The supermarket is any material storage point having a warehouse as its upstream or preceding operation and a customer as its downstream or subsequent operation. A customer buys a product, creating the need for a replacement, and the supermarket signals the warehouse to replace the quantity purchased. The warehouse in turn will signal the factory to produce more of the product.

An MRP-driven system initiates action in response to, and in anticipation of, demand. Where forecasts are inaccurate MRP makes product that is not required. Because of variability on the shop floor, lead times used to plan work in MRP are seldom accurate. This may lead to delay in completion of orders. Further, if problems disrupt the line, the upstream workstations (before the disruption) continue to make product (the push system), and inventory accumulates. MRP does not have an easy means of dynamically adjusting its schedule to variability.

A kanban or pull system has the following strengths:

- It initiates action in response to actual demand and dynamically adjusts itself to the variability.

- When there is a major disruption, since the work-in-process inventory (kanbans) is limited, the line shuts down for want of material.

- It focuses attention on correcting the problem, since the line shuts down.

- With limited inventory, quality problems are more manageable, both in locating the problem and in rectifying it.

- When a defect is detected, any operator can shut the line down by stopping work at the defective operation.

- By limiting the work-in-process inventory, kanban control ensures that throughput time is optimized.

- Kanbans clearly establish the link between inventory and lead time and workers see the value of lead time reduction.

- Production workers feel more in control of their area as they own and control the work in process (not thrust upon them by a preceding operation).

- Because work-in-process inventory is minimized, inventory investment is also minimized.

Kanbans provide a visible, simple, inexpensive means of controlling the shop floor.

The weaknesses of kanbans and the pull system have already been described in the section on just-in-time, but they are worth repeating briefly. The pull system cannot deal with:

- Engineered or made-to-order products
- Highly variable demands (frequent changes of product and small order quantities)

- Highly variable processes
- Complex products (having a large number of bill-of-material levels)
- Looking ahead, as it has very little planning capability

These limitations should be kept in mind.

Most of these drawbacks of the kanban system are strengths of the MRP II's planning system. It seems obvious that there is need to combine the two systems and achieve the synergy they provide. This is done through a hybrid planning and execution system (described later).

Using Different Kanban Systems

Process for Using the Two-Card Kanban System

The supermarket model can be used to describe the two-card kanban process.

1. All supermarket product (or lots) on the shelves will have a move or withdrawal kanban card attached to it. A move card will contain information on product description, the quantity it represents, and the route it moves in.

2. When the customer checks out, the move cards are removed and put in a kanban box *M*.

3. At regular intervals move cards are collected and sent to the warehouse, in the order that the supermarket wants replacement. This authorizes replacement parts to be picked.

4. Every warehouse product (or lot) has a production card attached to it. A production card will contain information on product description, the quantity it represents and what is required to make the part.

5. On picking the parts, the production cards are *replaced* by the move cards. The production cards are put in a kanban box *P*.

6. The parts with their move cards are sent to the supermarket.

This completes a loop from customer picking a part, to the checking out and collecting the move card, to replenishing stock from the warehouse to the supermarket using the move cards and releasing production cards.

7. The production cards are collected at regular intervals and sent to the factory, where they are the authority for making more parts. The sequence of cards determines the priority of the parts.

8. Raw material at the factory is picked using the production cards. Move cards on the raw material are removed and authorize ordering more raw material.

9. Parts are made according to the instruction on the production cards, and when made the production cards are attached to the parts.

10. The parts with their production cards are sent to the warehouse.

11. Material is ordered from a supplier with a move card.

It should be understood that the warehouse is eliminated wherever possible, and the signal is sent directly to the factory.

The whole process may be seen in Figure 9.11.

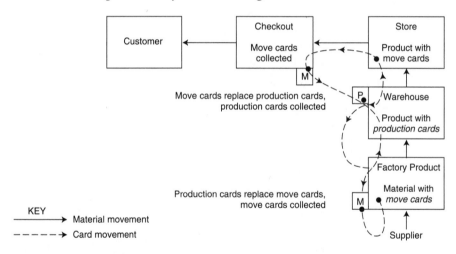

Figure 9.11. Supermarket Model of Two-Card Kanban

Signal Kanban (see Figure 9.12)

A signal kanban is used for lot production. The signal kanban is located at a position on a rack of parts, like a reorder point. When the signal is reached, the kanban authorizes production similar to the production kanban, except that the signal kanban authorizes a lot size (quantity) of product to be made. The location of the signal is at a point that has sufficient parts below it to satisfy demand, while the signaled lot is being made. A material withdrawal kanban may also be attached with a signal kanban, to authorize withdrawal of material. Thus the signal kanban functions as a single kanban and satisfies both move and production.

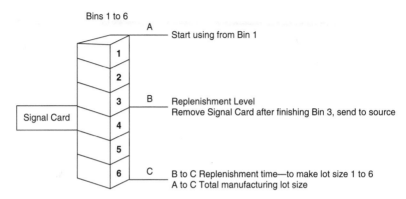

Figure 9.12. Lot Type Signal Card Kanban

Combined Kanbans (see Figure 9.13)

In an assembly process, the move and production kanbans can be combined and used to regulate the flow of work between any two consecutive workstations or operations. The kanban can take any form. One of the easiest and most effective kanbans is a taped area on the work bench or floor, equal in size to the product, located after a workstation. When the downstream workstation (B) takes a product from the kanban space, the upstream workstation (A) is authorized to pull a part from its upstream workstation, work on it, and fill the empty kanban space with a completed part. A whole process can be linked in this manner. The technique is simple, highly visible and effective. The number of kanbans between operations will depend on the throughput rate, the operation time and the variability.

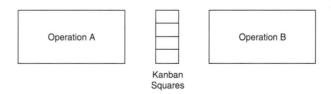

Figure 9.13. Combined Kanban System
Single Kanban: Operation A fills the 4 kanban squares (taped on the table) with parts (one per square). When Operation B pulls a part from a kanban square, Operation A makes another part and fills the empty square. Operation A can work only if there is a square empty.

Single-Loop Kanbans

Where there is a lot of variability in the operation times, it may not be possible or advisable to determine kanbans for individual operations. Rather than provide too many or too few kanbans between operations and adversely affect the throughput, it is better to provide kanbans for the entire process and locate them at the start of the process. By doing this the variability of individual operations can be accommodated without having to balance the line.

Example: The line in Figure 9.14 shows 8 kanbans (cards) for a process with 5 operations, an incoming-material kanban post, and a finished-product kanban post. *Up to 8 kanbans can accumulate at any operation or the kanban posts before the process shuts down.* This is the elegance of this method, as individual operations do not have to be balanced, and variability can be accommodated. The example shows two kanbans are available for starting product and await the completion of Operation 1 to pull incoming material. When Operation 1 pulls material, another kanban for incoming material becomes available to the previous process. Operations 2, 4, and 5 are idle and await product from the preceding operations. Operation 5 will continue to build product and ship to the product-kanban post, until all 8 kanbans are accumulated and the whole process is stopped for want of kanbans. Every operation can accumulate up to 8 kanbans before shutting down the line. This way the line can adjust to the variations of individual operations, while ensuring overall control at 8 parts in work in process.

The rules of kanban are observed and no product is worked on without a kanban. No product is moved until it is pulled by the subsequent

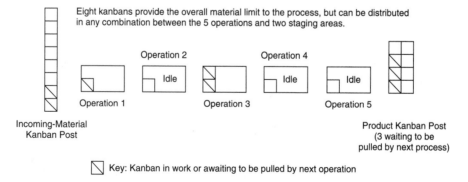

Figure 9.14.   Single-Loop Kanbans Serving an Entire Process

operation. When product is pulled from the product post (the last step) by the next process, a kanban is released and becomes available for pulling incoming material by Operation 1.

Kanban Process (Rules of Kanbans—see Appendix 9.5)

1. A downstream operation always pulls work from an upstream operation. Withdrawal must always be less than or equal to the number of unattached kanbans available.

2. Production must always be authorized by an available kanban to replace product pulled by a downstream operation. Production must not exceed the quantity authorized by the kanban. The production sequence must be in the order that the kanbans were delivered.

3. There should be no movement or production without a kanban.

4. When cards or other discrete kanbans are used, they must always be attached to product.

5. Kanbans must always be attached to quality product. If defective products are discovered, the line should be stopped and the cause of the defect ascertained.

6. The numbers of kanbans should be gradually and continuously reduced until the process becomes inefficient. Reducing kanbans minimizes inventory, improves throughput, and exposes inefficient processes.

7. Small fluctuations in demand ($\pm 10\%$) should be met by increasing or decreasing the working hours, *not* by increasing or decreasing the kanbans.

8. If there are large changes in demand, cycle times have to be recalculated and the number of workers changed. The number of kanbans may also have to be changed.

9. If a bottleneck operation is starved for work, temporarily increase the number of kanbans, labeling them as "emergency" kanbans.

10. Where batch processing of parts is unavoidable (such as in oven curing), consideration must be given to adding a buffer to the process.

Calculation of Kanbans

Kanbans are calculated one of two ways:

1. Where a constant quantity of product is withdrawn at varying intervals

Number of kanbans

$$= \frac{\text{average daily demand} \times \text{lead-time days} (1 + \text{safety factor})}{\text{container capacity}},$$

where lead time = processing time + waiting time + transport time
processing time = time interval between starting a work order (job)
and its completion
safety factor = depends on process variability (stable process = 0.1,
variable process 0.2 to 0.4)

This is the formula commonly used in a factory. The ideal of JIT is to work with one-piece production and transportation.

Example: If 10 units are to be made per day, and a product with much variability takes 5 days to build:
Number of kanbans = $10 \times 5 \times 1.3 = 65$
(lot size / container capacity = 1).
As the process stabilizes, the safety factor can be reduced to 1.1, giving a new number of:
Number of kanbans = $10 \times 5 \times 1.1 = 55$.

2. Where a variable quantity of product is provided at constant intervals

Number of kanbans

$$= \frac{\text{average daily demand} \times (2 + \text{transit delay})(1 + \text{safety factor})}{\text{container capacity} \times \text{deliveries per day}}$$

The transit delay equals the number of deliveries between supplier receipt and return of card. The number 2 is added because this is the minimum number of cards that must be with the supplier at any one time (one being returned and one being picked up).

The safety factor equals the additional days (hours) of stock kept in the store, and depends on supplier variability, such as unreliable suppliers, product with unpredictable quality, long lead time, and so on.

It may be seen that the new factor is

$$\frac{2 + \text{transit delay}}{\text{deliveries per day}} = \text{lead time.}$$

Example: A supplier fills and returns a kanban on the fourth day after it is picked up. He makes 2 deliveries a day. Then the lead time to fill the kanban is (2 + 6) / 2 = 4 days.

This type of kanban calculation is used when working with suppliers. Suppliers make trips at fixed times and fill varying quantities of kanbans.

Reacting to Change

As was mentioned in the section on uniform scheduling, every attempt should be made to freeze a schedule for two weeks to a month. This will keep the number of kanbans stable for that time period. However, when the demand increases significantly (+10%), there may be justification for increasing the number of kanbans. Before the number of kanbans are increased, however, every attempt should first be made to reduce the lead time. The use of overtime or temporary workers may be resorted to if the increasing demand is temporary. Where demand is reduced or where there are reductions in lead time, the number of kanbans must be reduced.

Dos and Don'ts

DOs

1. Do take time to train the operators and all management personnel in how the kanban system works.
2. Do set the production line uniform schedule in consultation with sales and marketing. They must commit to maintaining this schedule.
3. Do ensure that the final pull is the customer—that is, build to demand only.
4. Do ensure that the shop workers understand the working and value of kanbans and adhere to the kanban disciplines.
5. Do take time to lay out the shop floor and plan the kanban technique to complement this layout. Ensure that the kanbans and kanban collection points are clearly marked and understood.
6. Do plan the kanban system around small lot sizes and mixed model production.
7. Do measure the effects of kanbans on reduction in lead time.

DON'Ts

8. Don't use the regular kanban system for highly variable operations. If kanbans are used, they should be the single-loop type.
9. Don't allow product to be built without a kanban. This will require constant vigilance.
10. Don't increase the number of kanbans. On the contrary, seek to reduce the number of kanbans as a measure of the process learning and continuous improvement. The exception to this rule is if the bottleneck operation is starved for work, in which case emergency kanbans may be issued.
11. Don't forget that the kanban system represents a major cultural change in how most factories conduct business. Be prepared to spend a long time in institutionalizing the change and be persistent.

Constraints, Bottlenecks, and Buffer Management

Purpose and Description

The purpose of constraint, bottleneck, and buffer management is to reduce process lead time and improve throughput. Having laid out the shop floor, improved the method design, and introduced kanban control, it is essential that an understanding of operational priorities is developed and used. Which operation should first be improved and then closely monitored to ensure that lead time is minimized? Improvements in layout, method design, and processing must be initiated at the bottleneck.

A *constraint* is an operation that restricts the flow of product. Every process has a constraint—it is the slowest operation in the process. Where the throughput of the constraint is less than the (market) demand, the constraint becomes a *bottleneck*. (That is, the bottleneck cannot supply the market demand.) Time lost at a bottleneck cannot be recovered. To ensure that a bottleneck is never idle for want of work, a *buffer* is provided before the bottleneck. The management of a process is the management of its constraints and bottlenecks.

Discussion

Dr. Eliyahu M. Goldratt focused awareness on the need to manage a process by managing constraints and bottlenecks and much of this section is developed from his writings (1992). A balanced line is a myth. This is because variability is endemic to all operations and most operations are interdependent. Since all operations in a process are interdependent, an operation's time fluctuations tend to be passed on to the next operation, getting progressively larger downstream. This pattern is unstable and will give rise to a wave-like behavior in the internal transfer of work in process, with operations having periods of overload being followed by idle periods.

Goldratt illustrates this phenomenon by using the analogy of marching boy scouts. The scouts are making a product—the product is the roadway over which the troop has completed marching. Work in process is the roadway spread between first and last scout as they are marching. Lead or cycle time is the amount of time consumed by the entire troop to cover a section of the roadway. The objective of the march is for the troop to cover as much roadway as possible. The alternate running and walking of the scouts is analogous to the wave pattern in a plant. Gaps are formed because of different walking paces of the individual scouts (operation times), and large gaps are created toward the last scouts (downstream operations). How can a balanced troop of boy scouts be created? Attempting to set each scout (operation) to march at a precalculated rate will only lead to unbalance because of inherent variability in the capability of each scout. This in turn will lead to the scouts being spread out over a large distance (large work-in-process inventory) and the whole troop will take longer to cover a set distance.

How then is the objective of traveling the maximum distance by the whole troop achieved? By shouting at stragglers—expediting? By trying to get everyone to march at the same pace—quotas? An experienced scout master will recognize that the troop's pace is determined by one or two slow scouts (bottlenecks), and he will put these scouts in front to set the pace (drumbeat). He will also put the fastest marchers at the back to close gaps that are created. Finally, he will keep close watch to ensure that the gaps do not become so large as to create unmanageable waves upon closing the gap.

Goldratt formalized the unbalanced factory phenomenon into a system called the *drum-buffer-rope system.* In addition to the concept of pacing the line by the bottleneck (drumbeat), and making sure there are no large gaps or accumulation of inventory (rope), it is necessary to provide material before the bottleneck (buffer) to protect it against being idle for want of material to process.

Goldratt also stresses the need to ensure that actions are taken to optimize the whole process or factory (the global optimum), and not an individual operation or department (the local optimum). This approach further reinforces the need to concentrate on the bottleneck as it controls the entire process.

Figure 9.15 shows a process for a product with an assembly bottleneck operation of 10 pieces per hour. Each of the two parts making up the assembly have a buffer before the bottleneck operation.

Process for Maximizing Output from Bottleneck Operations (see Appendix 9.6)

1. Understand the requirements or market demand. This will establish the rate of product delivery. If there are no bottlenecks, this will be the *drumbeat.* If there is a bottleneck, the capacity of the bottleneck will determine the drumbeat.

2. Identify the bottleneck operation(s). Queues consistently form in front of a bottleneck operation. Asking the operators is another good

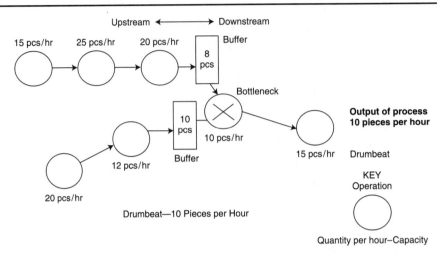

Figure 9.15. Process Showing the Constraint, Buffer, and Drumbeat

means of quickly finding a bottleneck. A bottleneck can also be calculated by finding the highest load/capacity ratio over a period of time (a ratio over 1 indicates a bottleneck).

3. Attempt to open the bottleneck by applying process and method improvements. If the bottleneck is removed, go back and determine the new bottleneck and repeat the process. Machines or equipment at the bottleneck should always be upgraded first.

4. If the process can be re-engineered and the product flow can be changed, ensure that the bottleneck is located as close to the start of the process as possible. This should also apply to new process designs. Earlier operations have less scheduling (arrival) variability and can be controlled better.

5. Place operations with the most capacity at the end of the process. This rule should be kept in mind while determining priorities for increasing capacity. Always start at the end of the process and work back to the beginning. This arrangement minimizes work-in-process inventory.

6. Arrange to release work to the shop based on work completed at the bottleneck. This will ensure that excess inventory will not build up. This is called the *rope*. Using kanbans is a good way to ensure this.

7. Fully utilize the capacity of the bottleneck. This is critical. It may be necessary to overmanage the bottleneck—by providing additional shifts, additional workers, and constant supervision. Staff the bottleneck with the best operators. Have cross-trained operators available. Ensure that the equipment, if any, is under a total productive maintenance program.

8. Inspect material before a bottleneck, thereby ensuring that defective material does not waste precious capacity.

9. Ensure that the bottleneck operation is in control and that its yield is close to 100 percent. If not, focus quality improvement efforts on the operation.

10. Attempt to reroute product to other work centers where capacity is available, particularly where the additional demand is infrequent.

11. Provide a *buffer* in front of the bottleneck operation consisting of material waiting to be worked on. This way, if there is a disruption in any of the operations before the bottleneck, the buffer will ensure that the bottleneck is not idled.

12. Keep a chart of the buffer size to determine if the buffer size is correct. Any person making a withdrawal should note the level of the buffer. This way, over a few months, a record of the range of fluctuations of the buffer is available. The amount of buffer stock should be based on the variability of the process and can best be determined by observing the level below which the buffer almost never falls. As a general rule, $\frac{1}{2}$ to $\frac{1}{3}$ the work in process in a well-run factory should be in front of the bottleneck.

13. Monitor the performance of the bottleneck. Measure throughput and a detailed breakdown of the bottleneck's operation time utilization.

14. Where setup time cannot be reduced, load the largest possible batch size on the bottleneck (in keeping with the other demands). As the operation is being completed, break this batch size into smaller lots and transfer them to the next operations.

   ■ Example: A specialized furnace having a long operation time should be loaded with a large lot size. This lot should be split and sent to subsequent operations in smaller lots.

15. Look at all the constraints in determining the sequence of operations. Very often the linear calculation of lead time can be misleading. Look at setup, run times, and batch sizes, and then prioritize operations sequence.

16. Beware of wandering or volume bottlenecks. These are not really bottlenecks, as their capacity exceeds average demand. However, due to a large batch of work reaching them in a short time, they may not be able to stay current. This is another reason for a buffer before a downstream bottleneck.

Dos and Don'ts

DOs

1. Do remember that *an hour lost at a bottleneck is an hour lost across the entire system.* This is the most significant concept of managing discrete manufacturing.
2. Do remember that bottlenecks govern both throughput and inventory.
3. Do manage a process by managing the full capacity of the bottleneck.
4. Do maintain tight control over all activities of the bottleneck.
5. Do constantly work on improving the bottleneck operation.
6. Do use smaller transfer batch sizes after the bottleneck, if the output is spread over time.
7. Do use kanbans where possible to balance flow (see the section on kanban control).
8. Do work on increasing the capacity of downstream operations after the bottleneck, as this will reduce the lead time.

DON'Ts

9. Don't increase production at a non-bottleneck operation. This will only lead to an increase in work-in-process inventory.
10. Don't compromise quality in an effort to improve the throughput of a bottleneck.
11. Don't try to achieve balance between all operations. Balance flow not capacity.
12. Don't spend too much time working on upstream non-bottleneck operations. Remember that time saved on a non-bottleneck upstream operation will not increase the throughput.

Lot Sizes and Lead Time

Purpose and Description

Processing large lot sizes is the greatest single cause of long lead times. Reducing lot sizes leads to direct reduction in work in process inventory and lead time.

In discrete manufacturing processes, it is commonplace to group units into lots and process the lot through each operation. Large lots exact enormous penalties by creating larger work-in-process inventory and longer lead times.

Discussion

It can be mathematically shown that processing large lots cause the single largest adverse effect on variability. Large lot sizes tie up equipment and increase work-in-process inventory, and this in turn increases the lead time. The Japanese understood this phenomenon and JIT stressed work-in-process and lot-size reduction as critical objectives. J. D. Little, in studying queuing in the 1960s developed a relationship between work-in-process inventory and lead time. Note the similarity between the formula and the kanban calculation! For any process:

Little's Law: WIP inventory = rate of product arrivals × lead time

Goldratt and Fox in *The Race* (1986) vividly show the effects of batch size on average inventory and lead time. In the sketch below the lead time to complete an order of 1,000 units is halved and the average inventory reduced significantly. (See Figures 9.16 and 9.17 of low- and high-level inventory manufacturing.)

Lot sizes affect manufacturing in many ways:

- Large lot sizes can have a negative effect on quality. In the case of small lot sizes, a defect can be detected within one cycle or less of the product and rework is restricted to the parts produced in that cycle. Further the defect is corrected sooner and proper parts are made, rather than having to rework a large defective lot.

- Small lot sizes subjected to engineering change allow improvements or redesigns to be affected early in the manufacturing cycle, and fewer parts need rework.

- Small lot sizes reduce the average on-hand inventory and this translates into improved cash flow.

- Smaller lot sizes need less space and equipment.

- Small lot sizes and shorter lead times permit forecasts to be made later and increase the probability of being more accurate and as a corollary, manufacturing can firm orders later.

- In many cases a small lot size with a short lead time will allow a product to be made to a specific customer order.

Summary: Work-in-process inventory is proportional to lot sizes, and production lead times are proportional to work-in-process inventory. Therefore production lead time is proportional to lot size.

Graphical Example of the Effect of Lot Size:

Lot Size: 1,000 units (Figure 9.17)

Transfer Lot Size: 200 units (Figure 9.16)

Time to complete: 2,450 hours

Time to complete: 490 hours

Total manufacturing time = (0.75 + 0.1 + 1.0 + 0.1 + 0.5) = 2.45 hours.

For 1,000 units: 1,000 × 2.45 = 2,450

For 200 units: 200 × 2.45 = 490

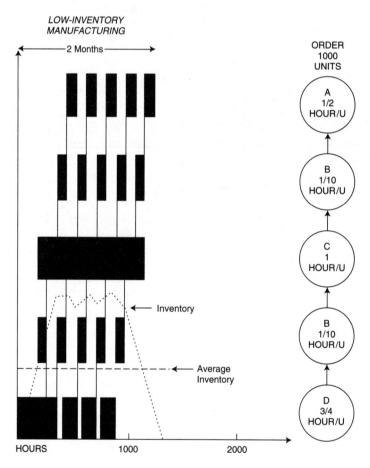

Figure 9.16. Small Lot size
Source: Reprinted with permission from *The Race,* Goldratt and Fox (North River Press, Inc.). Copyright © 1986.

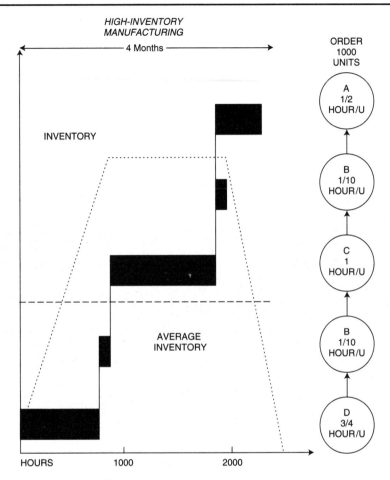

Figure 9.17.   Large Lot Size
Source: Reprinted with permission from *The Race,* Goldratt and Fox (North River Press, Inc.). Copyright © 1986.

Process for Lot-Size Reduction

1. Determine whether there is a large order or a steady demand schedule of products to be made in weekly or daily quantities, and make different products in succession (mixed mode) each taking about the same time (see section on uniform scheduling). If there is an irregular order schedule, then the lot size is determined consistent with known setup constraints.
2. Identify the process bottleneck.

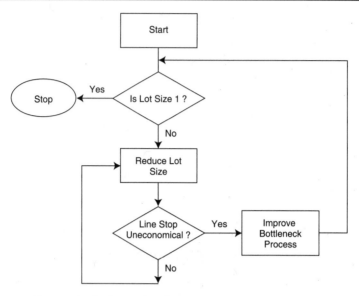

Figure 9.18.   Process for Reducing Lot Size

3. Develop a lot-size reduction strategy. The schematic in Figure 9.18 is a good logic diagram.
4. Reduce the lot size by 10 to 20%.
5. Observe the effect on the line. If there is no effect, reduce the lot size by another 10 percent. If the line slows down, or if insufficient product is made, increase the lot size a little.

    Example of empirical reductions: From EOQ formula:

    $$\text{EOQ lot size} = K \times \sqrt{\text{setup time}}$$

    If setup time is 4 hours and a lot size of 200 pieces is run, then

    $$200 = K \times \sqrt{4} = K \times 2, \text{ or}$$
    $$K = 100,$$

    reducing the setup time to 1 hour. Lot size $= 100 \times \sqrt{1} = 100$, or half the earlier lot size.

6. Improve the process at the bottleneck. Some approaches are:

    ■ Improving the layout and use of tools and fixtures
    ■ Reducing the setup time

- Improving the quality yield of good parts
- Providing for flexible workers
- Improving operator efficiency

7. Check that the line productivity has improved. If not, continue to improve the bottleneck process. If the productivity has improved, go back and reduce the lot size by about 10 percent.
8. Repeat Steps 3 through 6.
9. Ensure that shop layout and material movement is improved as reducing the lot size will increase the number of lots made and hence the movement of material.
10. Observe when the process time increases as the setup time has become dominant.
11. Reduce the setup time (see next section). Start the whole cycle over again from Step 3.
12. Use a transfer batch where it is not possible to reduce lot sizes at the bottleneck operation, or where a large lot size has to be used (processing equipment requirement like a furnace).

Dos and Don'ts

DOs

1. Do remember that reducing lot size is the most effective means of reducing variability and improving lead time (increasing throughput).
2. Do work on the bottleneck process to allow further lot size reductions to be made.

3. Do be aware that as lot size is reduced, setup time will become a critical factor and limit, if not increase, processing time.

DON'Ts

4. Don't neglect to improve process layout and transportation.

Transfer Lots and Lead Time

Description and Purpose

Traditionally one of the ways of selecting a lot size has been to determine the trade-off between inventory carrying costs and the ordering or setup costs. Where a bottleneck operation has considerable setup time that cannot be easily reduced, then, in order to maximize the productivity of the bottleneck, there is need to run a large lot size. In order to reconcile producing a large lot size at the bottleneck with the universal rule of working with small lot sizes, a variable batch size or transfer batch is used. The transfer batch allows for a large lot to be processed at the bottleneck, but splits this lot as it is being processed and transfers the smaller split lots to the subsequent operations. Thus a lot of 1,000 pieces may be processed through the bottleneck, and as 100 pieces are completed they are moved to the next operation, and all subsequent lots are of 100 pieces. The purpose of transfer lots is to reduce lead time without sacrificing throughput.

Discussion

Shigeo Shingo, in *Non-Stock Production* (1987), describes the effect of lot size on delay, which he calls *lot delays*. He says, "Lot delay, a very important concept in non-stock production, is missing in the production philosophy of Europe and the United States!" Process delays such as queuing, waiting, movement, and so on, are noticeable, but lot delays are hidden. The effect is compounded by the old concept that large lot sizes were desirable for spreading the effect of setup time. This can best be illustrated by an example.

Table 9.4. Effect of Setup Time on Lot Process Time

Setup Time	Lot Size	Operation Time	Apparent Operation Time
2 hours	10	1 minute	1 min. + [(2 × 60)/10] = 13 mins.
2 hours	180	1 minute	1 min + [(2 × 60)/180] = 1.6 mins.

It would appear that the operation time per piece has been reduced by 11.4 minutes. Unfortunately, lot sizes increase operation cycle time or lead times.

Using the same example, assume that there are seven operation steps, each taking 1 minute per piece. A comparison of processing the whole lot of 180 pieces to processing a lot of 10 pieces each is shown below:

Operation time per piece = 1 minute (t). Number of operations 7 (n)

If a lot of 180 pieces is processed, it will take 3 hours (T) per operation. Thus, through seven operations:

total lead time $L = n \times T = 7 \times 3 = 21$ hrs

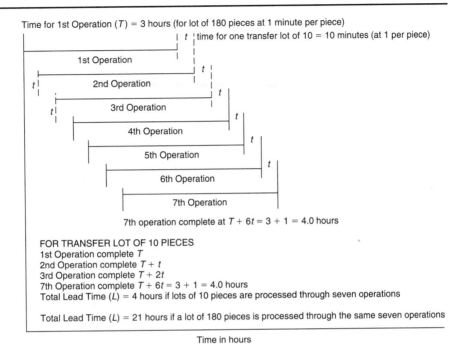

Figure 9.19.   Effect of Transfer Lots on Lead Time

If lots of 10 are split and transferred, each split lot will take 10 minutes per operation,

$$\text{Transfer lot lead time } L = T + (n - 1)t$$
$$= 3 \text{ hours} + (7 - 1) \times 10 \text{ minutes} = 4.0 \text{ hours}$$

as only the incremental time for one lot per operation is increased (see above).

Reduction in lead time is $30 - 4.0 = 26.0$ hrs, or about 87 percent.

If a single piece is transferred after every operation, each piece will take 1 minute per operation,

$$\text{Transfer lot lead time } L = T + (n - 1)t$$
$$= 3 \text{ hours} + (7 - 1) \times 1 \text{ minute} = 3 \text{ hours 6 minutes}$$

Reduction in lead time is $30 - 3.1 = 26.9$ hours, or about 90 percent.

It is clearly advantageous to reduce the transfer lot size.

Setting the Lot Size

As the run time to setup time ratio reduces (as the lot size is reduced), there will come a point when further reduction of lot size will increase the processing lead time and reduce the throughput. This is because setup time will become dominant and time will be spent setting up and not running product. This is shown in Figure 9.20.

It may be seen that the lot size can be reduced up to a point after which throughput time or lead time starts to increase. Furthermore, it may be seen that the point at which this increase occurs depends on the *utilization ratio* (capacity used). If utilization is low, lower lot size to setup ratios can be worked too, leading to reduced lead time. Higher utilization ratios lead to less capacity being available, and this in turn does not allow lower lot size to setup ratios, so the lot size cannot be reduced too much without starting up the curve and increasing lead time (see the graph in Figure 9.20). At all utilization ratios the lot size is reduced to a point after which the setup has to be reduced to maintain lead time. This is the preferred way. Much of the Japanese stress on reduction of setup time stems from their thrust on reducing lead time by reducing lot sizes.

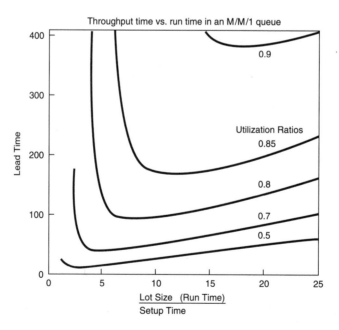

Figure 9.20.   Effect of Setup Time and Lot Size on Throughput
Source: Reprinted with permission. Copyright © AT&T 1980.

Process for Optimizing Lot Size with Setup

1. Identify the bottleneck operations.
2. Reduce the lot size of work passing through these operations using the process described above for reducing lot sizes.
3. Determine the minimum lot size. This is when the lead time shows signs of increasing, with a little variation in product or process.
4. Reduce the setup time and repeat Steps 1 to 3.
5. Where the setup time cannot be reduced further and the operation is still a constraint, process the entire lot through the constraint, but do not wait for the entire lot to be completed. Use a transfer lot to achieve minimum lead time.

Setup Reduction

Purpose and Description

The primary purpose of setup reduction is to reduce lead time by manufacturing smaller lots. An equally strong benefit is that a reduced setup provides manufacturing flexibility to make exactly what is required by a customer, and an ability to respond quickly to changes in demand. Finally with small lots work-in-process inventory is minimized.

Setup time is the time required for a specific machine, resource, work center, or line to convert from the production of the last good piece of a previous job to the first good piece of the current job. Setup time is also the time spent between successive pieces of the same job, when the machine requires adjustment. All setup time is technically nonvalue-added time—a necessary evil!?

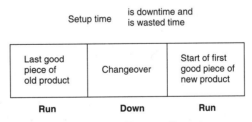

Figure 9.21. Schematic Showing a Setup Process Overview

Discussion

In the section on lot sizes and lead-time reduction, it was determined that the lot size could only be reduced to a point after which the setup time would become restrictive and result in an increase of lead time. In order to reduce the lot size and lead time further, the setup time has to be reduced. Shigeo Shingo is the pioneer most associated with the understanding of the effects of setup and has led the effort to reduce the setup time. Around 1950 he developed a setup reduction system called *SMED* (*single-minute exchange of die*), which meant that all setups should be reduced to less than 10 minutes—in single digits! Shingo says, "Improvement of setup operations has traditionally been ignored or given only cursory attention. However, my development of single-minute exchange of die (SMED) has brought about dramatic improvements." The statement is absolutely true. Shingo goes on to add, "I believe the just-in-time method, which is at the core of the Toyota production system, would most likely not have been developed unless the SMED system was available" (Shingo, 1987)—a formidable pronouncement.

All setup has two components: *internal setup* and *external setup*. *Internal setup* is that part of the setup operation that must be performed with the machine or resource stopped—such as removing a die in use. *External setup* is that part of the setup operation that can be performed while the machine or resource is running—such as preparing a new die, storing an old die, and so on. The essence of setup reduction is to reduce or eliminate internal setup time, usually by transferring it to external setup time.

Process for Setup Reduction

1. Identify the bottleneck or constraining operation: This is a starting point of the setup reduction plan (see the section on constraint management).

2. Identify all setup steps: Make out a detailed sheet listing *all* operations, from the time the last piece of a job is completed to the successful completion of the first piece of the new order and on the steps taken from one piece to the next in the same order. Use a process flow sheet or an operations activity chart to document this activity. It has been found that videotaping the process is an effective way of capturing all steps of the setup and has the additional advantage of allowing frequent review.

3. Separate internal setup from external setup: Internal setup requires the machine to be stopped; external setup can continue while the machine is in operation.

   1. Checklist. On the detailed operations sheet listed in Step 2 above, separate the internal setup steps from the external setup steps.

2. Function checks. Classify each step into parallel or sequential operations. In addition, determine the function performed in each step to be under:

- Locating and transporting materials and tools used in the setup
- Loading or unloading
- Alignment and adjustment
- Measurement
- Clamping

4. Convert internal setup steps to external setup:

1. Ensure that the next job and all jigs, fixtures, and tools required for the next job are at hand, conveniently located and easily accessible. The machine should not be stopped to look for them, nor should the operator wait until the job is done to look for tools for the next job.

2. Prepare in advance to combine or eliminate internal setup due to alignment, adjustment, measurement, and clamping. Some of the ways this can be done include:

- Working two setups. A second and identical setup fixture or die is used. The next piece or job is setup while the last piece is running. A quick means of unloading and loading is adopted (roller bearing track, swing arm, etc.) and the next piece is started.

- Setting up on same machine. Where the machine is large, a second setup can be made on the machine while the job is running. Here the machine arm or bed is moved to the second piece and work started.

3. Standardize the locating points, the means of location, the heights used and other dimensions that have to be adjusted before starting the next piece. This will ensure that no internal time is spent on these activities.

4. Use an intermediate setup where a complete parallel setup cannot be developed. This may include all the adjustments except the final setting.

5. Make sure that the jigs, fixtures, and tools used are appropriate and in good condition and time is not spent on struggling with clumsy clamping devices or nonfitting parts. This is particularly relevant in using dies (for stamping, presswork, etc.), where the die must be clean and well maintained to produce accurate work.

6. Attempt to eliminate the need for trial machining and test runs. This is a function of the effectiveness of the setup. Trials and tests can consume a large portion of the internal setup time.

5. Reduce or eliminate all internal and external setup activities:

   1. Ensure that loading and unloading of the job and fixture is done expeditiously. Storage of the fixtures and/or dies must be close to the point of use and there should be a rail or roller conveyor to move them off and on the machine.

   2. Eliminate adjustments. It has been estimated that adjustments constitute about 50 percent of the overall setup time. Some of the practices listed under paragraph 4 will help eliminate adjustments. Use accurate, calibrated tools for measuring and setting the job. Adjustments usually require a skilled operator, and so, once adjustments are eliminated, a quick setup can be performed by any operator.

   3. Develop simple and effective means of locating the job on the machine. There are numerous techniques of providing quick and accurate location:

      ■ Use a standardized base plate on which the job can be fastened as part of an external setup—particularly useful for dies. The base plate has a prelocated position on the machine.

      ■ Use locating pins with a chamfered top to ensure a quick location and a secure fit.

      ■ Use T bolts along a T slot of a machine bed.

      ■ A permanently set female V fixture on the machine bed can quickly locate all fixtures with the corresponding male V.

      ■ Use limit switches to locate the job and to adjust height or length.

   4. Improve clamping of the parts and fixtures on internal setup, after first moving as much of the clamping to external setup.

      ■ Use cut-off threads on bolt and holes to ensure minimum time and effort.

      ■ Use slotted bolt holes—Dahma holes—where the bolt is tightened in the slot and loosened in the hole. Use U-shaped washers instead of round washers.

Table 9.5 SMED System Outline

	Current *Identify all steps*	*Step 1* *Separate* *Internal–External*	*Step 2* *Convert* *Internal–External*	*Step 3* *Streamline External*
External		Checklists, machine functions	Advance preparation intermediate setup, standardization	● Improve transport ● Improve workplace
Internal				● Parallel operations ● Improve clamping ● Eliminate adjust

- Use off-centered hinged levers and cams for one stroke clamp and unclamp.

- Use pneumatic or hydraulic switches to clamp or unclamp parts and fixtures.

6. Conduct a videotape analysis of the process and improve it per the above steps.

7. Standardize the setup procedure. Look for a 75 percent reduction at no or low cost.

8. Ensure that the setup procedure is followed. Involve people.

Dos and Don'ts

DOs

1. Do start setup reduction at the bottleneck operations.

2. Do reduce lot sizes after reducing the setup. Remember that the real return on setup reduction is the ability to reduce lot sizes and lead time.

3. Do organize the workplace so that material, fixtures, and tools are conveniently located (good housekeeping) and easily identified (labeled and/or color coded).

4. Do make out a checklist. The format of the checklist might look like this:

Description of Step	Tool Used	Internal/External	Improvement

(continues)

Dos and Don'ts

DOs (*continued*)

5. Do focus on reducing internal setup first.
6. Do eliminate adjustment, which is around 40 to 50 percent of the setup.
7. Do develop one-touch clamping.
8. Do develop and use one-way fit— such as an audio cassette.
9. Do standardize the improved method and train the operator to ensure that

the improved method is understood and followed correctly.

DON'Ts

10. Don't use threads to fasten or clamp. If threads have to be used, have the threads cut to two surfaces.
11. Don't use hand tools, unless unavoidable.
12. Don't fasten if not required.

Examples of Setup Reduction (Courtesy Rick Frisby, Frisby Assoc.)

A. Separating Internal from External: Organization of Standard Procedures (Figure 9.22): The original method depicted was the current procedure the shop used prior to working with me to improve the process. The nuts, bolts, clamps, brackets, etc. were all stored in a box on a shelf behind the press. These parts were literally thrown together in the box. Nothing was

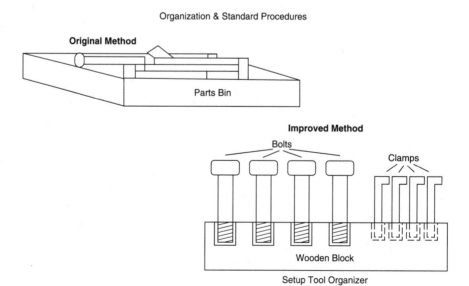

Figure 9.22. Separating Internal from External

marked. In order to complete a machine changeover, the setup operator had to rummage through the box, mixing, matching, and using trial and error to come up with a suitable set of clamping devices to complete the machine change. This was all accomplished during downtime and took from 15 to 25 minutes.

The improved method involved identifying all nuts, bolts, clamps, brackets, etc. in that box as well as determining exactly which parts were needed for each of the various machine changes. Each particular type part was stored together in its own bin. A wooden block holder was fashioned that could hold all the required parts for a particular changeover. Prior to the machine change, the setup operator would go to the bin during external run time, pick the required parts as listed on the setup instructions, and place them in position in the holder. This holder was then staged with the die and when the machine was shut down for changeover, all parts were in place and ready to be used. This eliminated 15 minutes from the internal downtime as well as minimized clamping time as all lengths and clamps were the proper type.

The total reduction realized from this simple procedural change was 50%!

B. Converting Internal to External: Advance Preparation, Preheated Die Changeover (Figure 9.23): In the original method, once the extrusion die was attached to the machine, the machine had to remain down for 20

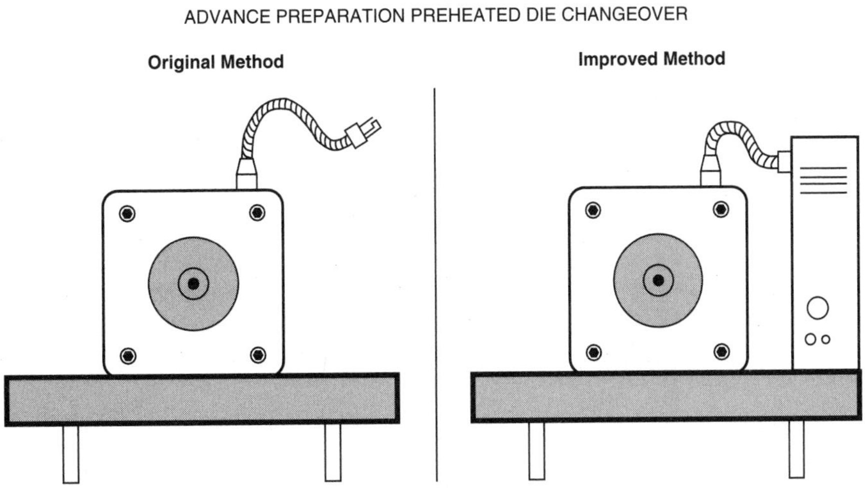

ADVANCE PREPARATION PREHEATED DIE CHANGEOVER

Original Method   **Improved Method**

Figure 9.23. Converting Internal to External

minutes while the heaters brought the die up to operating temperature. This was considered normal internal downtime.

In the improved method, this internal downtime was converted to external downtime by installing a bench next to the machine with a power unit to run the heating elements on the extrusion die. Approximately 20 minutes prior to die change, the new die was mounted on this bench and powered up, heating the die to operating temperature during external time. When it was time to change the die, the die was changed hot, eliminating the need for idling the machine for 20 minutes to bring the die up to temperature.

Additional benefits realized were that with the die hot, previous problems with removing and attaching the dies due to expansion and contraction problems associated with installing a cold die on a hot machine were all but eliminated. A 30% reduction in changeover time was realized by implementing this procedure and tooling!

C. Streamlining Internal Activities: Clamping Improvements, Wirebonder Magazine Loader (Figure 9.24): In the original method, each magazine loader was attached to the machine with six hex bolts. The setup operator had to loosen all six bolts, use one of the new magazines as a gauge by inserting and removing it from the loader, and hold the approximated position by hand while retightening the bolts. Often the loader would move, therefore, several remeasurements and retightening were needed to get a proper fit. This could take 5 to 7 minutes per loader to accomplish.

In the improved method, all magazines were measured and it was found that there were only 5 standard sizes. The 5 standard positions were measured on the loader and the bolts were replaced by tab/slot one-turn clamping devices. Each were color coded. When the operator needed

CLAMPING IMPROVEMENTS
WIREBONDER MAGAZINE LOADER CHANGEOVER

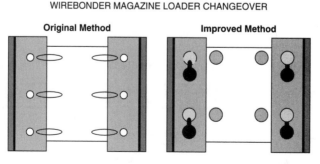

Figure 9.24. Streamlining Internal Activities

to change to a new magazine type, it was now a matter of checking the color code, pulling up and removing the loader brackets, and placing on and pushing down on the proper color coded tabs to lock the brackets in the required position. This method takes approximately 2 seconds, a 99% plus reduction in changeover time with minimal investment in retooling!

Hybrid Shop Floor Control Systems

Purpose and Description

The purpose of the hybrid manufacturing system is to reduce manufacturing lead time and increase throughput.

Many companies have implemented hybrid systems to run the factory combining the planning capabilities of MRP II with the execution capabilities of JIT, particularly the kanban pull system. This innovation is particularly beneficial where the product is complex and the process has a lot of variability.

Discussion

From the section on shop floor control with MRP II it was seen that MRP is good at planning—particularly for complex, multi-level products—since the computerized MRP program logic links and calculates all of the bill of material dependencies. MRP, however, cannot adjust lead times dynamically to meet shop floor variability and this makes its schedules unreliable. MRP II software is expensive and its implementation is time consuming. On the other hand, just-in-time with kanban control is good at shop floor execution, as it dynamically adjusts to shop floor variability. Kanban systems are inexpensive and relatively easy to implement. Kanban control, however, has difficulty in dealing with complex products

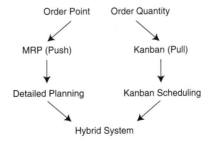

Figure 9.25. Development of the Hybrid Shop Floor Control System

and its planning capability is limited. It would seem natural to combine the strengths of the two to develop an optimum manufacturing system. To go a step further, the kanban execution capability of the shop floor can be improved by building the principles of constraint management into the hybrid system.

The key to integrating MRP and kanban lies in understanding the product characteristics and the production process.

Process for Installing a Hybrid Control System

1. Determine the characteristics of the product and the production process. These may be classified from high-volume, repetitive with little lead-time variability at one end to low-volume, nonrepetitive with high lead-time variability at the other end (see Figures 9.26 and 9.27).

2. For a high-volume, fairly simple, repetitive product with stable lead times: establish a drumbeat, synchronize all operations, and use kanbans or pull to manage the shop floor (see uniform scheduling).

3. For a repetitive product subject to varying schedules and variable lead times, plan the material procurement with MRP and work the shop floor with operation-to-operation kanbans.

4. For low-volume, complex products with long lead times, use traditional MRP II to plan and execute.

5. For complex products with demand and lead-time variability, but with some volume, use MRP for order planning and release. Use MRP

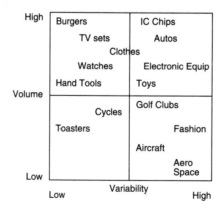

Figure 9.26. Examples of Product Characteristics

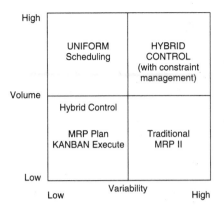

Figure 9.27. Scheduling for Different Product Characteristics

shop floor control for scheduling and tracking starts, moves, and yields. Link the kanban release to MRP's work-order system and then control the shop-to-shop movement with a single-loop kanbans. This is the essence of a hybrid control system. (See Figure 9.28 for an example of this system.)

6. For the hybrid system, identify the bottleneck and ensure that there is a buffer stock in front of it (see constraint management).

7. Measure lead-time performance under all combinations.

8. Review the kanban quantity and the lead times whenever the demand changes significantly.

9. Improve the performance of the bottleneck, continuously.

Dos and Don'ts

DOs

1. Do remember that MRP and kanban do not have to be mutually exclusive. They can be combined.
2. Do select a combination of MRP and kanban based on the product and process environment.
3. Do plan for excess capacity under conditions of high demand and lead-time variability.

DON'Ts

4. Don't allow inertia to become the bottleneck (Goldratt). In other words, do not become complacent after resolving the initial constraints. Improve continuously.

A Detailed Example of a Hybrid Manufacturing System

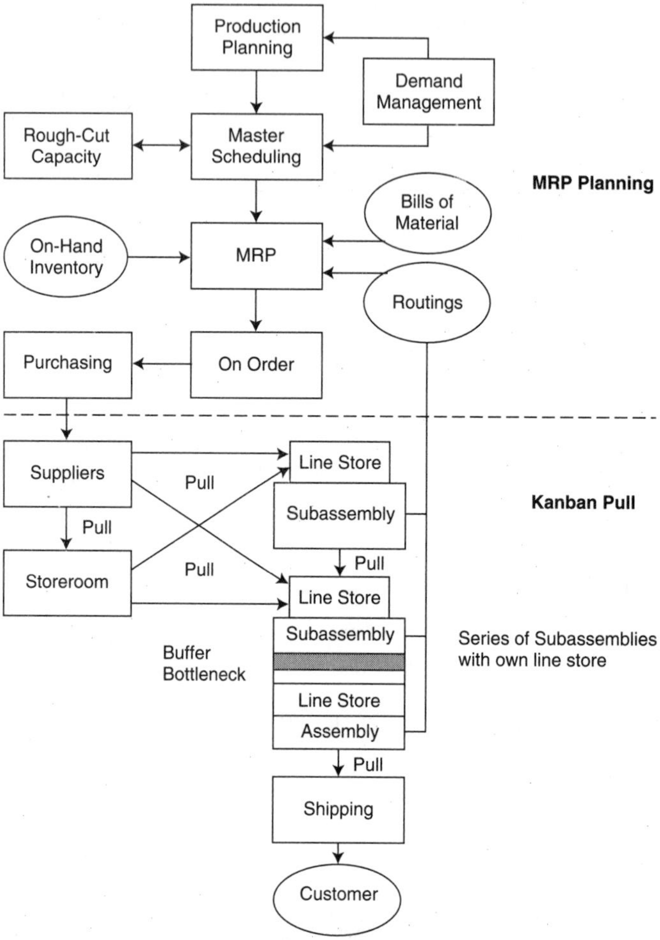

Figure 9.28. Hybrid Manufacturing System—MRP Planning and Kanban Execution

Background: The product is a high-tech assembly, progressively built through a series of subassemblies until it reaches the final assembly. The process is subject to considerable variation from operation to operation. The product type is stable and there is reasonable volume.

Process

    1. Manufacturing planning is done conventionally using MRP II. Forecasts and customer orders are entered into a master schedule, which

after being checked for rough-cut capacity are transmitted to MRP. On-hand inventory and on-order files are maintained, as are BOM and routing files.

2. Purchase orders are placed with suppliers based on requirements of MRP.

3. Shop orders are released based on the schedule. They create a transparent work-order tree, available at every subassembly level for tracking a specific subassembly.

4. The supplier sends in material based on a demand pull—the material planner electronically advising the supplier what to deliver and how much on a weekly basis, with a month's frozen demand.

5. Every subassembly is built in a separate area and has its own line store.

6. A line store contains all the parts needed to build the subassembly, including the previous subassembly. Parts are supplied to the line store by the supplier or by the storeroom. All parts are supplied based on a pull, either through a max-min report or a reorder card.

7. Each subassembly shop is allotted a finite number of kanbans corresponding to its lead time, variability, and demand. An available kanban is the authority to start a subassembly. A kanban remains attached to the subassembly while it is being built.

8. When a kanban is available in the subassembly line store, a subassembly can be started. A *start transaction* attaches the work order (created by MRP) to a specific subassembly serial number (input in the start transaction). This ensures that operators cannot build product that is not MRP scheduled.

9. A *select transaction* allows the operator to enter the parts picked from the line store that go to make up the subassembly. Many of the parts have serial numbers that are captured at this transaction. The select transaction validates that the operator has picked all the parts required for the subassembly.

10. The subassembly is built and the operation completion is recorded by a *move transaction*. Product cannot be moved unless all operations are complete.

11. On being completed, the subassembly is delivered to the next assembly line store. On being picked at the start of that assembly, the kan-

ban is released, goes back to its original line store, and becomes available to start the next subassembly.

12. If there is any breakdown, product will accumulate in front of the breakdown. No additional product (over the kanbans in each subassembly loop) will be built until the breakdown is cleared.

13. A bottleneck operation is identified and a buffer is kept in front of it. The level of the buffer is constantly monitored and never allowed to go to zero. If there is danger of the bottleneck operation being starved, additional kanbans will be released. This is the *only exception* to not allowing extra kanbans in the loop.

14. This whole hybrid process allows the factory to plan and control with MRP at an ordering and purchasing level, but uses a kanban pull system to regulate the flow of purchased materials and the movement of product on the shop floor. Shop floor lead-time variability is dynamically adjusted in real time by the rate of movement and the location of the kanbans.

15. If the factory falls behind schedule, the rate is increased by increasing the number of kanbans and the capacity (through overtime or additional flexible workers, for example) of the subassembly shop that is behind schedule.

Statistical Process Control (SPC) and Yield

Purpose and Description

When a process capable of making a product to specification is kept under control it will have a high yield. This in turn will minimize the lead time taken to satisfy a customer order.

Variability is a way of life, it cannot be eliminated. A control chart is a graph of statistically determined upper and lower control limits drawn on either side of a process average. *Statistical process control* is a means of tracking the variability in a process to make sure it operates within preset upper and lower limits. A process must meet the product's specification and variability. A process that meets these product needs and operates within limits will have a high yield. This will reduce the amount of rework and also reduce additional quantities that may have to be made to cover defective product.

Discussion

W. Edwards Deming has been credited with leading the Japanese quality revolution. Deming classified all process variations into two categories: *common causes*, which are random, systemic variations, developed by the sources of the process, namely groups of operators, machines or products; and *special causes*, which are produced by nonrandom variation within systems resulting from identifiable activities. These special causes must be eliminated before the process can be "in control" and tracked. Deming used *statistical process control*, a form of statistical probability, to distinguish acceptable variation from unacceptable variation (see Figure 9.29).

Random variation occurs within statistically determined limits, and as long as variation remains within these limits, a process is stable and in most cases in control. By collecting and plotting data on control charts, operators are able to monitor their process. If the trend of the data plots indicates the process is going out of control, or if the data show the process exceeding the control limits, both situations being created by special causes, the operator can take appropriate action to bring the process back into control or shut it down and get help. In either case, defective product should not be made. On the other hand, a process can be improved by eliminating common causes, since this leads to a reduction in variation and a narrowing of system limits.

Another and more generic way of looking at variation is to understand that there are two sources of process variability: *chance causes* and *assignable causes*. Chance causes are:

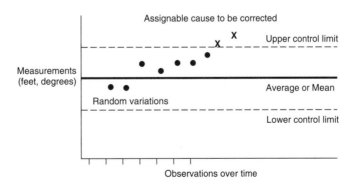

Figure 9.29. Typical Control Chart

- Always present
- Inherent in the nature of the process
- Neither identifiable nor removable
- Impossible to control

If only chance causes exist, a system is stable and predictable, and the pattern of variation can be used to predict future quality. Reductions to chance causes can be made by a fundamental change in the process. If a system is unstable and observations exceed or are trending to exceed prescribed limits, there are assignable causes of variation. Assignable causes are:

- Unpredictable, in occurrence time and frequency
- Identifiable and removable
- Capable of being controlled
- Require early action

Assignable causes can be detected, identified and eliminated.

Two types of control charts are commonly used: *attribute charts* and *variable charts*.

Attributes are based on characteristics that have a binary pass/fail option, like go/no go, or fits/does not fit. Attribute charts are often referred to as *p charts,* as they track the proportion or percent of defects p.

Example: The car paint is acceptable or unacceptable.

Variables are the actual measurements made on the process or product under study. Variable charts are referred to as $\bar{X}$ *bar* and *R charts*. On these charts the sample average $\bar{X}$ (pronounced "X-bar") and the range R are tracked concurrently.

Example: Machining a piece part.

(Details on how to set up and use control charts may be seen in Appendix 9.7.)

Control charts and sampling techniques may indicate problems, but they do not identify the cause of the problems. For determining the cause of problems other statistical techniques were developed, including Pareto analysis, fishbone or Ishikawa cause-and-effect diagrams, histograms, check sheets, scatter diagrams, and flowcharts.

The value of the control charts (for a process capable of making a part to specification) is in reducing or eliminating defective product and thereby increasing process yield (yield = good pieces/total pieces made).

Improvement in yield will reduce the number of pieces required to be made and this in turn will allow all orders to be completed in shorter lead times. A secondary benefit is the saving of not making defective product!

Generally control limits are set at three standard deviations (3σ), a standard deviation being a statistical measure to indicate the probability of occurrence. A control limit of $\pm 3\sigma$ denotes there is a probability of 99.7% (3000 parts in a million) that the sample being measured will be between the upper and lower limits.

Process Capability

Process capability determines the variability of the process and must be less than the requirement of the product being built. Described differently, the variability of the process must be less than the upper and lower tolerances of the product and should allow for some shifting of the process limits without exceeding the product specifications.

Process capability is expressed as a *Cp* or capability index.

$$Cp = \frac{\text{Product specification limits}}{\text{Process variability spread}} \quad \text{Must be greater than 1}$$

Cpk is similar to *Cp* with the shifting of the process being taken into account and included in the denominator as the *process variability spread*. *Cpk* is used in practice as processes will shift. Generally, product specification limits are determined by design or engineering, and the process variability is determined by a process engineer or product factors.

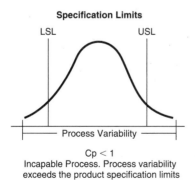
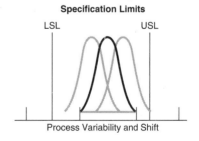

Specification Limits

LSL   USL

Process Variability

Cp < 1
Incapable Process. Process variability exceeds the product specification limits

Cpk < 1 Allows for no process shift

Specification Limits

LSL   USL

Process Variability and Shift

Cp > 1
Ideal Process. Process variability is within specification limits and can allow some shift

Cpk > 1 Allows for process shift as shown

Figure 9.30. Process Capability

Worker Involvement

A key to the success of an SPC program is worker training and involvement. The workers collecting data must own the system. It is essential that measurements are simple and can be easily taken. Before implementing the system it is essential that a good training program be put in place and the workers thoroughly understand the process. The out-of-control criteria must be constantly repeated until workers can quickly recognize the unnatural patterns they cause. There are software SPC programs that recognize the out-of-control criteria and signal the worker to stop.

Measurements may also be electronically collected. This is preferred since the process is not dependent on the operator to make inputs in a timely and accurate manner.

Yield

The primary purpose of SPC is to keep a process in control. As seen before, a process must be improved until it is capable of meeting product specifications before it can be used to make product efficiently. In other words, the process has less variability than the product specifications and hence makes an acceptable (usable) product. The *yield* of a process is the ratio of usable product to total product made. Where acceptable product is made during the first pass of a process, it is counted toward establishing a first-pass yield. Product that is reworked, and is then usable, counts as good product and it is counted in the final yield. Product that is scrapped or cannot be used is deducted from the total product made. High first-pass yield directly affects lead time, as there is no time spent on rework. Final yield also affects lead time favorably, as there is need to make more parts with high final yields.

$$\text{first-pass yield} = \frac{\text{total product accepted first time}}{\text{total product made}} \times 100 = \% \text{ yield}$$

$$\text{final yield} = \frac{\text{total product made—product cannot be used}}{\text{total product made}} \times 100 = \% \text{ yield}$$

Yield is captured as a scrap factor at the component level in MRP calculations and is considered in calculating quantity to be made.

Example: Order size 120 pieces and yield 60%. Parts to be made = 120/0.6 = 200.
Order size 120 pieces and yield 95%. Parts to be made = 120/0.9 = 126. Reduction = 74 pcs.

Process for an SPC Program

1. Document the process including a flowchart and measure the existing yield.

2. Clearly define the tolerance requirements for the product. Establish whether the process will use an attribute or variable chart. This will depend on the product characteristics, but whereever possible a variable chart should be used.

3. Select the measurements that define the acceptability of the product and determine the points in the process where these measurements will be taken.

4. Define how the measurements will be made and develop suitable instruments or gauges for making the measurements.

5. For the initial study, determine the sample size (keep the sample size to 4 to 5 for variable charts and 50 for p charts) and the number of samples (keep it to about 25 for both charts). On an ongoing basis, the number of measurements taken depend on the stability of the process, with more measurements needed for a process that is more variable.

6. Document the entire data collection process and train workers on measuring and collecting data.

7. Take the measurements and determine p and $\bar{p}$ or $\bar{X}$ and R. Calculate the UCL and LCL for the process. Generally these control limits are set at three standard deviations (3σ), which means that 99.7 percent of the product should be within the control limits. A measurement outside these control limits signals a problem (see Appendix 9.7).

8. Check that the process spread, including the shift is tighter than the product requirements ($Cpk > 1$).

9. If $Cpk < 1$, the process should be improved.

10. Plot the center line and the control limits on a graph.

11. Draft the rules defining the criteria for deciding that the process is going out of control. Train workers to understand and apply the criteria (see analysis of patterns below).

12. Plot the data from actual product under manufacture (p's for an attribute chart and X's and R's for a variable chart).

13. Interpret the results (see notes below).

14. Make a determination if the process is in control or out of control. Mark out-of-control points with **X.**

15. Use a team of quality engineers, process engineers and operators to determine the root cause and what corrective action should be taken when a process goes out of control. Ensure that the operator is involved in the analysis and understands what went wrong and how to prevent it recurring. Among the factors to be looked at when investigating a process out of control are changes in methods used, measurements used, the material, the machine's settings, the tooling, the operator, and the environment.

16. Iterate Steps 10 to 14 until the operators are familiar with the SPC process.

17. Compare the results produced by the SPC system with the quality of the product before the system was implemented.

Analysis of patterns: The pattern of the plotted points is first checked to determine whether the pattern is natural or unnatural.

Natural patterns: The primary characteristic of this pattern is that the points fluctuate at random, following no recognizable order. Natural patterns also tend to fluctuate somewhat symmetrically around the center line. Finally, the points seldom exceed the control limits.

Unnatural patterns: The primary characteristics of unnatural patterns is that the points do not fluctuate around the center line and there may be points outside or close to the control limits. Further an unnatural pattern has a distinct signature of trend or fluctuation.

Suggested rules for determining unnatural patterns:

- Rule 1: A single point falls outside the 3σ control limits.
- Rule 2: Two out of three successive points fall outside the 2σ control limits on the same side of the mean.
- Rule 3: Four out of five successive points fall outside the 1σ control limits on the same side of the mean.
- Rule 4: Eight successive points fall on same side of the mean.

In all cases mark the unnatural pattern with an **X** at the last point of the pattern. This helps to identify the unnatural patterns easily.

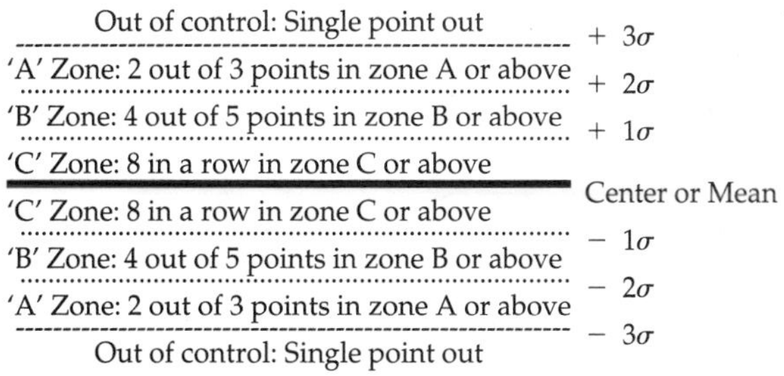

```
              Out of control: Single point out      + 3σ
--------------------------------------------------------
     'A' Zone: 2 out of 3 points in zone A or above  + 2σ
........................................................
     'B' Zone: 4 out of 5 points in zone B or above  + 1σ
--------------------------------------------------------
     'C' Zone: 8 in a row in zone C or above
                                                     Center or Mean
     'C' Zone: 8 in a row in zone C or above
........................................................
     'B' Zone: 4 out of 5 points in zone B or above  − 1σ
........................................................
     'A' Zone: 2 out of 3 points in zone A or above  − 2σ
--------------------------------------------------------
              Out of control: Single point out       − 3σ
```

Figure 9.31. Control Charts: Tests for Unnatural Patterns

Interpretation of charts:

- *Meaning of the* R *chart:* The R chart shows consistency. If the R chart is narrow the product is uniform and if the R chart is wide the product is not uniform. Machines in good condition and skilled operators make uniform products. If the process is out of control look for defective machines or untrained operators.

- *Meaning of the* X̄ *chart:* The X chart shows where the process is centered. If the center is not shifting, the X̄ chart is natural. If the X̄ chart shows a trend, the center is moving up or down. If the X̄ chart is erratic and out of control, something in the process is not functioning properly. Processes are centered by a machine setting or a process adjustment and a bias in the technique of the operator.
  If both the X̄ and R charts are out of control, look first for causes affecting the R chart.

- *Meaning of* p *chart:* Remember that the p chart shows the percentage defective or proportion of product classified as defective. When the pattern changes, it means that the proportion has changed. Such a change may signify that the percentage of bad product is increasing or decreasing and/or the criteria for determining defective product has changed. Both of these possibilities need to be checked. If the chart fluctuates erratically poorly trained operators or improper parts are the most common causes.

- *General:*

1. Look for trends. This indicates if the center is moving up or down.
2. Look to see whether fluctuations are becoming narrower or wider. These show the consistency of the process.

3. Check if there are obvious patterns, such as cycles or bunching. These indicate something in the process is wrong.

Applying SPC to Other Variables

The practice of SPC is to determine the acceptable limits of a process, monitor the process to ensure that it is in control, and take corrective action if it tends to or goes out of control (see Figure 9.31). It is primarily applied to product quality. There is no reason, however, why SPC cannot be applied to other variables that need to be controlled within acceptable limits, such as forecast accuracy, completion of orders on dates due, process lead time, and so on. In most cases, because the element to be controlled has a variable measurement, an $\bar{X}$ (X bar) and R chart will be used, upper and lower limits will be set up, and the element monitored in the same way that product quality is monitored.

Dos and Don'ts

DOs

1. Do have all operators properly trained to understand the value of SPC. Unless the process is adopted by them it cannot succeed.
2. Do present SPC as a clear and simple process and not a complicated statistical technique. Use graphical displays to help understanding.
3. Do select the appropriate type of chart for the type of data being measured.
4. Do ensure that when process defects are discovered, a root cause analysis is conducted, and the source of the defect discovered and corrected. Stop the line for out-of-control conditions.
5. Do keep good, accurate, clean records of the SPC measurements. Data must be kept in the exact same sequence they were recorded, or they are meaningless.

DON'Ts

6. Don't tinker with the process while the data are being gathered. The data should represent the process "as is."
7. Don't expect to achieve zero defects.
8. Don't let the fact that a process is operating in control reduce efforts to improve the process continuously.
9. Don't use the statistics as an end in themselves by using SPC charts for the sake of using charts. Do remember that the use of statistics is to monitor a process and prevent defects from occurring.

Poka-Yoke and Other Forms of Inspection

Purpose and Description

If the yield of a process is 100 percent, only the required number of parts need to be made, thereby minimizing the time needed to satisfy a customer order. In addition supplying customers with defect-free product is an absolute requirement. Finally, 100 percent yield minimizes the cost of product.

Poka-yoke (pronounced POH-kah YOH-kay), or foolproofing, ensures that errors cannot be made in the manufacturing process.

Discussion

The best way to deal with inspection is to have an error-free operation that eliminates the need for inspection! The least useful form of inspection is the conventional after-the-fact check by a second party or inspector. Such inspections add little value to the process, do not improve the quality of work, and lead to workplace friction and delay of product. What is even more ironic is that inspectors do not detect all defects. The reason is obvious. The source of quality is not the inspector. The source of quality is the operator, or the machine, or the process, or the product design. Defects can be reduced and quality improved only by correcting the source of the problem.

Management does not usually charter engineering or measure engineers on designing defects out of a process. Inspecting product by operators is desirable, but it is probably too little too late to make meaningful changes in the process. Focus should be first visited on the design of the product.

Separate quality control inspectors still exist in most manufacturing companies, as management does not have confidence in the operators to trust them checking their own work or each other's work. There is also a case for operators not being responsible enough to be entrusted with the task of monitoring their own work. The practice of using operators to check their own work must be planned for through education and training, so that the use of inspectors to check quality is eliminated. (See Appendix 9.8 for problems with inspection.)

The last section discussed statistical process control (SPC). The use of this technique by an operator to monitor a process is a considerable advance from the use of an external inspector.

Poka-yoke is a process means of ensuring zero defects. Poka-yoke is built into a process and attempts to prevent mistakes from occurring. The word in Japanese means mistake-proofing—not allowing errors to be made. The technique checks 100 percent of the parts and provides immediate feedback when conditions suggest that a defect may occur. Furthermore, where workers are empowered, each worker has the power *and the responsibility* to stop the line if a defective product is detected or if operations cannot be performed in accordance with the standard practice.

Shigeo Shingo (1986) tells the story of a process of assembling a switch having two buttons, an ON button and an OFF button. Each button has a spring under it. In the original process the operator forgot to install a spring every once in a while. Instead of trying to increase inspection or reprimand the operator or insist that he had to remember, Shingo devised a check into the process. The check consisted of a dish into which two springs would be put at the start of every assembly. If a spring remained in the dish at the end of the assembly, the operator realized that a spring was forgotten and the assembly corrected. Every attempt should also be made to incorporate poka-yoke into product design—audio and video cassettes are examples, as they can fit only one way.

Where poka-yoke cannot be applied there must be some other way of ensuring that 100 percent inspection is being conducted. There are two techniques for achieving this. One technique is to have all operators check their own work; another technique is to have each succeeding operator check the preceding operator's work. The biggest advantage of these two approaches is that feedback on defects is immediate.

In all cases of defects, a root cause analysis should be undertaken and the underlying cause of the defect identified and corrected. The simplest means of doing this is to ask Why five times. This usually gets to the core problem.

> Example: The grass is brown. Why? It is not being fertilized. Why? The owner does not have time to fertilize it. Why? He has other priorities. Why? He does not consider green grass to be of any importance.

Process for Inspecting Product

1. If a process uses conventional inspectors to check the quality of product, the practice should be reevaluated.

2. Develop a quality system involving the operator. This involves education and training. It also involves confidence and trust. Be prepared to spend time to develop this approach—at least a year.

3. Implement an SPC process (see the section on SPC).

4. Extend the SPC process to having the next operator check the previous operator's work. This practice also requires training and sensi-

tivity (to realize that it is the product and the defect that is examined and corrected, not the operator).

5. Consider shutting down a production line when defects are discovered. This will have a very convincing effect on an operator's attitude toward the quality of a product.

6. Start to adopt poka-yoke at the beginning of a line to ensure 100 percent product inspection and immediate feedback when defects occur. Poka-yoke consists of a detecting device and a signaling device. Among the type of detection techniques used to determine that a product is within specification are:

   - Contact devices such as limit switches, micro switches, liquid level relays, and so on.
   - Non-contact devices such as proximity switches, photoelectric switches, and various types of sensors—beam, fiber, area, vibration, and so on.
   - Gauges such as pressure, temperature, current, vibration, time, and counting gauges.

   Among the types of feedback or regulating systems used are:

   - Control methods that shut down the operation when defects occur.
   - Warning methods that use sound or light to alert workers that defects occur or are likely to occur.

7. Introduce poka-yoke if there is a need to ensure that all the parts of an operation have been completed or if a check is to be conducted on the completion of all the steps of an operation.

8. Ensure that there is a root cause analysis procedure to investigate and resolve the cause of the defect. Poka-yoke is not an end in itself and must be supported with a corrective action.

Examples of Poka-Yoke

- Prevention of a car accelerating forward when started—the car will not start unless the brake is depressed.
- A 3.5-inch floppy disk can be inserted only one way into the computer. The same check is maintained for an audio cassette and a VCR tape.
- Use color coding to distinguish between similar parts. For example, red is right and blue is left.

- Use templates with the outline of the correct orientation of the parts to be fitted.
- Control machine start-up with limit switches. If not operated correctly the machine will not start.
- *Have controls that are visible and/or that result in sound or light being activated.*

Continuous Improvement and Problem Solving

Purpose and Description

The purpose of *kaizen* is to focus on process improvement and hence reduce lead time and cost, and improve profitability.

The Japanese call continuous improvement *kaizen,* and in manufacturing, continuous improvement or kaizen seeks to find and reduce or eliminate waste in all operations and processes. The continuous improvement process is very similar to the problem-solving process.

Discussion

Masaaki Imai, who popularized the word *kaizen,* in a book of that name (1986), says that the most important difference between the Japanese and Western approaches to management is that the Japanese focus on process-oriented thinking and continuous improvement. The West emphasizes innovation and results-oriented thinking. The Japanese also stress the involvement of all personnel in the process, in contrast to a more individualized approach in the West. Today these distinctions are blurred, and most manufacturing operations look for continuous improvement of processes and involve the operator on the shop floor. One of the most effective ways of generating continuous improvement is by using the customer-supplier model.

In the model illustrated in Figure 9.32, customers are internal (the next process) and/or external customers. The objective of the model is to understand the requirements of the customer, and to use feedback to improve the process and satisfy the customer. In a manufacturing environment the model relates to successive operations, workstations, or work areas.

The story of the GM-Toyota joint venture NUMMI is an excellent example of the workforce being involved in designing and standardizing their work methods and then continuously improving these practices by an active suggestion program.

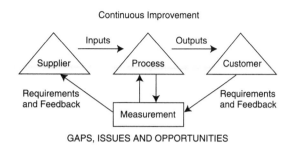

Figure 9.32.   Customer-Supplier Model

The key principles governing continuous improvement are:

1. Continuous improvement is driven by the need to satisfy a customer (and in the long term to improve profitability).
2. All work is part of a process.
3. All employees should be involved in the process.
4. Improvement is best achieved through teamwork.
5. Sustained improvement comes from changes in the process.
6. Continuous improvement never ends.

The formation, development and productivity of teams is not described in this section. The subject has been covered in Unit I of this book, under Employee Involvement. There are numerous generic continuous-improvement methodologies available for use. The process described below is a standard one.

Process for Continuous Improvement

1. Select the improvement area. A bottleneck or constraining operation should be chosen.

2. Define the objective of the improvement. Reduction in operation time or lead time are usually encompassing objectives, as lead-time reduction demands improvement in quality and results in reduction in cost.

3. Collect relevant data. Data should be accurate, timely, and easily available. Many improvement projects fail because of improper data requirements and difficulty in collecting data.

4. Analyze the data. Common techniques that may be used to help analysis are:

Graphs: line, bar, and pie graphs. Compare the use of the three:

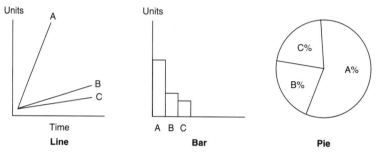

Line Graph	Bar Chart	Pie Chart
Use for comparing variations of units (Y axis) over time (X axis) of several sets of data. Shows changes and trends clearly.	Use for comparing small differences in quantity (size of bar). Can compare different categories with each other.	Useful for showing more than one level of stratification on the same chart, by showing proportion of each to the total. Equivalent to several bar charts.

Figure 9.33.   Comparison of Graphs

Histograms: Also called *frequency distributions,* histograms show the spread or distribution of data (height, weight, etc.). The number of occurrences (frequency) is plotted on the Y axis, while the groups or classes are plotted on the X axis. The histogram shows how the pattern of the data distribution about a mean—whether it is symmetric, skewed, has two points or modes, and so on. Always be suspicious of sudden and large changes in the data (bars).

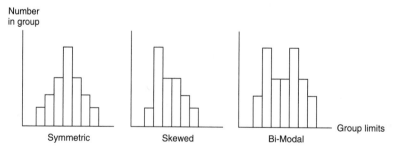

Figure 9.34.   Patterns of Histograms

Scatter Diagrams: These are similar to line diagrams except they have two variables and show the relationship of the variables. These diagrams are used to understand the effect one variable has on another related variable.

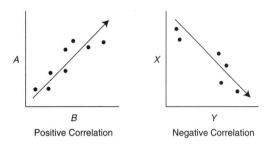

Positive Correlation   Negative Correlation

Figure 9.35. Examples of Scatter Diagrams

Flowcharts (see the section on product flow): Flowcharts are graphic representations of how a process flows.

Control Charts (see the section on statistical process control): Control charts are used to track the performance of a process over time. Control limits are specified and the process is monitored to ensure that it is within the limits and to initiate action if it is outside the limits.

Pareto Chart (see inventory ABC analysis, Chapter 10): A *Pareto chart* is a way of organizing data to show what the major factors are that affect the process being analyzed. It is also referred to as *ABC analysis* or the *80:20 rule*. The Pareto chart reveals the significant few (20 percent) factors or items that cause most (80 percent) of the effect. By establishing the significant few, priorities of execution can be established.

Fishbone Diagram (Ishikawa Diagram): The *fishbone diagram* is a cause-and-effect diagram that helps to identify causes or main contributors to an effect. The problem is written as the head of the fish and principal causes are identified as the main bones. Contributors to the main causes may be graphed as spines to the main bones.

The main categories may be generic (shown above) or specific to the problem (a specific machine may be shown). At each bone ask "why"

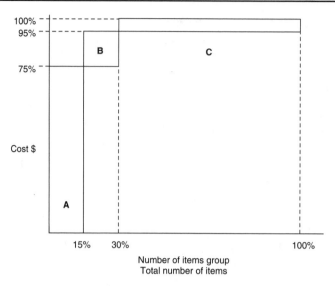

Figure 9.36. A Pareto Chart

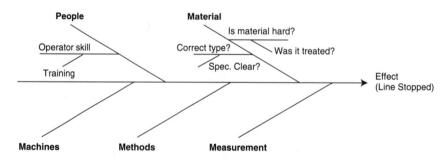

Figure 9.37. Fishbone or Ishikawa Diagram

to get a probable cause and then "why" again to get the probable secondary causes of this cause, which is now the effect, and so on.
For the above diagram the questioning may be as follows:

> Why was the line stopped? Defective material. Why was the material defective? Hard material. Why was the material hard? Not treated. Is the material the correct type? No. Was the specification clear? Yes.

The process of questioning developing grouped causes and effects should be continued until all likely reasons for the effect are identified.

5. Determine the root cause or main reason for the obstacle or problem. This is probably the most important step of the process. Correcting the root cause will lead to improvement or eliminate the problem. Techniques used are brainstorming, the fishbone diagram, interviewing, and experimentation.

6. Develop an action plan to test and implement the solution. This is an equally important part of the improvement process for two reasons. It confirms that the analysis is valid and it produces results. There is natural resistance to change and part of the implementation is the buy-in of the parties involved. Ensure that the action plan has dates and assigns responsibility to someone to perform the action.

7. Consolidate and standardize the improved process.

    Close monitoring and feedback should be conducted for some time after the new method is introduced, to make sure that it is followed and the gains are being realized. Document the process so that it can be followed by any operator.

Dos and Don'ts

DOs

1. Do encourage all personnel to look continuously for ways of improving existing processes. Provide incentives for suggestions and improvements.
2. Do start at constraining operations or work centers.
3. Do select the appropriate type and amount of data and ensure that they are easily available. Record how data are collected and who collects them.
4. Do remember that problem solving depends on meaningful, accurate, and timely data. Inaccurate data is worse than no data.
5. Do remember that "managing from data" avoids a lot of conflict and saves a lot of time.
6. Do involve the principal stakeholders—the operators, engineers, and production supervisors—in the analysis.
7. Do set up the improvement with authority to take corrective action at the lowest level.
8. Do determine the root cause if a problem is being solved. If appropriate, set up a small team of stakeholders to determine a solution.

DON'Ts

9. Don't get bogged down with large amounts of information that has limited value.
10. Don't neglect to roll out and communicate the new method with all concerned.
11. Don't fail to train the persons specifically involved in the changed process.
12. Don't forget to revise the documentation covering the process.
13. Don't neglect to consult with and involve the operators of the process.

Standardizing Operations

Purpose and Description

The purpose of standardizing an operation is to:

- Ensure that a job is consistently performed in the most effective manner
- To serve as an aid for future training.

Performing a job in the most effective manner will minimize the lead time and increase productivity.

After determining the most productive way of performing an operation, it is important that a permanent record be made of it. This record is called the *layout* or *standard practice*. Standards promote consistency and consistency is one of the major foundations of good quality. Standardization also helps in developing worker flexibility training.

Discussion

The development of permanent, standardized records is part of the practice of time-and-motion study and dates back to the days of Frederick Taylor (1911). The standard was the most efficient method for making a product and contained:

- A detailed sequence of operations
- The time and number of operators required for each operation
- The tools and gauges used in the process
- For a machining operation—details of feeds, speeds, tool settings and operation time.

Where standards are used as the basis of wage incentives, they have to be prepared particularly carefully. Standards also help train workers in different jobs and allow for the interchangeability of workers. Standards need to be carefully monitored and revised to reflect the continuous improvement that should be taking place.

The Japanese approach is to develop standard operations aimed at using a minimum number of operators in production. Toyota uses a standard worksheet that combines materials, workers, and machines. The standard worksheet includes three work elements—cycle time, operations routine, and quantity of work in process. These terms are defined as:

- *Cycle time* is the time between the completion of successive products.
- *Operations routine* is the sequence of work an operator follows (may differ from the product flow).
- Work in process is the minimum amount of inventory necessary to meet the demand.

As in other areas, the approach involves the operator and is focused on continuous improvement. Taiichi Ohno, the originator of the Toyota production system, says that the elimination of waste, the maintenance of high-production efficiency, the prevention of recurring defects, and the incorporation of workers' ideas are *all possible because of the inconspicuous standard worksheet*.

In the United States, the GM-Toyota joint venture NUMMI used the workforce to set work standards and then used the standardized work as the basis of continuous improvement and a precondition for learning.

With the development of imaging and graphical user interface (GUI) technology, it is only a matter of time before standards will be displayed on line on computer terminals. This will promote easy accessibility and ensure that the latest and valid version of a standard is used.

Process for Standardizing Operations

1. Ensure that the process has been developed and is the most effective way of making the product.

2. Determine time standards and operator requirements for all operations of the process. Each operation will have an individual standard practice or layout sheets that will contain detailed instruction on how to perform each step of each operation and include the tools and gauges used and the time taken. (See Figures 9.38a and 9.38b.)

3. Develop a process standard that describes the sequence of operations of the product and includes the layout of the process, a description of the equipment used, and any special instructions on material supply.

4. Make out a detailed job setup and operation sheet in the case of a machining operation (see Figure 9.39).

5. Specify the number of kanbans and the method of using them in the process standard, for processes under kanban control.

STANDARD JOB CONDITIONS

BASE RATE NO. 27112 CODE NO. —

DATE ________ STUDY NOS. 32906-32909 SYM. NO. — BINDER NO. 27

BLDG. 148A DEPT. No. 17 DIVISION Eastern OBSERVER Davis, W.T.

OPERATION Label 4-oz. Bottles Hardening Solution

SKETCH OF WORK PLACE

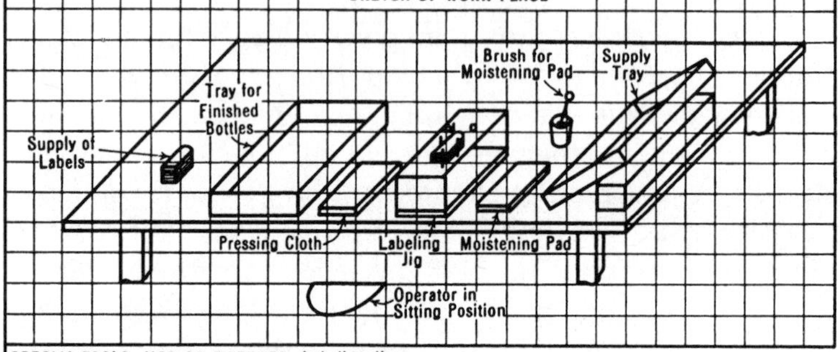

SPECIAL TOOLS, JIGS OR FIXTURES Labeling Jig

JOB ELEMENTS

1. Moisten pad with brush.	
2. Insert labels in jig.	
3. Procure bottle from supply tray, moisten bottle on moistening pad, label, using jig, press smooth on pressing cloth.	AUXILIARY
	Set up and clean up by handler or operator. No allowance in standard.
4. Dispose bottle to wooden tray.	
5. Upon completion of tray, make out ticket and place in tray as check against quality of labeling. Foreman can determine responsibility if labels are not up to standard.	Handler supplies bottles and disposes of finished tray.

AUDIT Production can be checked by order number. Foreman checks time turned in.

B118

Figure 9.38a. Example of Standardization: Standard job conditions form, size 8½ × 11 inches

Source: Reprinted with permission John Wiley & Sons. *Motion and Time Study.* Ralph M. Barnes, 1990.

```
┌────────────────────────────────────────────────────────────────────────┐
│                        GENERAL JOB CONDITIONS                            │
│                                                                          │
│  DATE OF ISSUE_____   BASE RATE NO.___27112___   CODE NO._____   │
│  BLDG.___148A___ DEPT.___No. 17___ DIVISION___Eastern___ OBSERVER___Davis, W.T.___│
│  TYPE OF OPERATION___Fill and Pack Bottles of Liquid                     │
└────────────────────────────────────────────────────────────────────────┘
```

LAYOUT OF OPERATION OR LOCALITY

Bottle Stock Room & Supplies — Bottle-Washing Machine — Bottle-Filling Apparatus — Solution Mixed on Floor Above. Bottles Filled by Gravity Flow

Packing Supplies

4 Stitch Cases — 3 Pack in Cases — 2 Pack in Cartons — 1 Label Bottles — Entrance

Shipping Room

First Floor Building 148 A

RANGE OF APPLICATION Unit designed for handling bottles of liquid product from 4-oz. to 32-oz. size.

DESCRIPTION OF STANDARD EQUIPMENT Balanced production line from supply room through to finished product in shipping room. Equipment consists of: bottle-washing machine No. 3712-A, bottle-filling apparatus No. 2192-O, battery of work places on long bench for labeling, packaging, and packing, and stitching machine No. 3127-C. Bottles handled in wooden trays to prevent accidents due to broken glass.

DESCRIPTION OF WORKING CONDITIONS Regular working hours 8-12, 1-5. Jobs performed in large airy room under daylight conditions. Artificial light available if necessary. Bottle washer wears rubber apron and gloves. Filling operator wears goggles, rubber apron and gloves, and cloth sleeves.

FLOW OF MATERIAL OR SUPPLIES Bottles supplied to washing machine from stock room. Washed bottles then moved to filling apparatus. Moved by truck from filling apparatus to labeling work place. Labeled bottles are then packed in cartons, cartons are packed in cases. Finished case is stitched on stitching machine, and then flows to shipping room. Packing supplies and labels are sent from supply room to position on work place.

B117

Figure 9.38b. Example of Standardization: General job conditions form, size 8½ × 11 inches

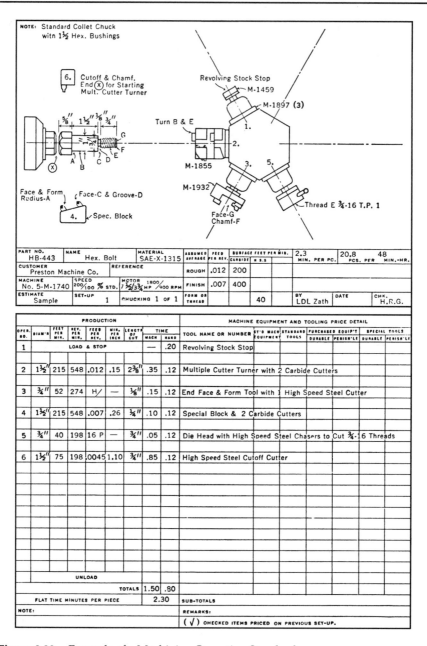

Figure 9.39. Example of a Machining Operation Standard
Source: Reprinted with permission John Wiley & Sons. *Motion and Time Study.* Ralph M. Barnes, 1990.

6. Specify points in the process where inspections will be conducted and detail the type of inspection required. A separate inspection standard practice sheet should be prepared.

7. Ensure that the standards to be used are prominently displayed.

8. Prepare an authorization chart of who can add, change, or delete any of the standard practice documents. These are quality control documents and must be carefully controlled to maintain their integrity.

9. Arrange to train all the shop personnel—supervisors and operators—since they will be working with and following the standards.

10. Encourage all personnel, particularly the operators, to make suggestions for improvement.

11. Review the standards on a regular basis with a view of making improvements.

12. Conduct periodic audits to ensure that the standardized procedures are being followed.

Dos and Don'ts

DOs

1. Do have formal, written, detailed standard documentation to cover the operations and processes of all product frequently manufactured.

2. Do elicit operator input and obtain operator involvement. Make operators responsible for the accuracy of the information in the standard.

3. Do realize that writing a standard is an exercise in communicating detail. It must be carefully and meticulously prepared.

4. Do seek to improve the standards continuously.

5. Do ensure only the current standard is in circulation.

DON'Ts

6. Don't neglect to set up formal controls (policy document) on who can add, change, and delete the standard.

Total Productive Maintenance (TPM)

Description and Purpose

The purpose of TPM is to prevent equipment from malfunctioning or breaking down, thereby increasing throughput and reducing lead time.

The APICS dictionary defines *total productive maintenance* as "preventive maintenance plus continuing efforts to adapt, modify, and refine equipment to increase flexibility, reduce material handling, and promote continuous flows. It is operator-oriented maintenance with the involvement of all qualified employees in all maintenance activities (see Figure 9.4)."

Discussion

The Japanese popularized TPM as productive maintenance carried out by all employees, and, like TQC (Total Quality Control), it has a companywide focus. The distinctive feature of TPM is that it involves operators with maintaining their own machines. The triple goal of TPM is no breakdowns, zero defects, and to have all equipment functioning optimally. Typically it would take about three years to implement fully an effective TPM program. In the United States, preventive maintenance (PM) is the equivalent of TPM, but in the past there has been no special effort at involving the operators in maintaining their equipment. Machines breaking down or being serviced periodically are a source of operational variability. Intermittent breakdowns not only introduce variability because work has to wait, but they also increase the average service (operational) time. Increasing the service time increases the utilization ratio and this in turn reduces the capacity available (the operation moves to a higher variability curve with attendant loss of capacity and/or lead time). Equipment failure or servicing has a dual adverse effect on lead times. From a variability viewpoint it is better to have planned shutdowns of short duration rather than unexpected, long breakdowns. To control the equipment maintenance process, the *mean time to repair* (*MTTR*) and the *mean time between failures* (*MTBF*) have to be understood and controlled.

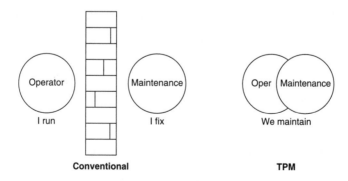

Figure 9.40. Attitudes in Conventional Maintenance and TPM

TPM seeks to achieve the following five objectives:

1. Maximize effectiveness (overall operating efficiency) of all equipment
2. Provide a system of comprehensive maintenance (preventive mainte-nance) for the life cycle of the equipment
3. Involve departments that plan, design, use, and maintain equip-ment—not only the maintenance group
4. Involve everyone—from management to production workers
5. Implement and sustain PM through autonomous small work teams

Maximizing Equipment Effectiveness

Process

1. Understand the six big losses. These are:
   Downtime:
   - Equipment failure—breakdowns
   - Setup and adjustment—exchange of dies, changing fixtures, and so on
   Speed losses:
   - Idling and minor stoppages—due to improper operation of equipment
   - Reduced speed—equipment not operating as designed
   Defects:
   - Process defects—quality defects
   - Reduced yield—product being scrapped

2. Define and measure **equipment effectiveness** to cover the six big losses. Set goals.

$$\text{Availability} = \frac{\text{operating time}}{\text{loading time}} = \frac{\text{loading time} - \text{downtime}}{\text{loading time}}$$

where loading time is total available time less planned maintenance time. *Downtime* includes all time except the planned maintenance when the machine cannot be operated.

Example: Assume a working shift is 8 hours or 480 minutes and planned maintenance is 20 minutes. Then loading time will be 460 minutes. If there were 20 minutes of breakdown, 30 minutes of setup, and 15 minutes of adjustment, then

$$\text{availability} = \frac{460 - 20 - 30 - 10}{460} = \frac{400}{460} \times 100 = 87\%.$$

Performance efficiency

$$= \text{net operation rate} \times \text{operating speed rate}$$

where the net operating rate measures the maintenance of a given speed over a given time, and operating speed rate is the ratio between designed speed and actual speed.

$$\text{Net operation rate} = \frac{\text{actual processing time}}{\text{operation time}}$$

$$= \frac{\text{processed amount} \times \text{actual cycle time.}}{\text{operation time}}$$

If 400 pieces are made at 0.9 minutes each, and the operation time is 400 minutes, then

$$\text{net operation rate} = \frac{400 \times 0.9}{400} \times 100 = 90\%$$

and

$$\text{operating speed rate} = \frac{\text{theoretical cycle time.}}{\text{actual cycle time}}$$

If the designed cycle time is 0.75 minutes and the actual time taken is 0.9 minutes, then

$$\text{operating speed rate} = \frac{0.75}{0.9} = 83\%.$$

Then

$$\text{performance efficiency} = \text{net operation rate}$$
$$\times \text{operating speed rate} = 0.9 \times 0.83 = 75\%.$$

Finally, if the quality yield is 95%, then

$$\text{overall equipment effectiveness} = \text{availability}$$
$$\times \text{performance efficiency}$$
$$\times \text{quality yield}$$
$$= 0.87 \times 0.75 \times 0.95 = 62\%.$$

Even though the availability is 97% and the yield is 95%, the overall effectiveness is 62%!

The goals for overall equipment effectiveness should be set at

$$\text{availability} = > 90\%, \text{performance efficiency}$$
$$= > 95\%, \text{and yield} = > 99\%.$$

Then

overall equipment effectiveness =
$$> 0.9 \times 0.95 \times 0.99 = > 85\%.$$

3. Develop a daily maintenance schedule. This should consist of cleaning, lubrication, conducting inspection, and performing adjustments as required. This should be performed by the operator or a team of operators.

4. Develop a preventive maintenance schedule. This should include periodic inspection and prescribed repairs and replacements. A preventive maintenance schedule is based on predicted wear and tear of the equipment, with oil and filter changes being the most common changes.

5. Put the TPM schedule on MRP using a replacement parts BOM. This will ensure that appropriate and adequate spare service parts are available. (The equipment manufacturer is the best advisor for developing this schedule.)

6. Plan regular courses for training operators and maintenance personnel on the use and care of the equipment. Refresher courses should be conducted periodically.

7. Institute an equipment management program.

TPM Implementation

Process

1. Obtain top management commitment. This must be vocal, visible, and continuous.

2. Promote the concepts. Understand that TPM represents a change from conventional maintenance. Many operators will resist it. There may be serious union issues.

3. Develop an implementation master plan with specific dates for introduction of the various phases of TPM.

4. Establish policies and goals. The policy should spell out the intention of the company to adopt TPM, that it is to have companywide applicability, and the expectations of the benefits it will confer. Goals may relate to the percent of overall equipment effectiveness, the reduction of failure rates and the improvement of the mean time between failures.

5. Train. Awareness training should be conducted with *all* employees. Specific team training should be conducted with the operators and maintenance personnel.

6. Improve equipment effectiveness (see above). This is the core of TPM.

7. Establish an operator maintenance program. Operators must provide some independent maintenance for their equipment and be responsible for some level of inspection. This is also probably the hardest change to make, particularly in companies that have operated conventionally for a long time and that have strong unions.

8. Set up a preventive maintenance program for the maintenance department. This program will depend on Step 7 above. The PM program must be equipment-specific and must detail inspections, checks, adjustments, and replacements that must be conducted at prescribed frequency or on the basis of equipment hours run. Obtain help and advice from the equipment manufacturer. Preventive maintenance schedules should be prepared monthly, and resources allocated to ensure that they are conducted. Detailed and accurate records of inspection and work done must be kept.

9. Conduct specific skill training on maintaining equipment for operators and maintenance people.

10. Develop an equipment management program. It has been estimated that 95 percent of the life-cycle cost of any equipment is determined at the time of design. Design staff should work with production engineering and maintenance personnel to ensure that all maintainability is given due consideration.

11. Implement TPM fully. Measure and ensure goals are being met.

Dos and Don'ts

DOs

1. Do implement a preventive maintenance system, with operator involvement and the advice of the equipment manufacturer.

2. Do be prepared for a long struggle to change the traditional separatist approach to maintenance.

3. Do understand the principal causes of machine trouble and prevent them from occurring:

(continues)

Dos and Don'ts

DOs (*continued*)

- Failure to maintain basics—cleaning, oiling, prescribed changes, and so on
- Failure to maintain operating conditions—temperature, humidity, and so on
- Lack of operator skills and/or maintenance skills
- Deterioration of machine through wear and tear

4. Do maintain detailed and accurate records of all maintenance performed, preferably on an appropriate preventive maintenance system.

5. Do ensure that an adequate supply of spare parts for maintenance are available. Seek advice from the manufacturer.

6. Do ensure that critical equipment (constraints) have a well-developed preventive maintenance schedule; that the mean time to repair (MTTR) is short, and the mean time between failures is long (MTBF). Measure both of these criteria on a regular basis.

7. Do integrate TPM with MRP. Treating the maintenance function as a product line within MRP is an effective means of planning and allocating maintenance resources.

DON'Ts

8. Don't exclude any personnel, particularly top management, from the TPM process.

Flexibility

Purpose and Description

Flexibility is a means of providing additional, on-the-spot capacity, by transferring resources to the operation that needs it. In doing so additional capacity is available when needed to deal with variability. Reducing variability reduces lead time and increases throughput.

In most manufacturing there is unpredictable product and process variability. Although continuous efforts should be made to reduce this variability, provision should also be made to deal with it and not let it adversely affect making product on time to meet a customer order. One way is to have flexibility. This flexibility comes primarily from multi-functional operators. It can also come from multi-purpose machines, quick setups, small lots, short lead times, and sufficient available capacity.

Discussion

In the section on shop layout and product flow it was suggested that a shop must be laid out according to a product's manufacturing sequence. It was also noted that in order to respond to varying demand, a U-cell arrangement of a product flow is most advantageous. It was seen that both these techniques led to reduced lead time and improved productivity. However, to implement these arrangements, workers must have multi-functional skills, and there should be sufficient machines to support some level of duplication that results from product-oriented layouts. Standardized working procedures help in rotating different operators through various jobs. The Japanese use the term *shojinka*, or flexibility, which translates into an ability to quickly alter the number of workers in each shop to respond to changes in demand. Toyota has a well-established program of job rotation covering supervisors and workers. The objective of this program is to have every worker master every kind of operation in every process in their shop! Toyota measures the multi-function rate as:

percent cross-trained
$$= \frac{\text{sum of all processes each worker has mastered.}}{\text{total number of processes} \times \text{number of workers}}$$

If there are 10 workers in a shop having 10 processes, and if all 10 of them can work 5 processes each, then

$$\text{the percent cross-trained} = \frac{10 \times 5}{10 \times 10} = 50 \text{ percent.}$$

Toyota was at 55 percent cross-trained in 1979 and have a 100 percent goal. See Appendix 9.9. How does one explain the phenomenon of such a high level of cross-training in Japan and so comparatively little in the United States? One of the reasons is the multiplicity of labor grades and levels in the United States and union contracts that do not allow easy interchangeability among grades. Another reason is management's inability to focus on and provide this need.

Where modifications to equipment are required to make different products, the changes should be understood, well rehearsed, and capable of being implemented in a short time. Chief among the changes required are setup changeover and it is essential that these are conducted in a SMED process (less than 10 minutes—see the section on setup reduction).

It was also pointed out that there should not be permanent material handling or transportation equipment (roller conveyors, car tracks, etc.), as these are difficult to change when the product or volume changes radically. They also provide a place for storing inventory!

Finally, the training and use of flexible workers has to be supported by the unions. It is often opposed as it is believed that flexible workers reduce the number of persons employed. This may be so, but it has to be viewed from the point of competitiveness also. It is not an easy task to convince the unions of this, but it is necessary.

Process for Providing Flexibility

1. Conduct awareness workshops in which the concepts and advantages of developing multi-functional skills are stressed. Seek buy-in from all personnel and the union.

2. Identify the operation's constraints and bottlenecks.

3. Ensure that sufficient operators are trained to execute the functions of the constraining operation. The number trained will depend on the stability of the workforce, but a good rule of thumb is to have at least one and a half times the number required as trained operators.

4. Publicize (large board display is one way) the operator's portfolio of skills.

5. Attempt to tie the number of required skills an operator has to his or her wage level. This may be a union issue, but if implemented it can provide strong motivation to develop multi-functional training.

6. Develop a job rotation program for all operators, particularly by having them work on constraining operations. Display the planned job rotation by operator on a monthly calendar.

7. Conduct a review of the machinery needed at the constraints. If the constraint is machinery limited (as distinct from manpower), additional machinery should be planned for. Where special purpose machinery is being used, it is essential that less sophisticated machinery be provided as additions or backup. (An example is in the use of large, high-volume furnaces and small lot furnaces as backup.)

8. Adopt a product focused layout using cells in the form of U layouts. This will allow the shop to take advantage of the multi-skilled worker (see the section on shop layout).

9. Consider including supervisors and managers in the job rotation plan.

10. Measure the effect of the flexibility developed on reduction in lead time and on time shipment of customer orders.

11. Provide a document covering a step-by-step description of the process. This should be used to confirm that the correct method is being followed and also to train new operators.

Dos and Don'ts

DOs

1. Do understand that the primary purpose of developing flexibility is to deal with variability. Start with constraints and bottlenecks.
2. Do remember that a major advantage of operator cross-training is to make operators aware of the total production process and improve their motivation and productivity by giving them variety and challenge at work.
3. Do strive to get buy-in for the process, particularly from the union, who may perceive it as an attempt to reduce the number of jobs.
4. Do make a commitment that personnel will not be reduced as a result of worker flexibility.
5. Do develop specific programs to put excess workers through when demand is reduced. Knowledge training, team training, learning new skills, and becoming computer literate are all activities that will improve long-term productivity and keep workers motivated.
6. Do capture the processes with clear, detailed work instructions.

Reduction of Manufacturing Lead Time: Summary

Description and Purpose

The reduction of manufacturing lead time decreases the manufacturing cost, demands quality work, and makes product available when required. Reduction of lead time is one of the fundamental measures of manufacturing effectiveness. JIT is sometimes described as a lead-time reduction program.

Lead time is the span of time required to perform a process or a series of operations (APICS). *Lead time* is used synonymously with *cycle time,* which is usually used to describe smaller time spans, such as a single operation. Manufacturing lead time starts on receipt of an order by the factory and ends when the product is shipped to a finished goods store or to a customer. The primary purpose of manufacturing is to make product economically and to satisfy customers. The reduction of manufacturing lead time is the most effective means of achieving this goal.

Discussion

This section attempts to summarize and integrate all the sections covered in this chapter on lead-time reduction and throughput improvement. There is some intentional repetition.

In discrete manufacturing, the lead time of an item is typically composed of the following elements:

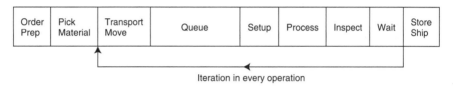

Figure 9.41. Elements of Lead Time

In many applications, a product spends only 10 to 20 percent of its lead time in being processed (that is, value-added time). The rest of the lead time is spent in being moved, set up, inspected, waiting to move, and in a queue in front of the next operation—all nonvalue-added time activities. If this analysis of lead time included defective product or rework of product, the value-added time may be even smaller. Of all the nonvalue-added time elements, *queue time* is usually the longest. *Value-added time* may be thought of as any time spent on actually transforming a product toward its final configuration.

There is an index to track the effectiveness of an entire process:

Manufacturing cycle efficiency

$$= \frac{\text{value-added time (processing time)}}{\text{total manufacturing lead time}}$$

This index contains nonvalue-added elements that are unavoidable, since product has to be ordered, or moved, or shipped. Another and fairer way of assessing the efficiency of the lead time is to measure the times butt to butt. *Butt-to-butt time* is the time a single piece of product takes to be made, if processed through its entire manufacturing cycle without any queuing or waiting. The times butt to butt will determine what percent of time a product is being worked on.

$$\text{times butt to butt} = \frac{\text{actual lead time}}{\text{butt-to-butt time}}$$

Example: *2 times butt to butt* means that a product is in operation 50 percent of the total lead time.
4 times butt to butt means that a product is in operation 25 percent of the total lead time.

The times butt to butt is a good measure of the comparative efficiency. The IC industry was at $\times 2$ with the benchmark at about $\times 1.5$.

The order preparation time, which may include preparing drawings or specifications, can be very large and can consume much of a product's total lead time. It should be monitored.

Queues

In the section on manufacturing relationships, the formation of queues was discussed, but this element of lead time is so important that it needs to be examined in more detail. For any stable operation, if the service (processing) time is less than the time between arrivals of jobs (utilization ratio $= < 1.0$), capacity is available to deal with every new job that arrives, there is no waiting, and the lead time is equal to the service time.

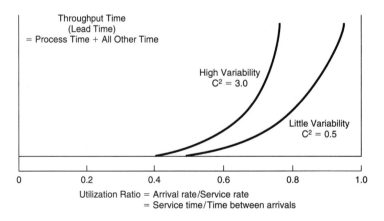

Figure 9.42. Lead Time versus Capacity (Utilization) versus Variability

As the utilization ratio approaches 1.0, any increase in the service or processing time or any decrease in the time between jobs will lead to jobs having to wait (in queue) to be worked upon. The increases and decreases in service and inter-arrival times are referred to as *variability*. If there are large variabilities, queues may start to form at low levels of capacity utilization, or if a resource is already at a high-utilization level, then large queues will form even at lower variability. In most cases small changes in service time and inter-arrival time will lead to large changes in waiting time and lead time.

The following table summarizes the effect of variability on capacity and lead time:

	Effect on Capacity	*Effect on Lead Time*
Low Variability	High level of Capacity used	Lead times minimized
High Variability	If High level of Capacity used	Large increases in lead time
High Variability	If Low level of Capacity used	Lead times minimized

Note: Minimizing variability is one way to optimize the use of resources while still meeting customer demand.

Note: Controlling and reducing variability is critical to improving lead time.

Practitioners should be aware of the variability of their key operations. One sure clue is to observe where queues form. The operations in front of which queues form consistently are most likely candidates for having high variability.

(For a calculation of variability and waiting time in queues, see Appendix 9.10.)

Manufacturing Execution Elements and Lead-Time Reduction

The sections covered under Manufacturing Execution were selected based on their bearing on variability and hence their effect on lead time. The elements affecting lead time are integrated as seen in Figure 9.43.

Arrival Variability

Ideally a process should receive product one at a time in predictable and equal intervals. Further, the time between the arrival of jobs or lots should be more than the time taken to process the lot, so there is no waiting.

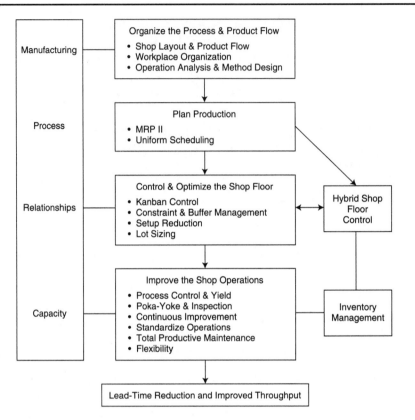

Figure 9.43. Elements of Lead-Time Reduction

In order to help control arrival variability, the following elements were covered:

- Manufacturing planning sections (master scheduling, MRP, CRP)
- MRP II shop floor control
- Uniform scheduling and cycle time
- Kanban control
- Constraints, bottlenecks, and buffer management
- Hybrid system shop floor control
- Lot sizes and transfer lots
- Setup reduction

All these activities are intended to provide consistent, predictable, small size lot of product arriving at a workstation. All of these activities are designed to reduce inter-arrival variability.

Service or Process Variability

An operation should be performed economically and in the least possible time. Furthermore, it should have a 100 percent yield so that there is no scrap or rework. Continuous efforts must be made to reduce the operation time. In order to provide this, the following elements were covered:

- Workplace organization
- Operation analysis and method design
- Process control and yield
- Poka-yoke and other forms of inspection
- Continuous improvement
- Standardizing operations
- Flexibility
- Total productive maintenance

All these activities are intended to provide consistent, predictable, minimum processing time on product arriving at a workstation. All of these activities are designed to reduce service variability.

Other Process Improvements

These consist of improvements to transportation; inspection; and ordering, storing and picking material.

Transportation Time: Considerable time can be spent on movement of material in a shop. In order to ensure that this element of lead time is minimized, the following element was covered:

- Shop layout and product flow

Inspection Time: Concepts on reducing inspection time were covered in the sections dealing with improving the quality of the process. While the major time gain in process improvement is realized from improving the yield of the process, time can be saved by rationalizing the inspection process itself, including the elimination of inspection as a separate operation. The reduction of this element of lead time was covered under:

- Process control and yield
- Poka-yoke and other forms of inspection

Ordering, Storing, and Picking Material: These elements of lead time can add considerably to the total lead time and should be minimized. Detailed coverage of these activities may be seen in the chapter on inventory control.

Technological Improvements

The time taken to actually manufacture is dependent on the variability and dependability of the process, which has been dealt with in this chapter. The time taken also depends on the technology of the process and the design of the product. Both these factors can affect the lead time appreciably. Design for manufacturability was covered in Unit I of the book. Process technology is out of the scope of this book, but should be in harmony with the techniques to reduce variability in the manufacturing process.

Other Significant Advantages of Reduced Lead Time

In addition to satisfying customer orders at an optimum cost, reduced lead time results in the following advantages:

- Less capital and better cash flow.

    This is a result of reduced inventory. Less storage space is required and less material is bought at one time. There is less loss due to product spoilage and deterioration.

- Improved quality.

    A focus on reducing lead time drives improvement in yield (minimal rejections). Further reduced lead time provides quicker feedback on defects and reduces rework.

- Improved market response.

    As a result of shorter lead times the customer orders can be made when they are received, eliminating or reducing the need for demand forecasting.

- Less rework due to engineering change.

    Shorter lead times generate less work in process and hence less rework due to change.

Summarized Process for Reducing Lead Time

J. D. Little, in studying queuing in the 1960s, developed a relationship between inventory and lead time. For any process:

Little's Law: WIP inventory = Rate of product arrivals × lead time (waiting time + service time)

1. Ensure that the products have a layout that is mostly unidirectional and that movement is minimized.

2. Eliminate, wherever possible, inspection as a separate step and intermediate storage of product.

3. Build only what is sold, in small lot sizes, covering not more than a week's demand (dependent on setup).

4. Limit work-in-process inventory. A pull system using kanbans is an effective means of ensuring this. Limit order release to what the constraining work centers can handle.

5. Identify the constraints and bottlenecks of the process.

6. Improve the constraint work centers by workplace organization and method improvements.

7. Understand the arrival (scheduling) and service (operation) variability of the constraining operations. Attempt to document and quantify the variability.

8. Reduce the variability of the constraints. Small reductions in service time lead to large reductions in lead time.
   Actions to take, in order of effectiveness, are:
   - Reduce the lot size until setup becomes the constraint.
   - Reduce the setup to enable further reduction of the lot size.
   - Use a transfer lot whenever possible (as small as possible).
   - Ensure that the equipment of the constraint is trouble-free.

9. Buffer the constraint. The constraint must always have material to work on.

10. Provide flexibility at the constraint by having multi-skilled operators.

11. Constantly improve the throughput (flow through) of the constraint by continuous improvement processes.

12. Measure and document the manufacturing lead time to confirm that improvements are being made. Analyze deviations and take corrective action.

Dos and Don'ts

DOs

1. Do concentrate on improving the whole process flow first.

2. Do ensure that the constraint is fully utilized.

3. Do keep work-in-process inventory down to a minimum.

4. Do understand the relationship between utilization, variability and lead time.

5. Do reduce lot sizes. It provides immediate benefits.

DON'Ts

6. Don't try to improve every operation. Focus on the constraint.

7. Don't neglect the order preparation time.

8. Don't plan to operate above an 80 percent level of capacity (utilization), unless the process is very stable.

9. Don't neglect to measure and document the lead time of the process to confirm that improvements are being made.

Appendix 9.1: Example of the Use of a Flowchart and a Process Chart

Present Method ☒	PROCESS CHART	
Proposed Method ☐		
SUBJECT CHARTED ___Requisition for small tools___	DATE _____	
Chart begins at supervisor's desk and ends at typist's desk in	CHART BY J. C. H.	
purchasing department	CHART NO. R 136	
DEPARTMENT ___Research laboratory___	SHEET NO. 1 OF 1	

DIST. IN FEET	TIME IN MINS.	CHART SYMBOLS	PROCESS DESCRIPTION
		●⇨☐D▽	Requisition written by supervisor (one copy)
		○⇨☐D▽	On supervisor's desk (awaiting messenger)
65		○⇨☐D▽	By messenger to superintendent's secretary
		○⇨☐D▽	On secretary's desk (awaiting typing)
		●⇨☐D▽	Requisition typed (original requisition copied)
15		○⇨☐D▽	By secretary to superintendent
		○⇨☐D▽	On superintendent's desk (awaiting approval)
		○⇨■D▽	Examined and approved by superintendent
		○⇨☐D▽	On superintendent's desk (awaiting messenger)
20		○⇨☐D▽	To purchasing department
		○⇨☐D▽	On purchasing agent's desk (awaiting approval)
		○⇨■D▽	Examined and approved
		○⇨☐D▽	On purchasing agent's desk (awaiting messenger)
5		○⇨☐D▽	To typist's desk
		○⇨☐D▽	On typist's desk (awaiting typing of purchase order)
		●⇨☐D▽	Purchase order typed
		○⇨☐D▽	On typist's desk (awaiting transfer to main office)
		○⇨☐D▽	
		○⇨☐D▽	
		○⇨☐D▽	
		○⇨☐D▽	
		○⇨☐D▽	
		○⇨☐D▽	
		○⇨☐D▽	
105		3 4 2 8	Total

Present Method ☐ Proposed Method ☒			PROCESS CHART			
SUBJECT CHARTED ___ Requisition for small tools				DATE _____		
Chart begins at supervisor's desk and ends at purchasing agent's desk				CHART BY J. C. H.		
				CHART NO. R 149		
DEPARTMENT ___ Research laboratory				SHEET NO. 1 OF 1		

DIST. IN FEET	TIME IN MINS.	CHART SYMBOLS	PROCESS DESCRIPTION
		●⇨☐D▽	Purchase order written in triplicate by supervisor
		○⇨☐D▽	On supervisor's desk (awaiting messenger)
75		○⇨☐D▽	By messenger to purchasing agent
		○⇨☐D▽	On purchasing agent's desk (awaiting approval)
		○⇨■D▽	Examined and approved by purchasing agent
		○⇨☐D▽	On purchasing agent's desk (awaiting transfer to main office)
		○⇨☐D▽	
		○⇨☐D▽	
		○⇨☐D▽	
		○⇨☐D▽	
		○⇨☐D▽	
		○⇨☐D▽	
		○⇨☐D▽	
		○⇨☐D▽	
		○⇨☐D▽	
		○⇨☐D▽	
		○⇨☐D▽	
		○⇨☐D▽	
		○⇨☐D▽	

			SUMMARY			
				PRESENT METHOD	PROPOSED METHOD	DIFFER- ENCE
			Operations ○	3	1	2
			Transportations ⇨	4	1	3
			Inspections ☐	2	1	1
			Delays D	8	3	5
			Distance Traveled in Feet	105	75	30
75		1 1 1 3	Total			

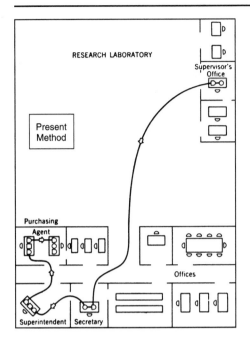

Flow Diagram of Office Procedure
Requisition written by supervisor, typed
by secretary, approved by superintendent
and purchasing agent, then purchase
order typed by secretary

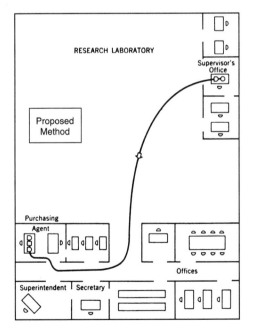

Flow Diagram of Office Procedure
Requisition written in triplicate by supervisor approved by purchasing agent

Source: Reprinted with permission John Wiley & Sons. *Motion and Time Study.*
Ralph M. Barnes, 1990.

Appendix 9.2: Operation Charting

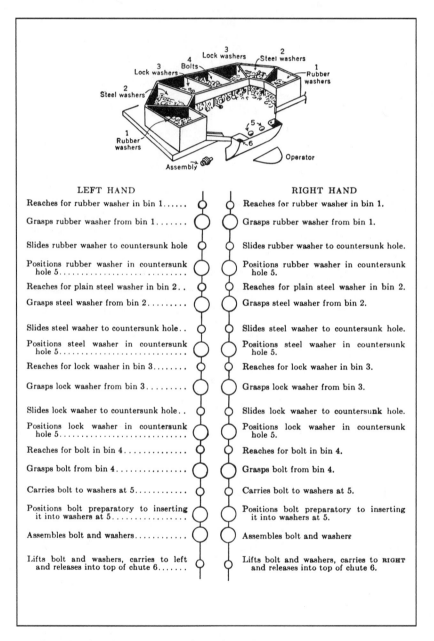

LEFT HAND	RIGHT HAND
Reaches for rubber washer in bin 1	Reaches for rubber washer in bin 1.
Grasps rubber washer from bin 1	Grasps rubber washer from bin 1.
Slides rubber washer to countersunk hole	Slides rubber washer to countersunk hole.
Positions rubber washer in countersunk hole 5	Positions rubber washer in countersunk hole 5.
Reaches for plain steel washer in bin 2 . .	Reaches for plain steel washer in bin 2.
Grasps steel washer from bin 2	Grasps steel washer from bin 2.
Slides steel washer to countersunk hole . .	Slides steel washer to countersunk hole.
Positions steel washer in countersunk hole 5	Positions steel washer in countersunk hole 5.
Reaches for lock washer in bin 3	Reaches for lock washer in bin 3.
Grasps lock washer from bin 3	Grasps lock washer from bin 3.
Slides lock washer to countersunk hole . .	Slides lock washer to countersunk hole.
Positions lock washer in countersunk hole 5	Positions lock washer in countersunk hole 5.
Reaches for bolt in bin 4	Reaches for bolt in bin 4.
Grasps bolt from bin 4	Grasps bolt from bin 4.
Carries bolt to washers at 5	Carries bolt to washers at 5.
Positions bolt preparatory to inserting it into washers at 5	Positions bolt preparatory to inserting it into washers at 5.
Assembles bolt and washers	Assembles bolt and washers
Lifts bolt and washers, carries to left and releases into top of chute 6	Lifts bolt and washers, carries to RIGHT and releases into top of chute 6.

Source: (Reprinted with permission John Wiley & Sons, Inc. Copyright © 1968. *Motion and Time Study,* Ralph M. Barnes, 6th edition).

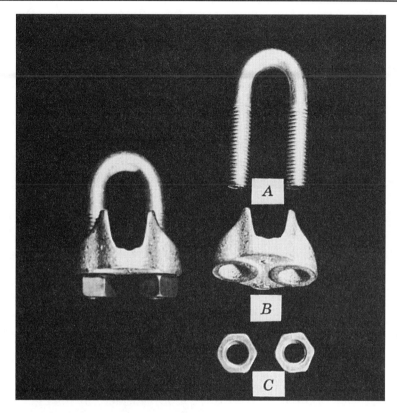

Rope clip assembly: *A*, U bolt; *B*, casting; *C*, nuts.

Appendix 9.3: Check List for Fundamental Hand Motions

Check List for Select

1. Is the layout such as to eliminate searching for articles?
2. Can tools and materials be standardized?
3. Are parts and materials properly labeled?
4. Can better arrangements be made to facilitate or eliminate select?
5. Are common parts interchangeable?
6. Are parts and materials mixed?
7. Is lighting satisfactory?

8. Can parts be pre-positioned during preceding operation?
9. Can color be used to facilitate selecting parts?

Check List for Grasp

1. Is it possible to grasp more than one object at a time?
2. Can objects be slid instead of carried?
3. Can tools or parts be pre-positioned for easy grasp?
4. Can a special screwdriver, socket, wrench, or combination tool be used?
5. Can a vacuum, magnet, rubber finger tip, or other devices be used to advantage?
6. Is the article transferred from one hand to another?
7. Does the design of the jig or fixture permit an easy grasp in removing the part?

Check List for Transport Empty or Transport Loaded

1. Can either of these motions be eliminated entirely?
2. Is the distance traveled the best one?
3. Are the proper means used—hand, tweezers, conveyors, etc.?
4. Are the correct members (and muscles) of the body used—fingers, forearm, shoulder, etc.?
5. Can a chute or conveyor be used?
6. Can "transports" be effected more satisfactorily in larger units?
7. Can transport be performed with foot-operated devices?
8. Is transport slowed up because of a delicate position following it?
9. Can transport be eliminated by providing additional small tools and locating them near the point of use?
10. Are parts that are used most frequently located near the point of use?
11. Are proper trays or bins used, and is the operation laid out correctly?
12. Are the preceding and following operations properly related to this one?
13. Is it possible to eliminate abrupt changes in direction? Can barriers be eliminated?
14. For the weight of material moved, is the fastest member of the body used?
15. Are there any body movements that can be eliminated?
16. Can arm movements be made simultaneously, symmetrically, and in opposite directions?

17. Can the object be slid instead of carried?
18. Are the eye movements properly coordinated with the hand motions?

Check List for Hold

1. Can a vice, clamp, clip, vacuum, hook, rack, fixture, or other mechanical device be used?
2. Can an adhesive or friction be used?
3. Can a stop be used to eliminate hold?
4. When hold cannot be eliminated, can arm rests be provided?

Check List for Release Load

1. Can motion be eliminated?
2. Can a drop delivery be used?
3. Can the release be made in transit?
4. Is a careful release load necessary? Can this be avoided?
5. Can an ejector (mechanical, air, gravity) be used?
6. Are the material bins of proper design?
7. At the end of the release load, is the hand or the transportation means in the most advantageous position for the next motion?
8. Can a conveyor be used?

Check List for Position

1. Is positioning necessary?
2. Can tolerances be increased?
3. Can square edges be eliminated?
4. Can a guide, funnel, bushing, gauge, stop, swinging bracket, locating pin, spring, drift, recess, key, pilot on screw, or chamfer be used?
5. Can arm rests be used to steady the hands and reduce the positioning time?
6. Has the object been grasped for easiest positioning?
7. Can a foot-operated collet be used?

Check List for Pre-position

1. Can the object be pre-positioned in transit?
2. Can the tool be balanced so as to keep the handle in upright position?
3. Can a holding device be made to keep the tool handle in proper position?

4. Can tools be suspended?
5. Can tools be stored in proper location to work?
6. Can a guide be used?
7. Can the design of an article be made so that all sides are alike?
8. Can a magazine feed be used?
9. Can a stacking device be used?
10. Can a rotating fixture be used?

Check List for Inspect

1. Can inspect be eliminated or overlapped with another operation?
2. Can multiple gauges or tests be used?
3. Can a pressure, vibration, hardness, or flash test be used?
4. Can the intensity of illumination be increased or the light sources rearranged to reduce the inspection time?
5. Can a machine inspection replace a visual inspection?
6. Can the operator use spectacles to advantage?

Check List for Assemble, Disassemble, and Use

1. Can a jig or fixture be used?
2. Can an automatic device or machine be used?
3. Can the assembly be made in multiple? Or can the processing be done in multiple?
4. Can a more efficient tool be used?
5. Can stops be used?
6. Can other work be done while machine is making cut?
7. Should a power tool be used?
8. Can a cam or air-operated fixture be used?

Source: (Reprinted with permission John Wiley & Sons, Inc. Copyright © 1968. *Motion and Time Study,* Ralph M. Barnes, 6th edition).

Appendix 9.4: Input-Output Control

Description

Work should be planned and controlled at all work centers, but it is critical that this control be exercised at:

- The gateway or first work center
- The constraining or bottleneck work center

To minimize work in process and optimize throughput and reduce lead times, it is essential that the load released to a work center (input) is linked to the production of the work center (output). This linkage is done through an *input-output control report*. The output is based on the capacity of the work center, and the input is regulated to maintain the use of this capacity. The report may be formatted as below:

Work Center: Milling machines

Week number	11	12	13	14	15
*Input					
Planned	900	900	900	900	900
Actual	850	875	825	900	1000
Cum. Dev.	−50	−75	−150	−150	−50
*Output					
Planned	900	900	900	1000	1000
Actual	900	950	900	950	1000
Cum. Dev.	0	50	50	0	0
*Queue					
Planned	1000	1000	1000	900	800
Actual Start 1000	950	875	800	750	750

Calculations are performed as follows:
Cumulative Deviation (Cum. Dev.) = Cum. Dev. of Prior Period + (Actual − Planned)
Week 11: Cum. Dev. = 850 − 900 = −50 and Week 12: Cum. Dev. = (−50) + 875 − 900 = −75
Queue = Starting queue + Input − Output (separately for planned and actual).
Week 11: Actual Queue = 1000 + 850 − 900 = 950 and Week 12 Actual Queue = 950 + 875 − 950 = 875

The input-output report is a very useful tool. In addition to determining how much work to release and to controlling queues, the report provides an early warning on the need for corrective action if queues are growing or being depleted too quickly.

Appendix 9.5: Kanban Rules (as Developed by Toyota)

Rule 1. A downstream operation (subsequent operation) should withdraw (pull) necessary quantities of product from the upstream or preceding operation only as and when authorized by a kanban.

Rule 2. The upstream or preceding operation should make only the quantity withdrawn by the subsequent operation.

Rule 3. Never send defective products to a subsequent operation.

Rule 4. The number of kanbans should be minimized.

Rule 5. Kanbans should be used to adapt to small variations in demand.

Rule 6. The whole process must be stabilized and rationalized.

Appendix 9.6: Theory of Constraints

Theory of Constraints

A constraint is anything that prevents a system from achieving a higher performance relative to its goal.

Step 1: Identify the system's constraint(s).
Step 2: Decide how to exploit the system's constraint(s).
Step 3: Subordinate everything else to the decisions made in Step 2.

Step 4: Elevate the system's constraint(s).

Step 5: If, in the previous steps the constraint has been broken, go back to Step 1, but do not allow inertia to cause a system's constraint. (This warns against complacency after the first problem has been solved) (Goldratt 1990.)

Global Rules

Another set of operational guidelines from Eli Goldratt on bottlenecks and constraints are:

1. Balance flow not capacity.

2. The level of utilization of a non-bottleneck is not determined by its own potential but by some other constraint in the system (the throughput of the bottleneck).

3. *Utilization* and *activation of a resource* are not synonymous.

4. An hour lost at a bottleneck is an hour lost for the total system.

5. An hour saved at a non-bottleneck is just a mirage.

6. Bottlenecks govern both throughput and inventories.

7. The transfer batch may not and many times should not be equal to the process batch.

8. The process batch should be variable not fixed.

9. Schedules should be established by looking at all of the constraints *simultaneously*. Lead times are the result of multiple interactions of all the activities of a schedule and cannot be predetermined.

Motto
The sum of local optimums
is not equal to
the global optimum.
(Goldratt and Fox 1986)

Appendix 9.7: Determining Control Limits for Attribute and Variable Charts

Variable Control Charts

These charts use variable data measured in the process. Two statistical variable readings are tracked: $\bar{X}$, the sample average and R, the sample range. The charts are known as $\bar{X}$ (X bar) and R charts. The $\bar{X}$ chart shows the average performance of the process, and the R chart shows the variability of the process. In general, the $\bar{X}$ chart is process dependent and the R chart is operator dependent.

X = an individual reading
$\bar{X}$ (X bar) = average of a group of X's
$\bar{\bar{X}}$ (X double bar) = average of $\bar{X}$ bars.
$\bar{R}$ (R bar) = average of R's

The formulas used to calculate the X chart control limits are:

$$\text{UCL} = \bar{\bar{X}} + A_2 \bar{R}$$
$$\text{LCL} = \bar{\bar{X}} + A_2 \bar{R}$$

The formulas used to calculate the R chart control limits are:

$$\text{UCL} = D_4 \bar{R}$$
$$\text{LCL} = D_3 \bar{R}$$

The values of A_2, D_3, and D_4 are constants that depend on the number of samples taken for a particular reading. These values can be found in the tables in many statistical books.

Process for *X* and *R* Charts

1. Decide on the sample size (n) to be used (about 5 is usually adequate).

2. Obtain a series of sample measurements (about 20 samples) each sample having n measurements.

3. Compute $\bar{X}$ for each sample and take the average of the $\bar{X}$'s (X double bar). This is the center line of the X chart.

4. Compute R for each sample and take the average of the $\bar{R}$'s (R bar). This is the center line of the R chart.

5. Calculate the UCL and LCL for the X and R charts using the above formula. Control limits are usually set at $\pm 3\sigma$, which means that 99.7 percent of the sample means should fall between the control limits. Plot these control limits on their respective graphs using appropriate scales, with the R chart being plotted below the $\bar{x}$ chart.

6. Plot the successive values of X and R on the chart, marking those outside the control limits with **X.**

7. Interpret the chart and determine if the process is in control. Take action to correct out-of-control conditions.

> Example: Driving to work every day. A work week of 5 days may be considered as a sample. The $\bar{X}$ will track the average of the time taken (process average) for the days of the week. The R will track the difference in time between high and low X's. The $\bar{\bar{X}}$ is the average of several weeks' samples, and the $\bar{R}$ is the average range of these samples. Using a standard chart, get the A and D values and calculate the UCL and the LCL.

Attributes Control Charts

These charts use data that is binary in nature and show characteristics that pass or fail requirements. No measurements are required. It is only necessary to count the number of pieces that are defective and convert it into a percentage. To construct an attributes chart, the average fraction defective, $\bar{p}$, must be known or estimated and expressed as under:

n = number of units in a sample

p = fraction defective in a sample = $\dfrac{\text{number of defective units in a sample}}{\text{sample size } (n)}$

$\bar{p}$ = average fraction defective in a series of samples

$\bar{p}$ = $\dfrac{\text{total number of defective units from all samples}}{\text{number of samples} \times \text{sample size (total number inspected)}}$

Process

1. Obtain a number of samples. A convenient sample size is 50 pieces, and about 20 to 25 samples should be taken.
2. Count the number of defectives in each sample and calculate p for each sample.

3. Calculate $\bar{p}$ (the average fraction defective). This is the center line for the $\bar{p}$ chart. Show as a solid line.
4. Calculate the upper and lower control limits for the control chart. Generally 3σ control limits (99.7 percent) level is used. So:

$$\text{UCL} = \bar{p} + 3\sqrt{\frac{\bar{p}(1 - \bar{p})}{n}}$$

$$\text{LCL} = \bar{p} - 3\sqrt{\frac{\bar{p}(1 - \bar{p})}{n}}$$

5. Plot the UCL and LCL lines on a chart as dotted lines using an appropriate scale.
6. Plot on this chart the successive values of p.
7. Mark out-of-control points with **X** using instability rules.
8. Interpret the chart. Is the percentage of bad product increasing or decreasing? Are there erratic spikes?
9. Take appropriate action to improve the process.

Example: The paint of cars is checked to determine whether it is "OK" or "Not OK." Last week the results were:

Day	Number of cars sampled	Number defective
Monday	30	6
Tuesday	30	3
Wednesday	30	2
Thursday	30	4
Friday	30	9

This week on Monday there were 6 defects and 9 on Tuesday. Is the process in control?

$$\bar{p} \text{ (for last week)} = \frac{\sum x}{\sum n} = \frac{24}{150} = 0.16$$

$$\sigma p = \sqrt{\frac{\bar{p}(1 - \bar{p})}{n}} = \sqrt{\frac{0.16(1 - 0.16)}{30}} = 0.067$$

$$\text{UCL} = \bar{p} + 3\sigma p = 0.16 + 3\,(0.067) = +\,0.361$$

$$\text{LCL} = \bar{p} - 3\sigma p = 0.16 - 3\,(0.067) = -\,0.041$$

Since defectives cannot be < 0, the LCL will be set at 0.

Monday's defective is $p = \dfrac{6}{30} = 0.20$. Tuesday's defective is

$p = \dfrac{9}{30} = 0.30$.

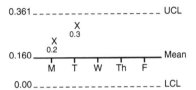

The process is exhibiting a trend toward going out of control.

Appendix 9.8: Problems with Inspection

- After the fact
- Even 100 percentage will not find *all* defects
- Some bad parts will get to the customer
- Adversarial "Big Brother" situation created
- Abdication of operator responsibility
- Reactive, not proactive inspection
- Controls but does not improve process
- Adds time and cost to the process

Appendix 9.9: Advantages of Job Rotation (Toyota Production System)

1. An ability to quickly increase or decrease the number of workers in a shop to match an increased or decreased demand.
2. Workers get less tired and are more attentive, leading to a reduction in factory accidents.
3. Working relationships between workers improve as feelings of work discrimination are eliminated.

4. Senior workers get to teach their skills to the younger workers and valuable expertise is not lost.
5. Since workers participate in all the processes of the shop, they feel a sense of ownership in safety, quality, cost, and production.
6. The system creates an environment for examining the process and encourages suggestions for improving the process.

Appendix 9.10: Calculation of Variability

To calculate the variability of an operation, the *standard deviation σ* of the operation must first be determined. The standard deviation is a measure showing the distribution around the mean, and of the frequency of occurrence of the different values of the subject being measured. Thus to find the σ of an operation's time, several values of the time taken to perform the operation should be recorded, and from these the standard deviation can be calculated.

Example: An operation is timed to take 3, 7, 4, 11, and 5 minutes respectively. Calculate the σ.

Individual Reading	*Average*	*(Individual Reading— Average)*	*(Individual Reading— Average)²*
3	6	−3	9
7	6	1	1
4	6	−2	4
11	6	5	25
5	6	−1	1
Average = 30/5 = 6			Total = 40

Then

$$\sigma = \sqrt{\frac{40}{5}} \quad \text{and variance} = \frac{40}{5} = 8.0$$

$$\text{variability} = c^2 \qquad = \frac{\text{variance}}{(\text{average})^2} = \frac{8.0}{36} = 0.22$$

By observing the intervals between arrivals of jobs or by noting the time taken to execute an operation, the arrival and service variability can be calculated.

Calculation of Average Waiting Time:

If

a = time between arrivals

and

s = time to process the job (service),

then

$u = \dfrac{s}{a}$ = utilization, and if

Ca = arrival variance

and

Cs = service variance, then let Q = length of queue (number of lots or jobs) or average inventory

and

l = interval (operational lead time) = waiting time + processing time

$$\text{Waiting or Delay time} = (s)\text{x}\left(\frac{u}{1-u}\right)\left[\left(\frac{Ca^2 + Cs^2}{2}\right)\right]$$

and $Q = \dfrac{1}{a} \times L$ = Little's Law

In inventory terms, Little's Law:

Inventory in Queue = Rate of arrivals $\times$ Operational Lead Time

10 Inventory Management

Summary of Chapter

This chapter covers specific methods for controlling inventory in the factory or warehouse. The chapter starts with an overview of inventory, followed by the *ABC technique*, an effective way of prioritizing approaches to controlling inventory. At the heart of all inventory management is *record accuracy*, and techniques for establishing accurate inventory records are described. This is followed by *storeroom management* including receiving, storing, and picking. Inventory planning is initiated by *material ordering*, which describes the general principles of ordering and explains the need for and the calculation of safety stock and economic order quantities. Specific techniques to *reduce inventory*, including slow moving, excess, and obsolescent inventory are described next. Some concepts on *material costs* are covered, followed by *performance measurement* criteria for inventory. The whole inventory control process is summarized in a section on an *inventory management program*.

A brief treatment of purchasing in the form of an overview followed by *supplier partnerships* and *best purchasing practices* closes this chapter.

The focus of the entire chapter is on managing inventory effectively.

Contents

- Overview and basic concepts
- ABC classification and application
- Inventory record accuracy
- Storeroom management
- Point-of-use storage
- Materials ordering and safety stock
- Customer service level determination and safety stock

- Inventory reduction
- Inventory costs
- Inventory performance measurement
- Inventory management program
- Purchasing
- Purchasing overview
- Make or buy decisions
- Supplier partnerships
- Best purchasing practices

Relationship of Inventory Management to Other Chapters in the Book

Inventory is planned and ordered through materials requirements planning (MRP). It is converted in the manufacturing process from raw material through work in process to finished goods. It is the means of satisfying a customer. Work-in-process inventory strongly influences the lead time of a process.

Overview and Basic Concepts

Purpose and Description

Inventory investment must be optimized for the best return, while maintaining a high level of customer service. By controlling the cost and the amount of inventory, a company decreases its cost of goods sold, increases its cash flow, and improves its return on investment.

Inventory used in manufacturing is material waiting to be worked on, being worked on, or waiting to be sold or distributed. Inventory also includes material that is not likely to be used because it is spoiled, obsolete, or in excess of need. Inventory is needed to build a product or deliver a product to a customer. Inventory is thus required. Inventory management consists of balancing the amount of inventory carried against the customer service required. Usually a higher customer service level requires more inventory. Unfortunately, inventory is also used to cover problems, such as poor forecasts, factory machine breakdowns, quality defects, and so on. Inventory costs money and excess or unused inventory means that money is tied up unproductively.

Discussion

There are many ways to look at inventory. Manufacturing views raw material and work-in-process inventory as product to be built and wants to have a lot available at all times, both as reassurance that there is work available and as protection against problems that may disrupt production. Sales and marketing want as much finished material as possible, available at all times, ready for immediate delivery. Finance and accounting see inventory as an investment, representing money that is tied up, and want inventory to be minimized or turned over rapidly.

Given these conflicting approaches, all of which have merit, making rational decisions on how to manage inventory is controversial and difficult. To help reduce the magnitude of the problem, it is essential that a company clearly defines its business objectives. (This is considered in the production planning chapter.) Inventory policies can then be formulated to support these objectives, and trade-off techniques can be used to resolve some of the conflicts. If, for example, a company decides that customer service is its primary goal, additional inventory in the form of safety stock must be stocked to cover fluctuations in sales and/or problems in supply of material.

Inventory is managed by planning for and producing the right quantity of material required to satisfy a customer order. Ideally the factory should make a product when a customer places an order. Since the time to make a product often exceeds the time a customer is willing to wait, to bridge the gap product requirements must be forecast. This also applies to a product that is sold off the shelf or from stock. The factory begins to make forecast product before the customer orders it and so covers the gap between the *factory make time* and the *customer want date.* The process of combining forecasts and customer orders and issuing factory orders while maintaining the appropriate time intervals is done dynamically through *master scheduling* and *materials requirements planning (MRP)*. (Both of these functions are covered in Unit II.)

Excess inventory can cover up many production and/or supplier problems, as the factory has additional material to work on and can ignore or postpone solving the problems. Inventory also directly affects the throughput time of product in a factory, and excess inventory leads to reduced throughput and increased lead times.

Classes of Inventory

Inventory is classified according to its condition during processing. There are four basic classes of inventory:

1. **Raw materials** are basic stock, such as steel, wood, plastic, and so on. They are usually used at the start of the manufacturing process (except for packaging material).

2. **Components** are parts or subassemblies used in assembling the product.

3. **Work-in-process (WIP)** consists of materials and/or components that are being worked on, or waiting to be worked on, in a factory. WIP uncouples or separates successive production steps and increases flexibility in operations.

4. **Finished goods** are completed products ready to be sold or delivered.

All of the four categories together are referred to as *full-stream inventory*.

Creation of Inventory and Types of Inventory

Finished goods inventory: This is generated by independent demand—forecasts, customer orders, service parts, and warehouse requirements—collected through and adjusted in the master schedule and/or the materials requirements plan. The materials requirements plan generates shop orders and purchase orders to satisfy the demand, thereby creating work in process and raw material inventory.

Cycle Stock: Inventory is also created by ordering and making product in *lot sizes*, when the product is manufactured or purchased in lots that exceed the specific quantity required. This inventory is also referred to as *cycle stock*.

Safety Stock: There are uncertainties in the type and quantities of the demand and there are variations in the ability of the factory or supplier to make sufficient product in time. To cover the fluctuations produced by these uncertainties and to ensure that the customer gets the order in time, safety stock is provided.

Transportation Inventory: Where a stream of product is required, inventory is needed to cover the time it takes to transport the product from factory or warehouse to customer.

Yield Coverage Stock: Inventory is produced by planning additional quantities to replace anticipated defective product. This amount of product replaced is called the *yield loss,* and so inventory is created to cover yield losses. A corollary to yield loss is inventory used to provide for *inspection time,* which may be considerable.

Anticipation Inventory: This inventory is often created for economic reasons, where pre-investment or anticipation inventory can provide savings by taking advantage of quantity discounts or to beat a price rise. Anticipation inventory is also produced to cover seasonal demand or smooth the factory's production schedule.

Strategic Inventory: This inventory is sometimes carried for strategic reasons such as scarce material and critically required material. Inventory may also be required to cover possible failures of supply in special situations such as strikes and political strife.

Summary

Whereas most of material planning and ordering is done through master scheduling and MRP, there are many reasons why inventory is required. It is essential that inventory is managed based on why it is required, and this understanding of the reason for the inventory should temper the recommendations of MRP. Materials requirements planning is an algorithm and does not cater to reasons why more or less inventory is required. MRP's recommendations must be reviewed and adjusted to meet the requirements of the business.

ABC Classification and Application

Purpose and Description

The *ABC technique* is one of the most useful and effective tools available to manufacturing personnel. It helps to determine priorities—where does one get the biggest "bang for the buck?" It is especially powerful in inventory management. A items should be carefully monitored and controlled, since they account for most of the inventory dollars. The technique is usually applied separately to raw materials, components, work-in-process, and finished goods.

In any group of inventory items, a small number (about 20 percent), will account for most of the total value (about 70 to 80 percent), of the group. The same principle holds true for cause and effect—a small number of causes are responsible for most of the effect. The small influential number are called A items, with B items being a slightly larger number (about 15 percent), and the balance being C items that, though making up 70 percent or more of the number, only contribute to 10 percent or less of the total effect. The A items are referred to as *the significant few,* while the C items are called the *trivial many.* The principle can be observed in daily life: 20 percent of the people in a country typically control 80 percent of the wealth, or 20 percent of the items on a family budget consume about 80 percent of the total budget. This effect is also called the *80:20 rule* or the *Pareto principle* (Pareto was an Italian who first discovered the relationship).

Discussion

Inventory can be classified into three categories as shown in Table 10.1. The principle is that a few items control a large amount of the total inventory value (*value* is defined as the cost of individual items times the quantity used). The percentages allocated depend on the type of product, and should be adjusted to ensure that only a small number of items are classified as A items, typically a number that can be tracked and controlled by a lead person or supervisor.

The ABC classification can be represented by a curve. Rules governing ABC items are shown in Table 10.2. It must be stressed that all items must be available when required, and a stock-out of a C item can be as harmful as that of an A item.

Table 10.1 Classification of A, B, and C Items

	A Items	B Items	C Items
Number of Items % (by #)	About 10 to 20%	About 10 to 20%	About 60 to 80%
Total Value % (by $)	About 60 to 80%	About 10 to 20%	About 10 to 20%
Cumulative Value % (by $)	About 60 to 80%	About 70 to 90%	100%

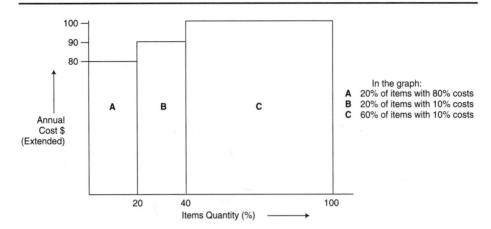

Figure 10.1. An Example of an ABC Curve

Table 10.2 Comparison of Controls on A, B, and C Items

Characteristic	A Items	B Items	C Items
Management	Very carefully tracked and controlled	Closely controlled	Floor stock, loose control
Inventory Accuracy	100% monthly cycle count Accounting for all errors	High accuracy Quarterly count Error investigation	Can have ±10% error No investigation of error
Delivery Lot Size	As small as possible—one week's requirement	Up to one month's requirements	Six to twelve months' requirements
Safety Stock	Minimum—less than one week	Moderate—two weeks	Liberal
Processing Priority	Highest priority	Medium priority	Low priority
Processing Lead Time	Should be minimum	Should be small	

There are other criteria that may be used to upgrade items into an A or B status, even though they do not qualify on a basis of total value. These include items that are critical to production, items with long lead time, items with short shelf life, material that can be stolen, material with erratic demand, and large bulky material. In all these cases these items are subject to the rules governing A items and require more frequent review and have to be controlled more carefully.

Process for Establishing A, B, and C Items

The classification of items into ABC categories is usually performed as follows (see Table 10.3):

1. List all items, their unit cost and their estimated consumption (forecast and/or actual demand) for some time period (usually a year). If the product is stable, historical consumption can be used.

2. Find the extended value (total value) of the item by multiplying the unit cost by the demand.

3. Sort the list in the total value sequence, with the highest value items on the top (Table 10.3, Column 6).

4. Add the total value in sequence to get a cumulative total value (Column 8).

5. Set a quantity of items at about 10 to 20 percent of the total number of items (Column 2). Starting from the top see how much of the cumulative total value is covered by these items. The cumulative value should be about 70 to 80 percent. Adjust the number of items to get about this value. *These are the A items.*

6. Take another 15 to 20 percent of the items and see that they cover about 15 to 20 percent of the cumulative total value, over and above the A item total value. The cumulative value will then be from 70 to 90 percent. *These are the B items.*

7. The balance are C items and should cover about 10 percent of the total extended value.

8. Review the list for items critical to production, or for characteristics such as erratic demand, long lead time, bulky size, and short shelf life. Upgrade them to A or B items, using good judgment.

The table shows that there are 6 A items, which are 15 percent by number and make up 73.6 percent of the cumulative total dollar value. There are likely to be about 10 B items covering 25 percent of the quantity and probably covering 15 percent of the cumulative value, leaving 24 (40 − 6 − 10 = 24) or 60 percent of the remaining C items with a cumulative value of 11.4 percent.

Table 10.3 Example of an ABC Table

1	2	3	4	5	6	7	8	9	10
Sequence No.	Item % Cumulative	Item Description	Unit Cost $	Demand #	Total Value $ Col. 4 × Col. 5	Item % Col. 6/ Total Col. 8 (5500)	Cumul. Value $ Add Col. 6	Cumul. % Col. 8/ Total Col. 8 (5500)	Classifi- cation A, B, or C
1	2.5		100.0	12	1200	21.8	1200	21.8	A
2	5.0		2.0	480	960	17.4	2160	39.2	A
3	7.5		30.0	24	720	13.2	2880	52.4	A
4	10.0		20.0	24	480	8.7	3360	61.1	A
5	12.5		5.0	72	360	6.5	3720	67.6	A
6	15.0		55.0	6	330	6.0	4050	73.6	B
nn									
nn									
40	100.0		0.1	480	48	0.9	5500	100	C

It should be noted that most MRP software calculates the ABC categories. The user is asked to specify break points, that is, what percent of the cumulative extended value should A, B, and C items cover. In Table 10.3, the break points will be A—73.6 percent, B—25 percent, and C—11.4 percent. A few iterations may be required to get break points that cover the right quantity of items.

Dos and Don'ts

DOs

1. Do classify all inventory into ABC items, even if all the information is not available. Start using the list to prioritize items to control.
2. Do have the order policy and the safety stock of A items reviewed by the inventory supervisor. Remember "the significant few."
3. Do use historical demand if future demand is not available.
4. Do select other As and Bs based on items being critical, having long procurement lead times, or having erratic demand, short shelf life, or being bulky.
5. Do keep the number of A items to a manageable quantity.

6. Do control C items in a simple manner. They should be ordered in bulk, reordered using a simple system such as a two-bin system (the empty bin generates an order), and have no daily record keeping. Remember "the trivial many."

DON'Ts

7. Don't mix raw materials and components with work in process and finished goods. They should be controlled with separate ABC classifications.
8. Don't restrict the ABC classification to inventory only, but apply it to customers, problems and any other cause-and-effect phenomena.

Inventory Record Accuracy

Inventory record accuracy covers inventory on-hand balance accuracy and order record accuracy.

Inventory On-Hand Balance Accuracy

Purpose and Description

Accurate inventory on-hand balances are required to:

- Make dependable delivery promises to customers.
- Order the right quantity of material required to meet the schedule.

- Minimize time and money spent in expediting emergency orders.
- Prevent overordering, leading to excess or obsolete inventory.
- Provide the basis of material requirements planning.
- Ensure that inventory value is correctly stated.

Inventory record accuracy covers the quantity and location of raw material in storerooms, of work in process in the factory, and of finished goods in warehouses.

Discussion

Storeroom and factory personnel often spend much time searching for parts to fill an important customer order. The parts are shown as available on an inventory status report, but they cannot be found. Every factory has an invisible monster that either devours materials or misplaces them! Maintaining accurate inventory records is probably one of the most underrated tasks in a factory. It has been estimated that the average inventory accuracy of the storerooms and warehouses in the United States is 70 percent. Only 7 parts in 10 have the right quantity and location on record.

What accounts for the failure to control this seemingly simple function? Perhaps it is the perception that the task is trivial and can be assigned to the lowest labor grade in the factory.

The entire problem is aggravated by some factories conducting an annual physical inventory. This exercise, usually directed by a public accounting firm, attempts to verify the inventory record count accuracy of the parts in a factory or warehouse (fully or partially). All factory operations are stopped for three or four days (including a weekend), and a large group of personnel are assigned to count all the material on record. Usually good controls are maintained to ensure that product is not counted twice, that all records are captured, and that random audit checks are made. The problem is that usually the counters are disinclined to count, and, as fatigue sets in, the accuracy of the count deteriorates. Further, many items are misidentified and this results in a double error; the original item is now wrong, and we have created a nonexistent item with a wrong location and quantity. At the end of the "physical," so many items have count discrepancies between "book" and "actual" that an attempt to reconcile all of them would lead to a long factory shutdown. For companies that classify their inventory into A, B, and C items (extended value), the problem is resolved by recounting the A and B items, and adjusting all other records to the count. Expect the record accuracy after a physical to be around 60 percent!

The best means of ensuring inventory record accuracy is through a cycle count program. Storeroom attendants count a few items every day, based on a predetermined frequency (usually on an ABC basis, A items being counted most frequently). Errors are investigated, reasons for error established and process improvements are instituted. It is essential that the causes of accuracy error are determined and the appropriate storeroom process is changed to eliminate the root cause of the error. If this is not done errors will recur, and inventory accuracy will not be achieved.

Process for Cycle Counting

1. Ensure that you have a "locked storeroom." This helps to ensure that only authorized personnel are allowed to receive, locate, or disburse material. This also ensures that information is processed at the same time as material is moved.

2. Establish a cycle count program. Clearly define that the purpose of the program is to determine and correct the root causes of inventory count and location errors.

3. Classify all storeroom inventory into A, B, and C groupings. Determine the frequency of count by group. This will depend on the people available. Typically, storeroom attendants will count 10 items per hour, and they should not count for more than 2 hours at a time. A plan based on counting 20 items per day per person is reasonable. A items are most frequently counted, and should be counted at least once a month, B items at least once in 3 months and C items at least once in 6 months. (See Table 10.4 for classification of A, B, and C items, with count frequencies and calculation of items to be counted.) Based on the frequency of count, the computer can be programmed to come up with a daily listing of items due to be cycle counted.

4. Determine the count error tolerance acceptable. Typically A items have no or very little error tolerance and C items can have up to ±5 percent. There are also location errors, where parts are not in the location specified or are in a location not specified.

5. Calculate how many items can be counted per day and how many persons are required to run the cycle count program.

Example for a warehouse with 800 parts:

Table 10.4 Typical Settings for A, B, and C Items and their Error Tolerance

Category	% Total Value $	No. of Items	Count Frequency	Tolerance
A	75%	80 (10%)	30 days	0%
B	15%	160 (20%)	90 days	±2%
C	10%	560 (70%)	180 days	±5%

Total counts per year = number of items × frequency of count
Total counts per year = (80 × 12) for As + (160 × 4) for Bs + (560 × 2) for Cs = 2,720

At 240 days per year: counts per day = $\dfrac{2,720}{240}$ = 11 items per day, or one person for about one-hour per day.

6. Run the cycle count program daily. Remember that error tolerances have been set for A, B, and C items. A computer-generated daily cycle count report should be prepared.

Table 10.5 Daily Cycle Count Report Format

Date	Part No.	Description	ABC	Location	Count	Book #	OK/Error Reason

7. Prepare a cycle count summary report summarizing accuracy.

Table 10.6 Cycle Count Summary Report Format

Items counted	Quantity Error	Location Error

$$\text{Inventory accuracy} = \frac{\text{items counted} - \text{errors}}{\text{items counted}} \times 100 = \%\ \text{accuracy}$$

within the accuracy of the tolerances permitted.

8. Analyze errors to determine their root cause. This is the principal reason for the cycle counting. Table 10.7 lists some of the common errors found in cycle counting and suggests how to resolve them.

Table 10.7 Error Analysis

Causes of Error	Corrective Action
Count errors	Train counters. If errors persist, change counters. Some people cannot count.
Timing errors	Count when no transactions occur, early morning or late evening. Check transaction log.
Items taken without authorization	Maintain locked store and enforce discipline for disbursement.
Incorrect count of receipts	Spot check count. Push back on receiving and suppliers.
Receipts not entered	Match moves to storeroom with receipts by storeroom.
Disbursement incorrect	Get counters to sign off disbursements.
Wrong transactions	Train attendants. Record initials of person making transaction. Watch for double entries.
Wrong identification	Use bar codes and wand identity.
Wrong book balances	Audit and correct BOM (bill of material) for back-flushed items. Allow only select persons to make adjustments.

9. After determining the cause of the error, adjust the on-hand balance to reflect the correct amount of inventory.

10. Set up a program to identify negative balances, and make an unscheduled cycle count check to determine the source of the error which caused the negative balance. A zero item balance is also a good time to do an unscheduled cycle count.

11. A cycle counting program never ends. It is a continuous process of maintaining inventory on-hand balance accuracy.

12. There is a school of cycle counting that counts items based on the frequency of their being picked. Items with high-pick frequency (once or twice a week) are cycle counted more frequently to ensure that the on-hand balances are accurate. This minimizes the risk of these items being out of stock while showing a balance.

Dos and Don'ts

DOs

1. Do establish the importance of inventory record accuracy and the storeroom function.

2. Do insist on good housekeeping (clean, neat, and everything in place). It helps to locate and identify material and it confirms the importance of the storeroom function.

3. Do set up a cycle counting program based on an ABC classification of inventory. Cycle count at a time when there are few, if any, transactions processed.

4. Do provide suitable training to the cycle counters. They need to understand the purpose of the cycle count program and its implications. They must recognize the importance of correctly identifying the item, its location, and its count.

5. Do make sure that the initials of the counter are recorded against the count.

6. Do make sure errors are investigated and root causes addressed. Initially the supervisor should conduct this activity and train storeroom personnel on how to do it. Always remember that the purpose of cycle counting is to find the cause of errors.

7. Do set an accuracy target (suggested is 98 or 99 percent). Prominently display cycle counting results. Give credit to cycle counters.

DON'Ts

8. Don't do a physical inventory. Fight it vigorously. Your cycle counting record accuracy should convince the auditors that your inventory records are accurate.

9. Don't correct your records without rechecking if you have to do a physical inventory.

10. Don't give all items the same importance in counting frequency and error tolerance.

11. Don't rate counters on cycle count results, but rather on cycle count accuracy. If individuals cannot count, change them. Do not discipline them.

12. Don't show the on-hand (book) balance to the cycle counters. This will tend to influence their count, and reduce their accuracy.

13. Don't apply the inventory accuracy program to the storeroom only. Whether material is stored at point of use or in work-in-process stores, it should be subjected to the same accuracy program.

14. Don't neglect to cycle count packing materials and other important non-production material.

Order Record Accuracy

Purpose and Description

If the order quantities are inaccurate or missing, MRP will not plan the right quantities and this may lead to customer orders not being delivered on time.

MRP takes the gross requirements for an item, subtracts its on-hand and on-order quantities, and establishes what is left to be ordered. Accuracy of the on-hand inventory and the on-order figures is thus essential. On-order covers orders for purchased parts and shop-made parts.

Discussion

The outputs of the MRP planning process are action messages to:

- place or release purchase orders and shop orders
- reschedule existing orders
- cancel existing orders

The action message will indicate what to order, how much (based on the ordering rules setup) and when to place the order. Delays or changes must be input, to enable MRP to plan with accurate information.

Process

1. Ensure that a complete and accurate BOM exists for all shop-made parts. This may mean conducting BOM audits on what is actually used. Parts omitted and parts with different usage quantity are likely sources of BOM errors.

2. Check the lead times used by the system with the actual lead times experienced. Lead times are the weakness of MRP, and they should be reviewed frequently to ensure they are up-to-date and accurate.

3. Check purchasing records to ensure that they are accurate and timely.

4. Run MRP daily. This ensures that the latest information is being worked with.

5. Run the action reports as often as they can be reviewed and acted upon.

6. Verify that the action messages of MRP are being acted upon.

7. Develop exception reports or error messages that alert the planner that an order is off schedule. Investigate all such exception messages, and determine whether the order can be supplied per schedule. In the event of the order being delayed, input the new promise date into MRP.

8. Make regular checks on receiving and incoming inspection to ensure that materials received are promptly and correctly entered into the system.

9. Ensure that scrap is entered promptly into the system. A decision has to be taken on how to treat rework and "on-hold for review." Usually such material is given a time frame to make a disposition.

10. Determine whether "return to supplier" represents return for credit (subtract inventory), or return for replacement (show inventory due).

11. In the event of a shortage because of the unexpected late receipt of an order, investigate the root cause for the delay.

12. Periodically purge the shop order file and the purchasing file of all old invalid orders.

Dos and Don'ts

DOs

1. Do make maintaining record accuracy a part of the company's culture.

2. Do insist on good housekeeping—neatness and orderliness promote accuracy.

3. Do ensure that there is accountability for maintaining the accuracy of records. Communicate this responsibility formally (see Table 10.8).

4. Do provide training in the use of the data control system.

DON'Ts

5. Don't be deceived by the apparent simplicity of maintaining record accuracy. It requires constant vigilance.

Responsibility for Record Accuracy

Table 10.8 provides a suggested charter of responsibility.

Table 10.8 Record Accuracy Responsibility Charter

Task or Data Record	Function or Group Responsible
Customer order status	order-entry personnel
Item master data	materials, engineering, purchasing and financial—each for their own data
Bills of material	product and/or manufacturing engineers
Routings	process and/or manufacturing engineers
Engineering changes	design or product engineers
Master schedule action report	master scheduler
MRP exception reports	inventory material planner
Purchase order status	buyer planner
Receiving and shipping information	shipping and receiving staff
Rejections, scrap, rework, return to supplier	material review board
Storeroom on-hand balances	storeroom attendants
Work-in-process inventory	individual operators
Cost information	accounting
Inventory policy	materials supervisor

Storeroom Management

Purpose and Description

Storage is required for four main reasons:

- To reduce transportation and production costs
- To reduce the problems of irregular supply and/or demand
- To help production and marketing
- To minimize handling and picking costs

Material has to be received, stored, picked and moved in the storeroom/warehouse.

Discussion

It has been estimated that the costs of warehousing and material handling constitute 12 to 14 percent of the total sales dollar of a product, hence they warrant careful consideration. If the demand for a firm's product were known and these products could be supplied immediately, there would be no need to store inventory. Neither of these assumptions is realistic. Companies use storage to cover fluctuations of supply and demand, to collect material for economic shipments, and to provide a production and marketing buffer. All these activities help to satisfy a customer order at an optimum cost.

It may be useful to make a distinction between the terms *storeroom* and *warehouse*. *Storerooms* usually house material in factories, whereas *warehouses* stock finished goods for customers or retail outlets. This section will focus on storerooms located in factories. The same principles of material handling apply to storerooms and warehouses. It is safe to assume that material records are maintained on a computerized system in all storerooms and warehouses.

Material handling covers the loading and unloading of material, moving the product to and from various storeroom locations, order picking, and delivering material to the production lines.

Today there are many sophisticated solutions to handle inventory effectively. These include automatic identification, radio frequency transmission, automatic guided vehicles (AGV's), and automatic storage and retrieval systems. Bar-coding and computerized record keeping are commonplace. As these are specialized subjects, no attempt will be made to describe or discuss their value and merits. However, a caveat must be issued—today's technology is making it possible to make and transport product faster, reducing the need for storage. Technology is also making product much smaller and more powerful so few components are replacing many. The future developments of a product should be understood before committing a large investment to storage or retrieval facilities.

Today *point of use* storage is becoming very popular. In this practice, each production facility in the factory has small sub-stores that stock the material required for the next few days. The line stores are replenished from a main storeroom—usually on a pull system.

Processes in Storeroom Management

Storage:

1. Ensure that all items in the store have a record of their quantity and a location. Locations can be fixed or random and usually have a *bin address*, which is a column or vertical reference and a row or horizontal reference.

   Example: Location A10 . . . A (vertical column) 10 (horizontal row).

Generally storerooms have random locations except those that handle a high-volume repetitive product.

2. Provide wide and deep storage bays for most general purpose products. See that high-volume products are handled with narrow high aisles and automated equipment. The aim is to achieve a balance between space utilization and material handling efficiency.

3. Provide racks that are adjustable to suit a variety of widths and heights of material.

4. Develop a reminder system for items that have a limited shelf life. The system should warn the storeroom that the items are coming to the end of their shelf life.

5. Store gases and other hazardous or flammable materials under special conditions, usually outside the building. Follow the Occupational Safety and Health Agency (OSHA) regulations for these products.

6. Reserve an area for consolidation of product to collect a truckload quantity.

7. Record all receipts, storage, and picks on a computerized management information system. This system must be integrated to the MRP system and the shop floor control system.

Material Handling

Material handling is a nonvalue-added process, so the objective is to minimize handling, thereby reducing time and costs.

1. Ensure that the largest possible load is handled to minimize the number of trips made. If necessary consolidate several loads before moving them.

2. Keep loads on pallets whenever possible. Use standard-size pallets that are 40 inches by 48 inches, as this allows two pallets to be placed side by side in a truck trailer or container.

3. Provide mechanical material handling equipment to meet the needs of the product and the volume handled. Equipment can be broadly classified into manual, power-assisted, and automatic, with manual being applied to low-volume, light product; power-assisted to heavier product, and automated equipment to high-volume, repetitive product.

4. Plan storage and pick to ensure that equipment does not run empty.

Order Picking

Order picking is the most important function of a storeroom, because generally it is labor intensive and time consuming.

1. Lay out a storeroom to allow efficient order picking.

    Repetitively picked and high-volume product must be stored in the same area, at a convenient height, located in the most accessible areas of the storeroom, and close to the shipping area.

    Bulky items must be stored at floor level on pallets, only a few near the shipping area.

2. Group parts that are likely to be ordered together in the same area, or in a uniflow picking sequence.

3. Plan the most efficient picking route if all locations are maintained on a computer. Combine and/or eliminate operations whenever possible.

4. Where large volumes of material are to be picked, the parts can be stored in storage bays, located away from the picking area, and picked from order picking bays. The picking bays are replenished from the storage bays during slack periods. Picking bays must be easily accessible.

5. Specialized equipment must be used to pick stock for very high-volume product such as groceries, fasteners, and so on.

Dos and Don'ts

DOs

1. Do understand the characteristics of the material being stored in terms of picking volume, size, shape, and ordering repetitiveness.
2. Do make decisions on material handling equipment based on material characteristics.
3. Do lay out the storeroom to achieve storage and picking efficiency.
4. Do use the computer to develop a daily picking plan that takes advantage of the layout.
5. Do ensure that the storeroom inventory is maintained on line, in real time, and that it is integrated with other factory ordering systems (MRP).
6. Do ensure that storeroom records are accurate (see the section on inventory accuracy).

DON'Ts

7. Don't react on an order-to-order basis. Organize picking documents. Batch customer orders for picking where possible.
8. Don't buy specialized equipment unless a high volume of repetitive product is assured for a long time.

Point-of-Use Storage

Description and Purpose

Storage at point of use generally reduces the time to build an order and it reassures the factory personnel that material is available and they can build to the schedule.

Rather than have the storeroom pick and prepare a material kit for every order, a sub-store or *line store* is set up in every production facility in a factory. The line store is stocked with about a week's required material.

Process for Point-of-Use Storage

1. Determine the items to be stored in a line store supporting a specific production shop. Generally this will cover all the materials, components and assemblies required to make the product in that particular facility.

2. Decide which items will be controlled and tracked and which items will be replenished on max-min basis. Generally, A and B items will be controlled, while C items will be floor or uncontrolled stock.

3. Set up a means of determining what materials are required to replenish the line stores and how much is to be provided.

   ■ For A and B type items (high-value items), a report can be generated that interrogates the orders scheduled in the forthcoming week and lists the parts required to build those orders (from the bill of material). The line store is replenished with those parts from the storeroom.

   ■ For C items, the parts can be stored in bins in the shop with a reorder card at an appropriate point in the bin. When the level of the bin reaches the reorder card, it is placed on a rack, picked up by a storeroom attendant, and the bin replenished. The two-bin system can also be used (see material ordering).

4. Pick the parts at the point of use when building an order. In the case of the controlled parts, a transaction has to be made to reduce the line store balance. No transaction is required for the uncontrolled parts.

5. Priority reports can be run that compare the available line store quantity with what is required per the schedule and assign a priority code.

Receiving, incoming inspection, and the storeroom can use this priority report to determine what to process on a day-to-day basis.

6. Develop a simple measurement to track the compliance of the line store.

7. Ensure that a storeroom attendant walks around the line store every day providing parts as required, replenishing bins, and picking up reorder cards.

Materials Ordering and Safety Stock

Purpose and Description

To satisfy a customer order, the right material must be available, in the right quantity, at the right time. A customer service policy determines the level to which a company plans to satisfy customer orders, and this policy greatly influences material ordering decisions. Material ordering must also optimize costs by using appropriate ordering rules. Today, fully satisfying a customer's order is paramount and inventory costs to meet the order are often a lesser consideration, particularly over the short term. In addition, there is considerable variability in the demand and supply of material. To protect against unpredictable variability, safety stock is kept. This helps to ensure that a customer order is satisfied.

There are two basic questions that are always asked when ordering material:

- How much to order? (lot size)
- When to order? (order points)

Material ordering covers the techniques used to determine the *quantity of the order* or *lot size*, and the *order timing* or *order point*.

There are three principal ordering systems:

- *Material requirements planning (MRP)*. MRP is a software algorithm that dynamically calculates how much to order and when to order (see MRP in Chapter 8).

- Order point techniques. The most common are

  - Two Bin
  - Visual Review
  - Order Point

- Periodic review technique

Table 10.9 Comparison of MRP and Order Point Ordering

Characteristic	MRP	Order Point
Scope of products ordered	Complete assembly and all components per the BOM.	Single part or component.
Interdependency ordering	Orders all parts by exploding BOM from top-level assembly and every subsequent level.	No capability to relate or order more than a single part.
Time-phased ordering	Backs off each part's lead time to determine when to order.	No time-phasing capability.
Scope of demand	Can plan any future demand. Can accept varying demand.	Cannot plan any future demand. Uses historical average demand. Cannot accept varying demand.
Adjustment to demand change	Automatically adjust to demand change in quantity or time.	Demand change has to be manually input.
Calculation of supply	Calculates and recalculates supply to meet demand or a pre-set lot size.	Cannot recalculate supply. No lot size calculation capability.
Action messages	Issues messages to order, reschedule, or cancel.	No message capability.

A common feature across all the above systems is the desired level of customer service. This target is one of the factors that determines the amount of safety stock maintained. Table 10.9 lists the important ordering characteristics and compares MRP with order point. The table shows that MRP should be the only ordering system used. Order point should be used only on simple, inexpensive items.

Material Requirements Planning (MRP)

See Chapter 8 for a detailed description of material requirements planning.

Order Point Techniques

Discussion

In determining when replenishment orders are to be placed, the desired customer service level must be balanced with the inventory investment and storage costs. In other words, a decision has to be made about how many customer orders are going to be filled and to what level. This decision has to be compared with the inventory investment needed to meet that service level. Another way of looking at this is to optimize the possibility of material not being available (a stockout) when a customer needs it. More inventory reduces the chances of a stockout, but increases the inventory investment. The availability of material and its inventory investment is also dependent on the accuracy of forecasts and the lead time of the items. Generally the more accurate the forecasts and the shorter the lead time, the lower the inventory investment required for the same level of customer service. Most reorder methods are based on one of the models described below.

Process

Two Bin: Two containers are used. The first, a primary container, is used daily and has a quantity usually equal to the order quantity of the item, which in case of C items is at least 3 months of expected consumption. The second container has a quantity of material that will be consumed during the time the primary bin is being replenished.

> Example: If 2,000 screws are consumed weekly (5 days), and it takes 2 days to replenish the stock of screws, then the second bin should have at least 800 screws, preferably 1,000 or 25 percent extra).

This second container is sealed when not in use, and opened only when the material in the primary container is consumed. On opening the second container, a signal or message is sent to the replenishing point (such as the storeroom, or supplier), to arrange to fill the primary container. The time it takes to use the items in the primary container should be noted to update the usage rate. This is a practical method of controlling low-value C items. Discipline is needed to ensure that a replenishment order is sent when the second container is opened.

Visual Review: The stock level is checked at regular intervals and ordered to restore the stock to a predetermined level. This method works well for controlling low-value items at point of use, when the replenishment lead time is short. Supermarkets use this method to restock their shelves. Like the two-bin system, inventory record keeping is eliminated. Again discipline has to be maintained to ensure that the periodic checks are made and that stock is put in the assigned place. Ideally in this process material should be replenished by the supplier, who comes around and fills the containers, a technique referred to as a *milk* or *bread run*.

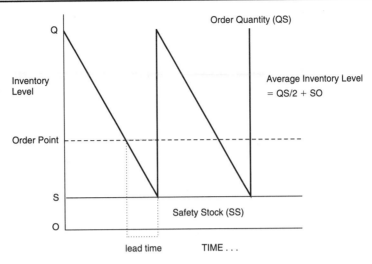

Figure 10.2. Reorder Point: Fixed-Order Quantity

Order Point (Fixed-Order Quantity): A reorder point is set at a stock level that will allow a replenishment order to be received before the material is consumed and a stockout occurs. The reorder point depends on the rate of consumption of the item, and its replenishment lead time (note the similarity to the second bin in the two-bin system). The key assumption is that the rate of consumption and the replenishment lead time are uniform and predictable. The process is illustrated by the sawtooth diagram (Figure 10.2). The problem with this method is that neither the consumption (demand) nor the replenishment is predictable or uniform, and this variability has to be covered by providing safety stock, as described below.

Periodic Review (Fixed Cycle)

Inventory records are reviewed periodically, usually once a week for important items, and once a month for unimportant C items. Sufficient material is ordered to restore the balance of the stock to a predetermined maximum level. The minimum level is enough to cover the consumption plus safety stock during the replenishment lead time (including a review interval). The technique is used where ordering costs are small (such as items on a blanket order), where a number of items are ordered from the same supplier, and for branch warehouse replenishment.

Dos and Don'ts (Order Points)

DOs

1. Do apply the simple order techniques of two-bin and visual review to frequent line replenishments of small, low-cost, high-volume items, so that no ordering or stocking records are necessary.

2. Do work the reorder techniques with supplier partners, so that they can replenish the items directly.

3. Do use the techniques to ensure that low-value, high-volume items are always available, and that they are ordered and replenished without risk of stockout.

DON'Ts

4. Don't use the order point techniques for material planning, customer orders varying in size, and timing and supplier orders varying in replenishment lead time. Understand these techniques cannot look out into future requirements.

5. Don't use the techniques for high-value A items that require accuracy of order timing and careful control of order quantities. Use MRP to order such material.

Customer Service Level Determination and Safety Stock

Discussion

The assumptions of predictable and constant customer demand and supplier replenishment are rarely experienced in actual operations. It is unreasonable to expect customers to buy to any predictable schedule. It is equally unreasonable to expect a replenishment order not to be subject to supply problems such as machine breakdowns, employee productivity variations, material shortages, and so on. Whenever the demand exceeds the assumed requirement (forecast), or the supply lags the expected deliveries, there will be a stockout, unless a safety stock is maintained (see the reorder point above).

To minimize the chances of a stockout, the reorder point has to contain two components:

1. A quantity to cover demand during the replenishment lead time, and

2. A safety stock to cover demand exceeding that forecast and/or supply not being received in time. See Figure 10.3. Point n is the normal level of consumption, with the normal arrival of the replenishment (supply) order. Point a is the level of the inventory when the demand exceeds that predicted, while point b is the level of inventory reached when

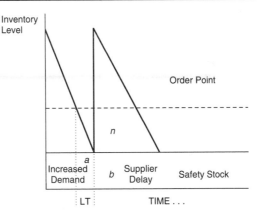

Figure 10.3. Safety Stock Protection Against Demand and Supply Variability

the replenishment order is delayed. In cases *a* and *b,* the safety stock has prevented a stockout.

Service Level

The question to be answered is: How much inventory should be provided as safety stock? To answer this question, a company must decide what level of service it will maintain. For a make-to-stock company, a service level of 95 percent means that the company plans to fill 95 percent of all units on all orders directly from stock. In today's environment, a 100 percent service level is not uncommon.

Safety Stock Assessment

Safety stock inventory is a function of:

- Replenishment lead time—measured in weeks or decimal months
- Forecast accuracy—as measured by MAD (mean absolute deviation) or SD (standard deviation). A standard deviation (SD) is 1.25 × MAD.
- Service level—measured in percent, (total items filled/total items ordered).

Safety stock is independent of the material ordering system used. (See Appendix 10.1 for more details on safety stock and relationship of MAD to standard deviation.)

Process for Determining Safety Stock (SS)

1. Set a desired level of service. For a *given time period,* this is:

$$\text{service level} = \frac{\text{number of orders delivered in time}}{\text{number of orders scheduled}}$$

Remember it should be at or close to 100 percent.

2. Determine the SF (safety factor) corresponding to the chosen level of service. (See common safety factors and levels of service, Table 10.11.)

3. Collect statistics on forecasts and the actual orders, to determine the nature of the forecast accuracy curve. Calculate the MAD as follows (normally 10 or more periods are used):

absolute deviation = forecast − demand (result has no sign)

Table 10.10 Example of Forecasts and Actual Demands

Period	Forecast	Demand	Absolute Deviation
1	1500	1450	50
2	1600	1550	50
3	1600	1650	50
4	1700	1850	150
5	1700	1800	100
			Total: 400

To calculate the MAD, divide the sum of the absolute deviations, by the number of periods, that is

$$\text{MAD} = \frac{400}{5} = 80$$

4. Calculate the safety stock from the MAD. (Assume equal lead times for the SS and MAD.)

In the above example, MAD = 80. Using a service level of 95 percent (MAD SF = 2.06),

SS = MAD × MAD SF
SS = 80 × 2.06 = 164.8 = 165

Table 10.11 shows the safety factors used for different service levels.

Table 10.11 Commonly Used Service Levels and their Safety Factors

Service Level %	Stockout Probability %	SD SF (Z value)	MAD SF (Z × 1.25)
90	10	1.28	1.60
95	5	1.65	2.06
98	2	2.05	2.56
99	1	2.33	2.91
99.86	0.14	3.0	3.75
99.99	0.01	4.0	5.0

5. As the service level increases, the amount of material required to be kept as safety stock increases quite rapidly as can be seen in the following example: MAD × MAD SF = SS.

Table 10.12 Example of Safety Stock Calculation

Service Level	MAD	Safety Factor	Safety Stock
95	80	2.06	165
99.9	80	5.0	400

6. If the safety stock lead time is other than that used to calculate the MAD (usually a month is used to measure and compare forecasts, and to calculate the MAD), multiply the safety stock formula by the square root of that lead time (which is used for the safety stock replenishment/manufacture).

Example: Where MAD is calculated on a monthly basis, and the lead time is 2 weeks (= half month), multiply the safety stock calculated by $\sqrt{0.5}$ (assuming 4 weeks per month), for a 95 percent service level.

$$\text{Safety Stock} = \text{MAD} \times K \times \sqrt{LT} \text{ where } K = \text{MAD SF}$$
$$= 80 \times 2.06 \times \sqrt{.5} = 116.$$

7. In event of expected delays, order early by one half the "worst-case delay."

Dos and Don'ts

DOs

1. Do remember that customer satisfaction is a paramount company objective, and delivering product on time is a key element in this objective.
2. Do remember that safety stock is a critical element in ensuring that product is available when unpredictable events occur.
3. Do carry safety stocks for items with erratic demand for service or replacement parts and for parts with uncertain supply.
4. Do calculate the order point from

order point
= expected demand during lead time
+ safety stock
$$OP = EDDLT + (MAD \times MAD\ SF) \times \sqrt{LT}$$

5. Do keep track of stockouts per item over time, to validate the performance of the safety stock calculations and to identify chronic offenders.
6. Do plan safety stock at the finished goods level whenever possible, instead of at the component level.
7. Do consider using an empirical rule of thumb to set the safety stock as a ratio of the expected demand during lead-time (EDDLT). This rule can be applied to finished goods or raw materials where MAD is not yet available. For a 98 percent service level, some commonly used multipliers and the category of material to which they are applied are given below:

	SS Multiplier of EDDLT		
Category of Item and Demand	*LT = 1 wk.*	*LT = 1 mo.*	*LT = 4 mo.*
C items	1.8	0.9	0.5
C items (irregular demand)	2.8	1.4	0.7
B items	1.2	0.6	0.3
B items (irregular demand)	1.8	0.9	0.5
A items	0.8	0.4	0.2
A items (irregular demand)	1.2	0.6	0.3

8. Do track forecast accuracy and provide feedback to the marketing and sales groups, so that they improve their forecast accuracy by paying more attention to it. If there were no forecast error, there would be no need for safety stock to cover demand variability!
9. Do understand and keep in mind that service level does not measure how long an item is out of stock, nor does it measure the effect on important customers.
10. Do try and establish service levels and safety stocks for families of items having common features.

DON'Ts

11. Don't set safety stocks blindly or forget about them once set. Review the amount set and adjust it based on actual consumption experience.

Inventory Reduction

Purpose and Description

Inventory is reduced to provide:

- Shorter lead times, as there is less time spent waiting in queues.
- Increased customer responsiveness as a result of the shorter lead time.
- Less impact of design changes because of faster manufacturing, and hence less obsolete inventory.
- More space available for productive use.
- Less cash tied up, improving return on investment.
- Earlier detection of quality problems, and hence less scrap or rework.
- Exposure of problems that may be hidden by excess inventory.

Inventory is required to build product. The material ordering section discussed how much is required and when it should be ordered. Because of supply and demand variability, inventory has to be balanced continuously. *Inventory should not be reduced if there is likely to be a negative effect on customer service.*

Discussion

Inventory control is probably one of the most talked about, but least acted upon functions in a factory. Inventory balances customer service in the control equation. When a company decides to improve its on-time delivery performance, it usually does so by increasing inventory. Inventory continues to climb until an opposite reaction sets in, usually initiated by finance, and inventories are cut by a percentage across the board. The cycle then repeats itself. The advent of just-in-time brought about a new consciousness of the levels and effects of inventory, particularly work-in-process inventory. The Japanese viewed inventory as evil. It covered up problems. By having excess inventory on hand, problems remained hidden, as a factory worked around them using the spare inventory. Where no excess inventory was available, there was nothing to fall back on. If a problem was encountered, it either got resolved or the line stopped.

There is the well-known analogy of the river and the rocks, in which the water in the river represents the inventory and the rocks represent factory problems. As the level of the water is lowered—that is, as the inventory is reduced—the problems of quality, maintenance, material, and so on, are exposed and the boat (production) is stopped. The problems have to be resolved, for production to continue. Many factories in the

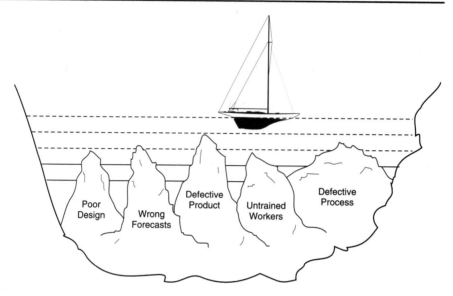

Figure 10.4. The River and the Rocks Analogy

United States have operated and generally continue to operate with con-
servatively safe amounts of inventory. Furthermore, in the United States
inventory is looked at from a cost point of view only, and not as a means of
problems solving. There is unwillingness to take the risk associated with
keeping a low inventory level, as a problem may disrupt production.

Process for Reducing Inventory

1. Set a target quantity of inventory, based on the inventory turns being
   achieved by the best company (the benchmark), in the industry group.
   The industry group can be determined from the SIC (Standard Industry Code), a classification issued by the government, grouping all like
   products in a common code. Publications such as *Manufacturing USA*
   published by Gale Research or *D & B* provide detailed comparative
   statistics.

   Example: If the industry leader has 15 turns, with a cost of sales of
   $20 million, then

$$\text{inventory target in \$} = \frac{\text{annual cost of sales}}{\text{turns}}$$

$$= \frac{20}{15} = \$1.33 \text{ million}$$

Table 10.13 Example of Setting Inventory Targets by A, B, C Categories

Category	% Cost of Sales	Annual Cost of Sales ($ million)	Target Turns	Inventory Target Cost ($ million)
A	75%	15	20	0.75 million
B	15%	3	10	0.30 million
C	10%	2	8	0.25 million
Total	100%	20	15.4 = (20/1.30)	1.30 million

2. Set separate targets for A, B, and C items, with As having the highest turns. Reconcile these individual targets with the overall target set in Item 1 above. Adjust the individual category target turns if necessary. This example is tabulated in Table 10.13.

    Example: If overall target set at 15.0 turns (benchmark from the best in the industry)
    and total annual cost of sales is $20.00 million,
    set A items at 75 percent, B items at 15 percent, and C items at 10 percent of cost of sales ($15, $3, and $2 million, respectively).
    Set target turns for each category, say A = 20, B = 10, and C = 8.
    Then, for A items, inventory target = 15/20 = $0.75 million, and B = 3/10 = $0.30 million, and C = 2/8 = $0.25 million, or total = 0.75 + 0.3 + 0.25 = 1.3 million.
    Reconcile overall turns target: $\dfrac{\text{Annual sales}}{\text{total inventory target}} = 20/1.3 = 15.4$
    turns. Adjust if necessary.

3. Target inventory can be the number of days of stock or the cost of material, and should be based on benchmarking or making an improvement on historical performance.

    Example: From the above inventory profile, overall target is approximately 15.0 turns.
    For a year of 240 working days, 15 turns works out to 240/15 = 16 days.
    15 turns also works out to $1.30 million.
    Thus a working day is equal to 1,300,000/16 = $81,250.
    The 15 turns target can be expressed as 16 days' inventory, with each day worth $81,250.

4. Develop a means of measuring the cost of raw material, work-in-process, and finished goods. Set separate targets for the three categories.

Raw material cost can be measured from the storeroom balances.

Work-in-process can be measured from a standard costing module or from a simple monthly input-output calculation: WIP $ = raw material $ issued + labor $ added − cost $ of product completed.

Finished goods cost can be measured from warehouse balances.

Example: To set targets for the inventory classification:

Set WIP days equal to manufacturing lead time (start to finish of order in the factory).

Set finished goods at a level calculated from the service level required (see material ordering section).

The remaining days are the raw material days' target.

5. Reduce and control amount of inventory by type of inventory. (See detailed inventory reduction checklist in Appendix 10.2.)

5.1 Raw Material
- Develop fast response arrangements with key suppliers—this will reduce the need for safety stocks.
- Improve scrap and/or yield factors, to reduce quantity of material ordered.
- Increase delivery frequency on A (high-value) items—monthly to weekly—so as to receive less inventory at a time.
- Arrange supplier delivery of material when needed, like a bread man routine (bins checked periodically and filled when necessary) for C items.
- Reduce (or eliminate) time from receiving to inspection to stores to shop. A dock-to-stock program achieves this objective by delivering material at point of use.
- Work with engineering to standardize raw materials and components.

5.2 Work-in-Process
- Reduce lot sizes. Arbitrarily reduce lot size in 10 to 20 percent increments and see the effect. Then reduce by another 25 percent and so on until the setup becomes the limitation.
- Reduce setup time as this limits the lot-size reduction.
- Schedule work to avoid queues thereby increasing throughput.
- Identify bottlenecks and overmanage them by ensuring that

they are always fully utilized, thus increasing throughput and reducing inventory.

- Improve shop reliability of delivery of the product.
- Reduce lead time—Work-in-process inventory is directly proportional to lead time (see Unit III Manufacturing Execution Overview for details).

5.3 Finished Goods

- Develop a good sales, customer, order entry, and manufacturing interface to make what is really needed when it is needed.
- Reduce safety stocks by improving forecasts and/or entering into customer partnerships.
- Shorten reaction time by shortening lead time from order entry to shipment.
- Migrate from build-to-stock to configure-to-order. Raw materials cost a lot less than finished goods.
- Reduce transportation time to distribution centers.
- Reduce the number of distribution centers and warehouses.

6. Assign specific ordering authority. A item orders must be the responsibility of a departmental manager or supervisor. The target turns for this inventory should be among the performance goals. Similarly, set inventory performance goals for all material planners, purchasing agents, and market forecasters.

Dos and Don'ts

DOs

1. Do develop a company culture that understands the effect of inventory in concealing problems.

2. Do ensure that the manufacturing, design, and process engineers understand that they have a direct and strong influence on the creation and reduction of inventory.

3. Do set inventory targets in units or pieces (based on dollars, which come from the target inventory turns), as

shop persons relate to units not dollars or turns.

4. Do measure and monitor. Reduction in inventory is achieved just by making the system aware that inventory levels are being watched.

5. Do understand that lead-time reduction is a key element in inventory control, and lot size is a key driver of lead time. Setup time in turn limits lot size. Reduce all three elements iteratively.

6. Do use an ABC approach to prioritize and focus inventory reduction plans.

(continues)

Dos and Don'ts

7. Do develop strong interactive partnerships with key customers and suppliers (see supplier partnerships and customer service). Obtaining information on what is required (customers) and being able to time the delivery of material (supply) will reduce inventory significantly.

DON'Ts

8. Don't use inventory as a substitute for problem solving.

9. Don't have an across-the-board percent inventory reduction program. Reduce inventory by type and by requirement.

10. Don't neglect inventory accuracy. Much excess inventory results from understated on-hand balances.

11. Don't neglect the importance of forecast accuracy. Forecasts drive inventory. Involve marketing and sales and make them accountable along with operations for the level of inventory.

Excess and/or Slow-Moving Inventory

One Definition: Inventory on hand in excess of one year's requirements or inventory on hand that has not moved in six months.

Some causes of excess and slow-moving inventory are:

1. Reduction in forecasts or demand from prior levels
2. Inaccurate forecasts and/or irregular supply, requiring large safety stocks to cover the uncertainty
3. Large batch sizes
4. Long setups, requiring high levels of work in process
5. Management pressure to keep operators busy

Obsolete Inventory

One Definition: All inventory items, purchased or produced, that have *not* been moved into production or sold within a 12-month period.

Obsolete inventory is caused by:

1. Product design or specification changes
2. Changes in methods of production
3. Excess purchase or production
4. Marketing decisions such as product discontinuation, product changes, and promotional strategy changes
5. Regulatory decisions

Process for Controlling Slow-Moving, Excess, and Obsolete Inventory

1. Define slow-moving, excess and obsolete material.
2. Ensure that there are reports that identify and measure slow-moving, excess, and obsolete inventory as defined.

3. Review these categories inventory once a quarter through a Materials Review Board.

   3.1 Reduce or eliminate all future forecasts for building/buying these products/materials.

   3.2. Determine the disposition of the materials

   - Where possible, return products to supplier, even at a discount.
   - Keep items for spares and service parts. Sell the items to customers of the product.
   - Rework items to a usable state.
   - Scrap all other materials.

4. With accounting, determine the write-off or write-down of the inventory value.

5. In all cases of obsolete, slow-moving, or excess material, determine causes and take preventive action to avoid recurrence.

Inventory Costs

Purpose and Description

Inventory must be managed to minimize the total cost incurred for an established level of service. The sale of inventory must generate income that exceeds all of the costs incurred in producing it and storing it.

Considerations in managing inventory are:

- Costs. Ordering costs, setup costs, and storage costs, and these costs have to be added to the cost of a product.
- Major costs associated with business lost because items are out of stock.
- Investment costs. To be added to other investments, to calculate the return on investment of capital.
- Cash flow. The timing of when and how much inventory is purchased influences the cash flow of a company.

Discussion

The understanding and treatment of inventory cost is essential to the effective management of inventory. With the emphasis on customer satisfaction, there is misconception that this means satisfying the customer at any cost. Obviously this is an oversimplification. Costs have to be controlled so that the company earns a profit. Controlling inventory related costs is one part of the solution. A total cost approach should be adopted when making inventory decisions.

In inventory cost calculations only *direct* or *out-of-pocket costs* should be considered, that is, those costs incurred because of a specific inventory decision. General overhead costs such as the cost of electricity, space, and so on, should not be allocated to inventory. These are *fixed* costs and spreading them on inventory can be very misleading. Similarly, costs that have been incurred in capital expenditure for equipment, land, and other resources are *sunk costs* and should not be included in inventory planning decisions. On the other hand, *opportunity costs* should be recognized, even though these are not recorded in accounting statements. The question should be asked: "What else could have been done with the money or space and what would have been the result?" This is the opportunity cost.

The problem is that many costs associated with inventory are difficult to quantify. For example:

- What is the cost of a stockout?
- What is the cost of a late delivery?
- What is the cost of a dissatisfied customer?

Yet these costs are directly dependent on the level of inventory. In calculating carrying costs and ordering costs, assumptions have to be made. Therefore these calculations are, at best, estimates.

Inventory costing has to be dealt with both at a macro or company level to assess the intangibles of customer satisfaction and lost sales, and at a micro or departmental level to deal with the actual cost of purchasing, storing and manufacturing parts.

Process for Establishing Inventory Costs

Four types of costs are relevant to inventory decisions.

1. Ordering Costs:

1.1 Establish the cost of placing a purchase order and a shop order. Costs associated with placing a purchase order include material planning and requisitioning, selecting a supplier, placing a purchase order, expediting, inspecting, delivering material to a storeroom, and finally processing and paying an invoice. Costs associated with placing a shop order include paperwork (material planning and ordering); and machine setup, startup, scrap, and any other one-time costs associated with the startup of a new order (training, drawings, etc.).

1.2 Assign values to the activities, being careful to include only direct costs—that is, only those costs that vary with the number of orders placed.

2. Carrying Costs:

2.1 Establish the types of expenses incurred because of inventory carried. These include obsolescence, shrinkage (pilferage, spoilage, damage), taxes, insurance, storage handling, and cost of capital.

2.2 Determine the direct cost of storage space and direct handling cost.

2.3 The largest element in the carrying cost is usually the cost of capital used to purchase materials and/or components. Usually the prime rate or the rate at which the company can borrow money is used in calculating the cost of carrying inventory.

2.4 Establish individual values for each type of cost and combine to determine a carrying cost.

3. Out of Stock Costs:

3.1 A stockout occurs when there is insufficient stock to fill an order. Determine whether the customer can wait for a backorder.

3.2 If there is a backorder, calculate the total cost. Include the ordering costs covered above, and any special charges incurred to obtain and ship the material in an emergency.

3.3 If the customer is unwilling to wait and the order is lost, there is the loss of profit of the direct order and the loss of customer goodwill. The latter is impossible to calculate, but some record should be kept to relate future orders to the stockout.

4. Capacity Related Costs:

4.1 Capacity associated costs include overtime, subcontracting, hiring, training, layoff, and idle time costs. These costs are incurred when it is necessary to increase or decrease capacity.

4.2 Ensure that the costs are actual out-of-pocket costs and not accounting costs. Also ensure that the costs are the result of a specific inventory decision, for example, to invest in an early buildup of inventory in anticipation of high demand or reduced supply, such as vacation shutdown.

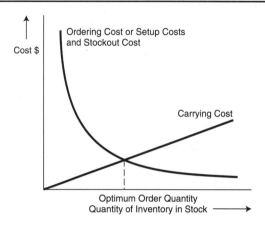

Figure 10.5.   General Concept of Optimizing Inventory Costs

Making an Inventory-Costing Decision

Conceptually the quantity of inventory to be ordered is a balance between ordering (or setup) costs and carrying costs. Figure 10.5 illustrates this. Having estimated the various inventory costs, it is necessary to trade off the carrying cost with the cost of getting inventory including the cost of not having inventory.

Dos and Don'ts

DOs

1. Do use material requirements planning to determine when and how much to order and intelligently follow the recommendations of the program.

2. Do understand that the largest and most tangible cost is the financing cost of the material held in inventory. Reducing the inventory level is the best method of controlling costs.

3. Do make sure that the costs applied are out-of-pocket costs and are directly associated with the inventory decision. Ask, "Can the cost be recovered if inventory is reduced, or avoided if inventory is not increased?"

4. Do remember that almost all the costs are estimates and should be treated as such.

5. Do remember that inventory-carrying cost is best used as a management variable to estimate the effect of customer service level on safety stock.

6. Do establish empirical costs for each broad cost category—that is, ordering costs, carrying costs, out-of-stock costs, and capacity-related costs.

> Example: cost for processing a purchase order: $30
> annual carrying costs: 20 percent of average inventory cost.

DON'Ts

7. Don't use inventory costs as the *only* input to decisions on customer service.

Inventory Performance Measurement

Purpose and Description

Inventory is primarily needed to:

- Maximize customer service
- Optimize investment costs

In today's competitive world, customer satisfaction is of paramount importance and is a strategic goal of most companies. Inventory investment must support the customer service strategy. There should be little conflict between the two goals and inventory management must be subordinate to and in support of customer service. To ensure that inventory is being managed to achieve these objectives, performance measurements are needed. Customer service and investment costs are monitored by numerous measurements, many of which are product and/or industry specific. Selection of the appropriate measurements is extremely important. As measurement drives behavior, properly selected measurements are one of the best ways of improving customer service and inventory investment.

Discussion

The two main types of business are make to stock and make to order. In a make-to-stock business, a company usually measures customer service as

$$\% \text{ demand filled} = \frac{\text{items available off the shelf}}{\text{total demand (items ordered) in same time period}} \times 100.$$

In a make-to-order business, the emphasis is on

$$\% \text{ orders shipped on time} = \frac{\text{orders shipped per customer request date}}{\text{orders scheduled to ship during same time period}} \times 100.$$

The problem in developing a suitable measure of customer service is that there are many factors involved, all of which complicate the basic issue of knowing if a customer is being satisfied. Some issues are:

- What effect do stockouts have on customer satisfaction?
- How can the effects of the length of time an item is out of stock be measured?

- How does providing preferential treatment to important customers or special customers affect the service measurements?
- How should shortage of critical items be measured?

All of these questions can lead to very complex measurements requiring more effort than they are worth. Effective measurements must be kept as simple as possible and should be analyzed for trends.

Care must be exercised that performance measures are not distorted to show favorable results. To avoid this, the goals set should be realistic and be accepted by the personnel who are responsible for the functions. However, in order to be competitive, a company's targets must match or exceed that of the best in the industry. These best-in-class standards can be determined through benchmarking.

Finally, a shipment schedule must be based on when the customer requires an item, not when the company can supply the item. If the customer's required date cannot be met, an acceptable alternative date must be mutually agreed to and scheduled.

Process

Customer Service Measurements:

1. Investigate customer service measurements best suited for the product and process environment. They can be a *percentage type* or an *absolute type.*

    Among the most popular percentage type measurements are:

    - Percent complete orders shipped on schedule
    This measurement does not account for the value or importance of an order and is affected by the number of line items per order. It is a strong indicator of customer satisfaction, as it represents how many customers receive their orders complete and on time.

    - Percent total units shipped on schedule
    This measurement, although recognizing the difference in size of the orders, does not deal with value or importance. This measurement has limited usefulness and is not recommended.

    - Percent dollar volume shipped on schedule
    This measurement can be biased by large value orders, but helps to measure the revenue generated by the company.

    - Percent order line items shipped on schedule

This measurement indicates the degree of individual item availability.

Among the most popular absolute measurements are:
- Order days out of stock

The days an order is out of stock does not recognize the value or importance of the individual orders involved.

- Total item days out of stock

This measurement is derived by multiplying the number of items out of stock by the number of days out of stock. This is a useful measurement for the efficiency of the ordering group.

- Total number and value of current backorders

- Overdue orders, aged by days overdue

Overdue orders listed by backlog value or by days overdue, clearly indicates how long backorders have been open and is a very powerful measure of customer satisfaction.

For these measures to be meaningful there must be some basis of comparison and benchmarking against best-in-class is extremely useful.

2. Select a measure of customer service. There is no "right" or "wrong" measurement, only what is appropriate for the business. The measurement should be confined to data that are available and the results measured should have a direct impact on productivity and profit.

3. Determine the desired level of customer service. Here an A, B, and C approach may be adopted, where important A items have a higher service level target than less important C items. The level of service should also be influenced by market expectations, and by the company's present performance and that of the competition. This approach is debatable and depends on the company's strategy. It should be implemented with care and sensitivity.

4. Determine the level of investment required to maintain the desired service level. This can be done by establishing the ordering size and the safety stock (see the material ordering section). Obtain concurrence for this level of investment.

5. Ensure that weekly and/or monthly measurements are circulated to all principal players, including production workers. Hold a regular

review analyzing reasons for shortfall in service and determining corrective action to be taken.

Process

Inventory Investment Measurement:

1. Determine the inventory investment.

   The absolute dollar value of inventory does not mean much, since both unit volume and prices are constantly changing. A 10 percent increase in inventory for a 20 percent increase in sales represents a relative reduction in inventory investment.

   Two common methods used to measure the relative size of inventory are:

   - Inventory turnover (turns)
   - Months of sale in inventory (I/S)

Both these measures indicate how quickly inventory moves through the production process. These measures do not correlate the effectiveness of inventory in maximizing customer service, nor do they indicate how much an inventory turn is worth in investment dollars.

   1.1 Inventory turnover. This is the relation of average inventory to sales at cost. For example, if the average inventory is $4 million and the annual cost of sales is $12 million, then the turns would be:

   $$\text{inventory turns} = \frac{\text{annual sales at cost}}{\text{average inventory cost}} = \frac{12}{4} = 3.0 \text{ turns}$$

   Inventory turns can be applied to historical or to future projections. The main problem is that when expressed as an across-the-board measure, it obscures the needs of different groups of material.

   1.2. Months of sales in inventory. This is the inverse of the turnover measure. It is the period average inventory related to the average month sales at cost. In the above example, the $4 million in inventory is compared to an average of $1 million of sales per month at cost. The ratio is:

   $$\text{months of sales in inventory} \frac{\text{average inventory cost}}{\text{annual sales at cost}/12} = \frac{4}{12/12} = 4.$$

This ratio is easier to relate to, as there is a better feel for available inventory in terms of months of sales. Some companies apply on-hand inventory month by month to future forecasts, to estimate how long current stock will take to be consumed.

2. Set a target quantity of inventory, based on the inventory turns being achieved by the best company (the benchmark), in the industry group. The industry group can be determined from the SIC (Standard Industry Code), a classification issued by the government, grouping all like products in a common code. Publications such as *Manufacturing USA* published by Gale Research or *D & B* provide detailed comparative statistics.

3. Develop an inventory budget. The budget has the projected absolute dollar value of inventory based on the forecast sales, the planned purchases and the planned production in dollars of cost. Obtain concurrence from finance, production, marketing, and sales. The inventory plan can be constructed as follows:

Table 10.14 Example of an Inventory Budget

Function/Month		Jan	Feb	Mar	April	May	June	Total
Sales @ cost $		20	10	10	30	30	20	120
Production $		20	10	10	10	10	10	70
Purchasing $		10	10	10	10	10	0	50
Inventory $	Start 60	70	80	90	80	70	60	Total: 450

Formula: End Inventory = Begin Inventory + Production + Purchasing − Sales

January end = 60 + 20 + 10 − 20 = 70 and February = 70 + 10 + 10 − 10 = 80

annualized cost of sales = 120 × 2 = 240 and average inventory = $\frac{450}{6}$ = 75

inventory turns = $\frac{\text{annualized cost of sales}}{\text{average inventory}}$ = $\frac{240}{75}$ = 3.2 turns

4. Track actual inventory, sales, purchases and production, and compare with budget. Investigate deviations. The main numbers to follow are the actual sales compared to forecast sales as they affect all other results. If sales confirm a trend, adjust the production and purchasing plans.

Dos and Don'ts

DOs

1. Do establish a means of measuring customer service, commensurate with the type of business. Keep it simple.
2. Do determine a target level of service.
3. Do establish an inventory level in dollars and inventory turns commensurate with the service goal.
4. Do try and get a benchmark of the service level and inventory turns the best-in-class manufacturers are achieving.
5. Do measure, analyze, and circulate. Ensure that deviations are understood and corrective actions are taken. Maintain history and graph on time scale.

DON'Ts

6. Don't try and calculate exact inventory target levels. A broad ABC approach is adequate. The key is the ability to react quickly to changing trends.
7. Don't treat the service level measurements as in an absolute sense. There are many factors that cannot be adequately represented. Look for trends.
8. Don't use measurements as a substitute for talking to customers and finding out firsthand how they feel.

Inventory Management Program

Purpose and Description

The purpose of an inventory management program is to establish inventory policies and oversee their execution. The major policies to be established are:

- Customer service level
- Inventory investment level
- Measurement of service and inventory performance

Inventory management is a technical function, with a specific body of knowledge containing rules, techniques, and policies. It requires a companywide approach with production, marketing, and finance developing integrated programs and shared objectives. Furthermore, the inventory objectives must be consistent with the business strategies of the company.

Discussion

Inventory is too important, volatile, and potentially disastrous an investment to be left to individual departments. It must be managed as part of a company strategy, with common objectives and goals. Perhaps the greatest difference between the U.S. and Japanese approach to manufacturing is the attitude to and control of inventory. The Japanese use inventory as a means of detecting problems and forcing problem resolution.

This is a high risk strategy requiring discipline in and control of the manufacturing operation. The U.S. approach has changed from making and selling what has been made, to making what is required to sustain a high level of customer service. In the United States, due to the greater diversity of customer demands, there is more variability in the process and this necessitates having more inventory to buffer the effect of the variability and to maintain a high level of customer service.

In today's environment, high customer service has become *the* most important objective of most companies and delivering product in time is a key element of this objective. Customer service is a competitive strategy in most companies and inventory must be used to support this strategy. Inventory cost must be optimized so that customer service is maximized at the lowest sustainable cost.

The two basic objectives of inventory are somewhat in conflict—increased service level generally requires an increased level of inventory investment. Therefore there must be planned inventory management. Companies that do not have an integrated customer service and inventory plan oscillate between excess inventory and poor customer service.

Process for Managing Inventory

1. State the company's strategic objective, with emphasis on *customer service and/or cost*. (In most companies this is clearly service.)
2. Establish a customer service policy. This should include:

   - Service level: establishing service level categories for different groups of products and/or different types of customers
   - Inventory investment: establishing investment limits by service category

3. Establish measurement criteria. (See the section on performance measurement.) These should include:

   - Types of service measurements
   - Types of inventory measurements

4. Establish a policy cost guide to enable calculation of operations/inventory trade-offs and ordering costs/inventory trade-offs. The guide should include:

   - Cost of money
   - Cost of storage, ordering, and setup

Develop and present the cost required to support different service levels.

5. Apply the ABC approach to service and inventory (see ABC section).

6. Use conservative forecasts based on actual experience, as forecasts are seldom accurate. Forecast accuracy needs to be recognized as a major inventory cost driver.

7. Set up an inventory/service standing committee consisting of finance, marketing, sales, operations, and materials management. This group must establish and monitor inventory and service goals by product group.

8. Conduct a monthly review of the service and inventory performance. Identify problems, determine the root cause of the problems, and decide on corrective action.

9. Document and publish monthly standing committee reviews. Include action directives specifying persons responsible and probable date of completion.

Dos and Don'ts

DOs

1. Do establish a clear priority between customer service goals and inventory levels. Lack of this leads to indecision and lack of focus.
2. Do get the general manager to maintain a personal interest in the use and control of inventory. This will establish the importance of managing inventory.
3. Do focus on improving forecast accuracy. It improves customer service and reduces inventory.
4. Do push to reduce lead times. This also has the dual effect of improving service and minimizing inventory.
5. Do make the operations and marketing managers jointly responsible for customer service and inventory. Establish common performance goals for both.

DON'Ts

6. Don't expect to have an effective customer service or inventory program, if all major departments have not been involved in establishing a clear company policy.
7. Don't look at inventory as a means of satisfying customers only. Remember the problem detection value of reduced inventory.
8. Don't impose an across the board arbitrary percent reduction of inventory. Inventory reduction should be product specific.

Purchasing

Because materials are becoming the dominant cost of a product, purchasing is becoming a critical function in manufacturing. There are many excellent books covering purchasing. The intention of this book is to briefly describe the function and some of its practices. Included are:

- Purchasing overview
- Make or buy decisions
- Supplier partnerships
- Best purchasing practices

Purchasing Overview

Purpose and Description

Purchasing covers all activities from receiving a requisition, through selecting and monitoring a supplier, to receiving materials at the required time, to specified quality, and in the right quantity.

Major categories of items purchased are: equipment; services; supplies; and manufacturing materials consisting of raw materials, components, and subassemblies. Purchasing plays a key role in attaining a company's objectives of delivery, quality, and cost.

Discussion

Today, in most companies, material costs constitute over 60 percent of the cost of goods sold. To be competitive a company must plan and manage its purchasing function effectively. Developing and maintaining supplier partnerships is becoming essential in establishing and sustaining a competitive advantage. Cost reduction of purchased material goes directly to the profit line, with a small percent of material cost reduction resulting in a large percent increase in earnings. Further, global competition has resulted in global sourcing of material and this puts more pressure on costs. Purchasing must be part of a company's competitive strategies and must contribute to achieving a company's goals.

Process

The traditional process followed by purchasing is described below:

1. Requisition received from user (generated by MRP). This authorizes the start of the purchasing process. The requisition must include:

- Identification of what is to be purchased. This is a detailed description of the material or a material specification, which can include drawings, physical and chemical specifications, and other details relevant to describing the product fully.

- Quantity to be purchased and a delivery schedule with special delivery instructions, if any.

- Account to be charged, with the approval of the appropriate authority.

2. Requests for quotation are sent to approved suppliers (established through a separate process). Based on their responses, a supplier is selected. This step may be iterative, having several cycles of clarification and negotiation. The requisitioner is often involved in this step, as they are able to provide answers to technical questions and help determine which supplier is most likely to meet the company's needs.

3. A purchase order is issued. This is a legal commitment that authorizes the supplier to produce or acquire the item, and binds the buyer to pay for it. In addition to having the details specified in the requisition, the PO contains price, discounts, terms of payment, and any special terms of purchase. It also contains shipping and billing instructions. Purchase orders and authorization are today frequently transmitted on EDI (electronic data interchange).

4. Open purchase orders are tracked to ensure delivery is made as committed and to take proactive action if there are technical and supply problems. The material planner must be informed, ahead of time, if the required delivery schedule will not be met. Where there are blanket orders or contracts, releases must be monitored. This function may be delegated to the planner or requisitioner.

5. Process receipts and handle problems of quantity and quality. Accept overshipments and undershipments to a preapproved tolerance. Ensure that quality problems are identified and the supplier informed. Defective material should be returned or reworked at the supplier's expense, in consultation with the supplier.

6. Maintain accurate and timely records of open purchase orders, receipts, required deliveries, supplier performance and cost. (This is most often done by a purchasing module in the MRP II system.)

Make or Buy Decisions

Purchasing should be party to deciding whether it is advantageous to make an item in-house or buy it from an outside company. Certain factors make outside purchase of material obvious, and these include: inability to make the part in-house, parts covered by patents, and a clear cost or technological advantage of the outside company. Where the decision is not obvious, many factors have to be considered, including:

- Capacity and capability—in-house and of the external supplier
- The relative quality levels
- The marginal costs of working the item in-house
- Lead times
- Cost/benefit analysis

It must be remembered that if there is internal capacity and personnel are not likely to be laid off, the marginal cost of producing an item is the cost of material only.

Dos and Don'ts

DOs

1. Do have a formal document describing the purchasing process and the requirements for complying with the process.
2. Do clearly define how items are to be specified in the requisition. Lack of definition causes delay, leads to defective material, and increases item cost. Return requisitions that do not comply.
3. Do ensure that the buyer is the only authority to order material or change a purchase order. The planner may interact with the supplier, but decisions must come through the buyer only.
4. Do maintain good supplier performance records.
5. Do benchmark the cost, quality, and delivery characteristics of major purchased items.
6. Do set up and develop a supplier partnership (see the next section).

Supplier Partnerships

Purpose and Description

In today's global environment, participative and productive relationships with key suppliers must be established to improve competitiveness. Furthermore, to sustain competitiveness, this relationship must be maintained and strengthened. A customer-supplier partnership is a mutually beneficial relationship built on trust, respect and commitment, requiring the sharing of knowledge and resources.

Discussion

Traditionally the buyer-supplier relationship was adversarial, with the buyer trying to gain the most favorable cost terms, often by pitting several suppliers against one another. In addition, neither party trusted each other and each tried to gain an advantage at the expense of the other. Most companies had a large number of suppliers and used them on a case-by-case basis. Three quotations for every purchase was the ground rule, followed by closed-door negotiation. The lowest quotation was usually awarded the order. Competition and the Japanese changed this approach. The Japanese developed long-term relationships with their key suppliers, and these partnerships achieved significant improvements in product quality, delivery, and cost, making both parties more competitive. Repeat business was awarded to the same suppliers, with new business being given to key suppliers. Most information was shared with the supplier, particularly forecasts, schedules, target pricing and "should costs." Key suppliers were involved at the design stage, and in continuously improving the product and its manufacturing process.

Today, supplier partnerships are an integral part of manufacturing. Companies and their suppliers become increasingly interdependent, and, as they work closely together, their relationship grows. The company (customer) shares information with the supplier, reduces or eliminates the nonvalue-added functions of checking and monitoring, and relies on the supplier to deliver a quality product in time. The supplier in turn depends on the customer for future business. Their mutual self-interest binds them and their continued association strengthens their partnership.

Process for Developing a Supplier Partnership

1. Review the supplier base for potential partners. Partnership should be sought with *key* suppliers only. Factors to be considered include:

   - Availability of competition for the requirement (less competition favors partnership)
   - Supplier management strengths (honesty, flexibility, openness, and so on)
   - Sufficient annual demand to warrant devoting company's resources to partnering
   - Long-term probable demand for the products

There should not be too many suppliers for the same types of product. Typically one or at the most two suppliers should be developed for each type of product or commodity.

2. Establish a partnership.
   Conduct interactive sessions with selected suppliers to communicate:

   - Objectives: mutual benefits; win-win relationship
   - Company policies and procedures
   - Roles and responsibilities of supplier and customer
   - An overview of the supplier partnership process
   - Seek supplier buy-in to participate in the partnership

3. Perform a quality systems audit.
   This should be conducted by a qualified quality organization (an ISO 9000 registrar, for example) and consists of a review of the supplier's process capability and quality control.

   - Ensure that the supplier understands the product and process requirements. These should be formally documented.

   - Conduct a technical audit of supplier facilities, to establish their capability and capacity to make product to required specification. The audit should also assess the suppliers process control and confirm that they should be able to manufacture consistently to the desired quality level.

   - Compile a corrective action program with the supplier to rectify any nonconformance detected in the audit.

   - If the audit is satisfactory (no major nonconformance), confer *conditional approved status.*

4. Develop a plan for the supplier to become responsible for material quality.

   - Over time and with proven supplier quality performance, there should be a plan to migrate progressively from:

   - inspection at the using company's facility, to

   - to inspection only at the supplier's facility, to

   - supplier self-certification of product.

- Establish on-site inspection criteria and an inspection plan for the products under review.

- Specify detailed requirements for documentation to be maintained; including control charts, maintenance of key dimensions, raw material certifications, and so on.

- Conduct periodic system audits to ensure that the supplier maintains his quality system.

5. Develop and implement a supplier rating program:

- Decide the criteria on which to rate suppliers—on-time delivery, product quality, and cost reduction are the usual criteria.
- Decide how to measure and rate the criteria.

Example: Delivery, quality and cost reduction are the criteria used to rate suppliers:

Delivery %. For all items: $\dfrac{\text{total scheduled order receipts}}{\text{total scheduled}} = \%$ on-time

Penalty 1% for every 1% on-time lost. (If on-time = 90%, 10% points lost on delivery.)

Quality %. Measured by: Defects found at inspection and in the using shop.

Penalty of 5% per defect type found per lot received. (If 3 defects found, 15% lost in quality.)

Cost reduction %. Measured by dollars saved.

Bonus / Penalty: 1% added for every $10,000 saved per year. If no savings for a year, 5% penalty.

The above rating is only an example and should be varied to meet the specific needs of the company.

- Determine a progressive level of supplier certification, commensurate with a supplier's performance improvement.

Typical certification levels are "Approved Supplier," "Certified Partner," and "Preferred Partner." Each succeeding level has higher-quality expectations, imposes less external audits and inspections, and relies more on the supplier's internal process controls. At a Certified-Partner status there should be no inspection of the product, and the supplier's quality system should be accepted totally. At a

Preferred-Partner level, supplier should expect to be given new business in addition to being given all their existing product supply.

■ Develop and document a supplier rating program to include all of the above considerations.

■ Have an introductory session with *all* principal suppliers to explain the intent and the substance of the supplier ratings. Accept appropriate supplier suggestions and change the rating program accordingly. Be careful to assure the suppliers that this is not a means of discrediting them. Stress the continuous improvement goal of supplier ratings.

6. Monitor, review, document, and publish supplier ratings.

■ Supplier rating must be reviewed at regular intervals (usually monthly or quarterly), and observations (favorable and unfavorable) communicated to the suppliers.

■ Provide an opportunity for receiving supplier feedback on the ratings and the process.

■ Have periodic face-to-face interactive sessions with primary suppliers. Share company future planning, including forecasts of expected business.

7. Upgrade suppliers based on performance.

■ Determine what level of rating or audit inspection success will change the status of a supplier and what advantages the status confers on the supplier.

Example:

To move from Approved Supplier to Certified Partner, a supplier must be rated 95 percent or more for 12 successive months. Once so certified, the supplier should not fall below 95 percent for more than 2 consecutive months. There will be no incoming inspection on the product of a Certified Partner. Furthermore, such a supplier will be given all repeat orders of his product.

8. Determine how and when to involve a supplier in the development of new designs and new methods of manufacture.

■ Certified Partners and Preferred Partners should always be involved in any redesign or process change involving their product.

9. Determine any necessary corrective action to be taken for suppliers not meeting required standards.

   ■ As in the case of upgrades, there should be levels of rating and audit inspection failures that will downgrade or disqualify the supplier. If possible, offer training and technical assistance to improve the supplier's performance.

10. *Always have an easy means of obtaining supplier feedback on their rating.* Give all such feedback due consideration. Remember that the intention of the partnership is to establish a synergistic relationship, which will lead to mutual benefit.

Dos and Don'ts

DOs

1. Do create a relationship first. Stress the mutual benefits of the program. Reassure the supplier that he is not going to be exploited. Without trust the program cannot work.

2. Do document the proposed process—even a draft will do. This should form the basis of discussion with the supplier. There should be a clear definition of roles, responsibilities, and expectations.

3. Do design the rating to reinforce the company's primary goals. If delivery is critical, give it a larger percent of the rating total.

4. Do develop ratings that are easy to use, easy to collect and easy to explain to the supplier.

5. Do use a multi-disciplined team to work the supplier partnership process. Include members from operations, materials, purchasing, engineering, and accounting.

6. Do reduce the supplier base to one or two suppliers for each major product line or commodity.

7. Do try to encourage and develop the secondary supplier.

8. Do make sure a good supplier benefits. If after working the program, all suppliers continue to be treated alike, there will be no incentive for performing well.

9. Do involve higher management on both sides in the process (assures commitment).

DON'Ts

10. Don't go through a supplier-partnering process *without gaining a measurable business advantage*—more customers, higher profits, less inventory, and so on. If you are not realizing an advantage, step back and review the program. Speak to your suppliers. Change appropriately and repeat the process.

11. Don't neglect to communicate with the suppliers on a regular basis, formally and informally. Communication must be clear and consistent. Sup-

(continues)

Dos and Don'ts

DON'Ts (*continued*)

pliers must understand and meet expectations. Just as good performance must be recognized and rewarded, poor performance must not be accepted.

12. Don't develop new suppliers without first trying hard to improve existing suppliers.

13. Don't expect all suppliers to reach the certified or preferred level. Most suppliers do not have the resources or the capability to be involved in design of product.

14. Don't forget that this program represents change. There will be explicit and implicit opposition, within the company and from the supplier. Persistence is the key.

Special Partnerships

Today special partnerships are becoming popular, particularly between large discount houses and their suppliers, where the supplier is responsible for stocking the shelves of the store (or maintaining a predetermined level of inventory). In such cases, the supplier may electronically monitor sales at the store checkout counter or the supplier may establish a warehouse and replenish the product periodically. This is the trend of the future.

Best Purchasing Practices

Purpose and Description

The purpose of these best practices is to ensure that a quality product is purchased, and delivered when required at the lowest price. Having established the importance of the purchasing function in influencing a company's success, and having described supplier partnerships as one of the best techniques of achieving purchasing effectiveness, a few other best purchasing practices need to be discussed. These are:

- Purchasing objectives
- Buyer planner
- Cost control approaches
- On-time delivery approach

Purchasing Objectives

It is essential to define the overall goals of a purchasing function. These may include the following objectives:

1. Provide an uninterrupted flow of the materials, supplies, and services required to operate the organization.
2. Keep inventory investment and loss to a minimum.
3. Locate, develop, and maintain top-class suppliers, by commodity.
4. Help to standardize as much material as possible.
5. Purchase materials and services at an optimum price.
6. Ensure highest quality of purchased product.
7. Establish a competitive advantage through a superior purchasing function.

Planner/Buyer

Traditionally, a material planner analyzes an MRP run and decides which planned orders to finalize. After getting the necessary approvals, he or she sends a requisition to a buyer who locates suitable suppliers, negotiates price, delivery and terms, selects a supplier, and issues a purchase order. During the negotiation, there may be several iterations of clarification between supplier, buyer, and planner. The planner is usually called upon to expedite material and he or she in turn contacts the buyer, who does the actual expediting with the supplier. Today many companies are combining these two functions into a single job, that of planner/buyer. Usually a planner/buyer is given responsibility for a specific product or commodity line of inventory, and with single-point accountability, the whole planning and purchasing process is improved. Furthermore, it takes only one person to make both an inventory and a buying decision and so the process is shortened.

Cost Control Approaches

Control of the cost of purchased materials is one of the critical functions of purchasing. Several techniques can be used to establish a buying cost advantage:

1. Share the items forecast with the suppliers, enabling them to book capacity and level load.

2. Issue a purchase order to cover a year's expected requirement, but establish a minimum quantity commitment, say consumption over the supplier's manufacturing lead time, for which the purchaser is liable. Base supply on releases. This helps the supplier to plan, without committing the purchaser to the full year's requirement. This also obtains volume purchase discounts.

3. Wherever possible, establish blanket orders in which shipment is made based on releases.

4. Involve key suppliers in the design of the item, and allow them to take advantage of their capability.

5. After appropriate checks, move from external inspection at the supplier's premises to quality system monitoring.

6. Concentrate on the A and B items.

7. Encourage the suppliers to look continuously for material and/or process simplifications. Reward them for cost-effective improvements by sharing the costs saved and by giving them preferential treatment for new designs.

8. Source globally.

On-Time Delivery Approaches

Along with quality and cost, managing on-time delivery is critical. Many techniques help to develop reliability in this element.

1. Share the master schedule with the supplier. If possible, have the supplier on-line with the master schedule and purchasing systems, so that he is aware of all changes in delivery schedules and can react to them.

2. Set up a demand pull in which a supplier is given a long-term schedule (at least a year), but is scheduled to deliver product at a weekly rate (kept constant for a month). Delivery is scheduled according to the short-term needs of the business and can fluctuate.

3. Encourage the supplier to keep a buffer stock of items where there is delivery variability.

4. Develop a supplier to a level where product can be shipped directly to the storeroom or the line (see the section on supplier partnership).

5. Measure on-time delivery and communicate the performance with the supplier. Insist on an explanation and a corrective action plan for repeated delivery failures.

6. Work on capturing an accurate supplier lead time for MRP. This will create credibility in the required material delivery dates. Avoid hedg-

ing by asking for items earlier than required on the assumption that they may be late.

7. Reward a supplier who has 100% on-time delivery.

8. Insist the suppliers inform your purchaser as soon as they know of problems with promised deliveries.

Appendix 10.1: Safety Stock

Assuming that demand during lead time (DDLT) is normally distributed, a safety factor (SF), corresponding to the desired level of service, is used with the standard deviation (SD) to determine the required safety stock (SS). The area under the normal curve on the right half or + side represents the probability of the demand being more than the average or expected demand during lead time (EDDLT), and can be read off as the Z value in any standard reference as the safety factor. This safety factor is set by the desired level of customer service. It may be noted that standard deviation was discussed to relate inventory control with statistical process control (SPC). In practice, MAD works equally well and it can be easily calculated. (1 standard deviation = 1.25 MAD).

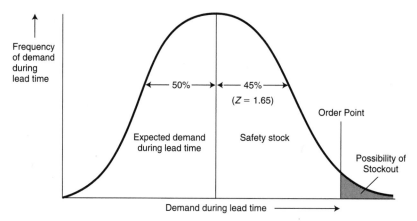

Appendix Figure 10.1. An Example of a Safety Stock Curve for 95% Service Level

To calculate safety stock:

$$OP = EDDLT + SS$$
$$SS = SD \times SF \ (Z \ value), \ or$$
$$SS = (MAD \times 1.25) \times SF \ (since \ SD = MAD \times 1.25) \ or$$
$$SS = MAD \times MAD \ SF \ from \ table \ below.$$

where
- OP = order point
- EDDLT = expected demand during lead time
- SS = safety stock
- SF = safety factor
- Z value = area under curve corresponding to service level (see a reference table)
- SD = standard deviation
- MAD = mean absolute deviation

Appendix Table 10.1 Commonly Used Safety Factors

Service Level %	Stockout Probability %	SD SF (Z value)	MAD SF (Z × 1.25)
90	10	1.28	1.60
95	5	1.65	2.06
98	2	2.05	2.56
99	1	2.33	2.91
99.86	0.14	3.0	3.75
99.99	0.01	4.0	5.0

Appendix 10.2: Inventory-Reduction Checklist

A. Identify and Attack *Unplanned* Inventories and Deviations:

1. Obsolete and rejected
   Arrange for salvage, discounted sale, disposal.
   Investigate why it occurred, how to avoid next time?

2. Slow moving
   Analyze reasons to minimize future occurrences.

3. Cumulative sales under forecast
   Analyze forecast for bias. Reduce if possible.
   Utilize forecast review team of marketing, finance, and operations.

B. Question Factors Influencing *Planned* Inventories:

1.0 Raw Materials (most points will apply to purchased finished goods):

1.1 Shorten incoming material ownership time, where F.O.B. supplier.
     Local versus overseas sources
     Air versus sea; rail versus truck
     Just-in-time delivery arrangements

1.2 Increase delivery frequency.
     Daily, weekly versus monthly on high-value items

1.3 Shorten inspection time allowances.
     No time allowance for waiting
     Prioritize items for inspection
     Adequate inspection capacity
     Eliminate inspection for qualified, certified vendors

1.4 Pressure suppliers for reliable delivery, quality.
     Eliminate allowances for delay
     Eliminate allowances for rejection

1.5 Determine high-usage protection in combination with work-in-process, finished goods safety stocks. (Avoid duplication.)

1.6 Coordinate order quantity sizes with usage requirements.
     Use period order quantity rather than EOQ

1.7 Reexamine supply assurance and other contingency inventories.
     Eliminate/reduce as causing factors diminish

1.8 Balance buying ahead for discounts or ahead of price increase with the resulting inventory costs.

1.9 Hedge commitment to new products based on risk analysis; delay final production and short lead time purchases wherever possible.

2.0 Work in Process:

2.1 Can aging, impounded or inspection times be shortened?

2.2 Reduce allowances for waiting time; prioritize; utilize slack principle (which order has most time available over time required) to prioritize final steps.

2.3 Lot size inventory
   Reduce changeover time/cost
   Is carrying cost of inventory valid?
   Overlap production stages where possible (see Chapter 9—transfer lots)

2.4 Determine high usage protection in conjunction with raw materials and finished goods.

3.0 Finished Goods:

3.1 Are contingency inventories (supply related safety stocks) still needed?

3.2 Reaction lead time—can it be shortened?
   Experienced coordinators
   Work in process or raw materials, safety stocks
   Shorten quarantine time
   Machine capacity available?
   Cross-train workers

3.3 Sales forecast deviation—safety stock
   Analyze actual sales for average deviation and bias and use in safety stock calculation
   Improve forecasting accuracy
   Consider service objectives and requirements
   Shorten reaction time

3.4 Reduce lot size inventory

3.5 Branch distribution centers
   Improve forecasting reliability
   Can intransit time be reduced?
   Can number of branches be reduced?
   Can pre-released products be quarantined en route?

Note: Many of the above suggestions require extra coordination time, which may be worthwhile on high-value items only. The ABC classification should be used to direct the effort to high-value items. Note also that reducing inventory may increase other types of cost, such as transportation, purchase price or manufacturing. The cost increase should not be greater than the period savings in inventory carrying costs.

IV Customer Service

11 Customer Service

Summary of Chapter

The chapter on customer service starts with *Basics* and an *Overview* covering the strategic and tactical importance of focusing on customer satisfaction. Customer service is *Defined* and *Measurements* of customer service are proposed. The chapter closes with *Management of Customer Service* and *Best Practices for Customer-Supplier relationships.*

Contents

- Customer service: overview
- Customer service: basics
- Customer service definition and measurement
- Price satisfaction
- Management of customer service
- Best practices for customer-supplier relationships
- Customer best practices

Relationship of Customer Service to Other Chapters in the Book

Customer satisfaction is the primary purpose of manufacturing. Parts are made to fulfill a customer order or in anticipation of a customer order. The effectiveness of the manufacturing infrastructure of quality, employee involvement and design; the information system; the master scheduling and material requirements planning; and, finally, all the execution techniques and practices, must be judged by their contribution to and the value they add to satisfying a customer at a profit to the company.

Customer Service: Overview

Excellent customer service is the desired result of the effective manufacturing planning and execution, and it is essential that it is defined and measured.

There has been a lack of specific information on *how* customer service can be improved using manufacturing planning and execution techniques. Many innovative approaches are being used in the field, but many of these have yet to find their way into the body of knowledge covering manufacturing.

It is hoped that this book will bridge this gap.

Customer Service: Basics

Description and Purpose

Total customer service is a series of complex relationships a company develops with the external world. It covers many facets dealing with value and perception of value. It also covers many functions such as advertising, promotions, supply chain management, and new products introduction. Briefly defined, it is meeting or exceeding a customer's expectations.

The purpose of manufacturing is to make a quality product in time to satisfy a customer order and meet or exceed customer expectations. The challenge of manufacturing is to achieve this objective while optimizing costs so that the entire operation is profitable. Customer service and customer satisfaction are the ultimate drivers of all industry, and this simple fact must not be lost sight of in the complex and hectic pace of daily manufacturing.

Executive Overview

Today customer service is a strategic issue. It is becoming the principal differentiator in establishing and maintaining a company's competitive advantage. Today industry's approach and attitude to the customer is epitomized by the slogan "The Customer Is King." The 1990s is the decade of the customer.

The issue is, Who can attract and retain the most customers? As markets become more global and accessible, competition becomes more intense. Brand names are not enough to differentiate a product and the price premium is being reduced. There is a push for products to become commodities. A simple diagram illustrates this phenomenon (Figure

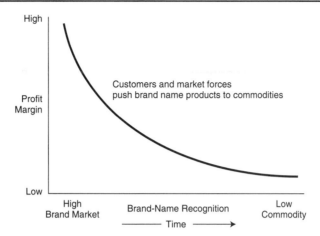

Figure 11.1. Product Recognition Movement

11.1). The balance of power has shifted from seller to buyer—the individual and the distributor. Large chains such as WalMart exert tremendous leverage on the manufacturer. Today it is not enough to design and build a good product and then use marketing to create a need and sell it. Customer needs (individually or collectively) have to be determined and then these needs have to be satisfied.

The quality movement increased the expectations of the customer and primary needs have a certain minimum level of acceptance. Both the individual customer and the industrial buyer have become more sophisticated and compare products and services before making a decision. For companies that meet all the basic quality expectations, the only remaining distinguishing characteristic becomes customer satisfaction. Thus customer service has become today's differentiator. A PIMS (a Massachusetts group) study has shown that a perception of high quality and service supports a higher price. Johnson and Johnson are one of the best examples of this axiom. There also is a growing body of evidence that links delivering a high quality of customer service with dramatic improvements in a company's profitability and market share.

Service: Focus on the Customer

What is customer service? Theodore Levitt suggested that customers do not buy a product only—they buy benefits. This is a very important distinction, as it focuses on the value that is perceived by a customer and not on the specific product or service provided. Customer service seeks to satisfy a bundle of tangible and intangible customer perceived needs, in the process of delivering a product. The primary tangible needs are delivery, price, reliability (quality), and product options. There may be intangible needs such as responsiveness and empathy.

It must also be stressed that customer service seeks to satisfy and retain a customer on a long-term basis. Research has shown that it costs more to attract new customers than to retain old ones; and what is more, retained customers buy more, cost less to serve, pay premium prices, and act as a reference for new customers. Retained customers are thus more valuable than new customers. With this in mind, superior customer service focuses not only on a single sales transaction but also on developing an ongoing relationship with a customer. Superior customer service results in satisfied customers.

Customer satisfaction has to be strategically integrated with other business objectives such as profitability, return on investment (primarily inventory levels), and market share. In today's environment, customer satisfaction is a company's key competitive strategy. Many companies evolve to a customer satisfaction strategy through a quality program such as TQM (see Chapter 1), as they seek to eliminate defects, customer complaints, and customer returns. It would appear that a customer satisfaction strategy is very closely linked with a total quality management strategy.

Some anecdotal observations on the influences of customer satisfaction are:

- Satisfied customers will tell an average of 5 others about their experience.
- Dissatisfied customers will tell an average of 10 others.
- Customers expect visible action within 48 hours.
- Customers will accept one hand-off to a specialist.

Process for Setting Up a Customer Service Environment

1. Establish customer satisfaction as a strategic objective and publicize it through a vision or quality statement.

2. Ensure that top management actively and visibly propagates and supports the establishment of customer satisfaction as a competitive strategy. This is probably the most critical requirement for success.

3. Establish communication channels or conduct a survey to determine the customer's needs and expectations.

4. Decide and communicate how customer service will be measured. The measurements must be aligned to the customer's needs.

5. Determine the performance of the best companies through benchmarking (see Chapter 1 for benchmarking).

6. Set and communicate targets commensurate with the best in the field.

7. Train the entire organization in a customer-service culture. Provide intensive training to those staff members who have direct contact with the customer.

8. Provide for excellence at the customer interfaces. Order entry, order inquiry, product information, and after-sales service are all critical junctions.

9. Delegate decision-making authority to the customer interface level.

10. Provide integrated systems support, so that accurate on-line information is available on inventory, backlog, and capacity to support on-line customer promises.

11. Monitor performance. Take corrective action. Ensure that performance is aligned to the customer service vision.

12. Reinforce the focus on customer satisfaction. Remember that a company's culture may have to change.

13. Repeat the cycle. Plan. Do. Check. Act.

14. Provide training periodically, as practices and techniques for providing superior customer service evolve.

15. Talk to actual customers and set up an easily accessible feedback loop.

Customer Service Definition and Measurement

Definition and Purpose

A set of primary customer needs related to manufacturing are delivery, price, reliability, and product options. A brief definition of each of these elements is appropriate.

- *Delivery.* A product is delivered or is available when required by a customer.

- *Price.* This is more than the sticker number. It includes whether the customer feels he or she has received value for the money they have spent.

- *Reliability.* The product performs as specified, consistently and dependably.

- *Options.* Customers have sufficient choices, within the product line, that provide an equivalent service to satisfy their needs.

Precise service measurements of these elements are required to monitor performance and determine if targets are being achieved.

Discussion

It is necessary to ascertain which elements contribute the most to customer satisfaction. This can be determined by asking customers what they consider important and then confirmed by monitoring their actual behavior. A listing of the relative importance or priority of the elements must also be compiled. For example, how will a customer choose between price and reliability? This evaluation has to be conducted separately for each product family and market. Based on the results of the evaluation, tentative objectives and measurements of customer service can be determined. It now becomes necessary to determine what level of performance is required for the elements a customer considers important. Here *benchmarking* has to be performed in which the performance of the best companies is determined and used as targets to be achieved or exceeded. (See Chapter 1 for the benchmarking process.)

Process for Measuring Customer Service

1. Decide on the customer service criteria that will be used to measure customer service.

2. Determine measurements for each service criteria selected. Ensure that data for these measurements are accurate and easily available.

   (Earlier it was discussed that there are four primary customer service needs, these being delivery, price, reliability, and options. This is by no means a definitive set, and other needs can be identified.) Using the above set of needs, the following set of measurements may be used:

   #### Delivery

   On-Time Order Delivery
   $$= \frac{\text{Number of orders delivered complete on the customer's required date}}{\text{Number of orders scheduled to be delivered in the same time period}}$$

Note: There is a distinction between the customer's required (need) date and the supplier's commit (promise date). Every effort must be made to accept and commit to the customer's required date. If this date cannot be met, another acceptable date must be negotiated. The customer should understand and accept the reason for his need date not being met.

Other measurements that track delivery may include:

- Order completeness or average number of shipments per order. (A scale may be used such as 100 percent for orders shipped completely, 95 percent for orders requiring two shipments, 90 percent for three shipments, and so on.)

- Orders delivered per promise date versus orders scheduled per promised date.

- Number of overdue orders and extent to which (time) they are overdue. (Overdue aged report.)

- Off-the-shelf availability—percent of items confirmed from available stock at the time of order entry.

- Order cycle time—total time from placement of order to delivery.

Price

- Price comparison—product price compared to competition
- Revenue ($) generated through sales booking and shipment

Reliability

- Percent product returned
- Warranty claims made—number and dollars expended
- Complaints received on product problems
- Service calls made

Options

- Number of models in a product group

3. Compile a composite service scorecard by customer. This should be done by identifying the most important service criteria, assigning an importance percent weight to them, and then multiplying the actual measurement by the weight.

An example of this may be as follows:

Table 11.1 Composite Customer Scorecard

Service Element	Performance Weight	Performance Level	Weighted Score
On-time Delivery	30%	95%	0.285
Order Completeness	20%	98%	0.196
% Product Returned (100 − %)	20%	98%	0.196
% of Complaints (100 − %)	20%	96%	0.192
% Overdue Order (100 − #)	10%	90%	0.09

Total Composite Score = 0.959

Calculation:

Weighted Score = Performance Weight × Performance Level

Weighted score for on-time delivery = 0.3 × 0.95 = 0.285. Similar calculations for all service elements

The sum of all the weighted scores is the composite score.

4. Set up statistical process control limits for each of the criteria being measured.

   Example: Lower limit for on-time delivery 90 percent, and order completeness 95 percent.

5. Monitor the process.

6. Take proactive corrective action when the process starts to deteriorate.

7. Continuously improve the existing process. Here the benchmarked best-in-class performance should be used as the goal to be met and exceeded.

8. Develop measurement to gauge responsiveness and empathy. Being soft-performance skills, they are more difficult to measure. Measurements may include:

Responsiveness

- Number of requests for improving delivery and number complied with
- Number of requests for delivering small lots more frequently, and number complied with
- Invoice accuracy and order accuracy

- Complaints received on "Not easy to do business with" (through occasional surveys)
- Number of inquiries on order status and time to respond to such inquiries

Empathy

- Number of customers retained—repeat purchases per year compared to total customer base
- Total number of customers and customer base growth

Dos and Don'ts

DOs

1. Do set up customer service measurements based on what customers consider important.
2. Do ensure that top management visibly and frequently reinforce the importance of customer satisfaction.
3. Do set targets based on the best-in-class.
4. Do ensure the measurements being used are communicated to and understood by the entire organization.
5. Do make a customer service composite target a part of management's incentive plan.

6. Do seek direct feedback from the customer.
7. Do make periodic checks to ensure that data used is accurate.
8. Do make important data easy to get.

DON'Ts

9. Don't establish measurements that require hard to get data.
10. Don't neglect to train and retrain all personnel in the customer satisfaction culture. Focus on personnel who have direct contact with a customer.

Price Satisfaction

There is a strong feeling that price drives customer satisfaction, and there is indeed truth to this perception. However, customer satisfaction is also driven by the quality and responsiveness of a company's goods and services. There is need to understand and balance both sides of this equation. Research by AT&T has provided some key insights (AT&T, 1995).

1. High price and high price satisfaction are possible, and price satisfaction can be changed by factors other than changing price.

2. Price change may or may not change price satisfaction, as satisfaction with price is relative to the price competitor's charge at a point in time. Simply lowering the price won't necessarily improve price satisfaction if the competition follows suit and keeps the price differential the same.

3. Customer satisfaction with prices is influenced by more than just initial purchase price. Besides the initial price needed to acquire a product or service, the full life-cycle set of costs for owning, operating, and using products and services influences a customer's overall satisfaction with price. These costs include such things as financing cost, warranty terms, and maintenance costs.

4. Customer satisfaction is also influenced by noncost drivers such as billing satisfaction and sales satisfaction. If customers perceive bills to be inaccurate and have difficulty in resolving this, they feel they have paid more than they should have. Sales knowledge and professionalism are key drivers, and customers should not feel that they were sold more than they needed.

5. Relative satisfaction with products and services affects price perception. A study of the telephone 800 services market was conducted comparing AT&T with MCI and Sprint. It showed that customers who perceived AT&T's products and services to be 50 to 80 percent better perceived the price to be only 10 percent higher. Customers who perceived the products and services to be no different from those of the competition perceived the prices to be 20 percent higher.

6. Perceived price and actual price differ, and perceived price can be managed. Studies have shown that perceptions of price differences between AT&T and the competition can be totally inaccurate. However, these misperceptions can be corrected through advertising and educational materials. Since the marketplace is dynamic, perceived and actual differences need continual monitoring and management.

Summary

Actual price, perceived price, and satisfaction with price are clearly different but related concepts. Satisfaction with price can be improved by factors other than the actual price itself. Indeed, the people and processes that deliver a product and services to customers have the biggest impact on price satisfaction.

Management of Customer Service

Description and Purpose

Providing outstanding customer service is a critical business strategy. In order to be effective it must be positively and continually managed. The purpose of providing superior customer service is to develop a competitive advantage.

This book has been structured and written to show *how to* provide superior customer service through manufacturing planning (Unit II) and manufacturing execution (Unit III). It may be worthwhile to review a schematic of the model of the book. Competitiveness is the common high level requirement for all companies. For manufacturing to be effective it must have a supporting infrastructure of design, quality, and people. A manufacturing information system provides the database for forecasting, master scheduling and materials planning—all activities required to schedule customer orders. The manufacturing information system also issues orders to purchase and/or make parts, and then tracks the progress of these orders. The shipment of quality product on time, through lead time management, results in customer satisfaction. Managing inventory is one of the cornerstones of being profitable (Figure 11.2).

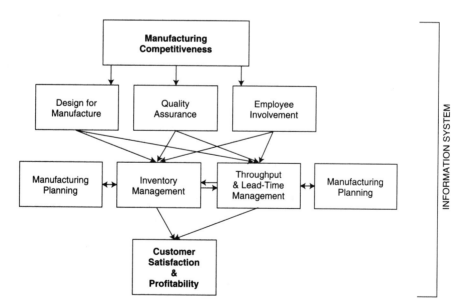

Figure 11.2.   A Schematic View of *Manufacturing for Survival*

Discussion

A typical customer-supplier relationship is illustrated below.

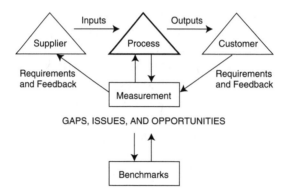

Figure 11.3. Customer-Supplier Model

Large manufacturing companies have complex environments with numerous interdependencies. Suppliers, carriers, warehouses, subcontractors, and factories are all linked in making different products, with often
unpredictable customer requirements of quantity and delivery. In such a
complicated network there is a great deal of variability or uncertainty—
suppliers do not deliver on time, supplied product is defective, factory
machines break down, customers change their minds, and so on. It is this
variability that causes delivery to be delayed and/or increased costs to
be incurred. There are two main methods of dealing with this variability:

1. Buffer the variability with inventory or with additional capacity.

2. Attempt to understand and control the sources of the variability.

Both of these methods must proceed simultaneously and continuously.
In Unit II (Manufacturing Planning) and in Unit III (Manufacturing
Execution Overview) manufacturing techniques to reduce variability
and practices to buffer variability were described.
In the next section on best practices for customers and suppliers,
the supplier-factory-customer relationships are reviewed and specific
practices to reduce the uncertainty of the process are discussed.
An illustration of the whole customer-manufacturer-supplier chain
showing the variabilities of the process is shown in Figure 11.4 below.

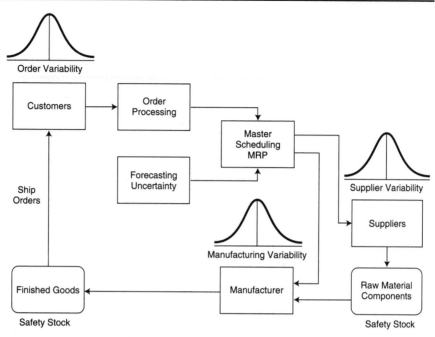

Figure 11.4. Customer-Supply Chain Relationship Showing Variability

Process for Managing Customer Service

1. Determine what criteria will be used to measure customer service based on a survey or understanding of what the customer considers important. (See the section on customer service definition measurement.)

2. Benchmark the current performance of each criteria selected.

3. Identify major gaps between the performance of the best-in-class competitors and the company's performance.

4. Determine the performance characteristics considered significant by the customer that the company can excel in. Be aware that a company need not excel in every service criteria. The analogy of the decathlon athlete should be used, where the champion excels in two or three events and is good in the rest.

5. Select performance criteria that are rated as being important by the customer and have major competitive gaps. Also select criteria in which the company has strength.

Example: Importance Rating Scale 1 (low) to 7 (high). The four criteria selected were:

Table 11.2 Example of Assessing Performance

Criteria	Importance	Measurement	Company's Performance	Benchmark Competitor	Distributor Requirements
Order-Cycle Time	6.4	Time (days)	10	5	8
Product Available	6.6	Fill Rate	58%	85%	90%
Promises Kept	6.7	On-time %	65%	80%	90%
Order Complete	6.5	Ships/Order	2	3	2 to 3

In this example the customer is the distributor. The largest gap is
in the Product Available and Promises Kept criteria. These are also
considered the most important. Clearly these have to be the two
areas that the company must improve. Even though the order-cycle
time is not considered as important, it results in failures in the
product availability and promises kept criteria and must be improved. The company is shipping more complete orders (ships per
order measures how many shipments are made to complete an
order). This is a strength that should be retained.

Criteria selected for improvement are order-cycle time, product availability, and promises kept.

6. Set targets based on the competitive benchmark or a requirement of
   the customer. (In the example, the targets set for the three criteria
   selected will be the distributor's requirements.)

7. Determine how to close the gap for the criteria selected. The performance of all three service criteria selected are affected by variability.
   There are three sources of variability: customer demand, supplier
   performance, and factory supply.

8. For each source of variability, list the causes of variability and the
   countermeasures that can be taken to minimize the variability.

   Example: Some of the causes and their countermeasures are listed
   in Table 11.3. This list will vary depending on the type of product
   and process used. It is not meant to be comprehensive. The whole
   of manufacturing planning and manufacturing execution deals

Table 11.3 Variability—Sources and Countermeasures

Criteria	*Cause of Variability*	*Countermeasures*
Product available, Promises kept	Customer: Irregular and unpredictable demand.	Partner with customer, joint agreements, booking capacity, increase safety stocks.
	Supplier: Late deliveries, unable to respond quickly, defective material.	Partnering, sharing forecast and schedule, having safety stocks, direct supply to customer, set up process for continuous improvement.
	Factory: Improper scheduling, machine breakdowns, bottlenecks, poor process flows, poor yields.	Improve manufacturing process and product planning and execution. Factory simulations for scheduling.
	Poor product design, poor manufacturability, no standardization.	Develop design concurrently with factory, customer, and supplier.
Order-cycle time	Customer places orders with no predictable required time.	Develop expectation of what order cycle is acceptable for families of products.
	Supplier: Does not meet quoted cycle time consistently.	Use more standard components, provide safety stock, improve processes.
	Factory: Does not meet planned lead time, cannot reduce lead time.	Reduce work in process, setup times and lot sizes, use kanbans, improve product flow.

with reduction in factory, supplier and customer variability. In Unit I, design variability is considered.

9. Develop a plan to reduce and/or eliminate the principal sources of variability affecting the selected service criteria.

10. Successful implementation is the hard part. Having goals and clear measurements will help the process. Establish an action plan with specific, time-bound, event-driven steps, each having clear accountability. Ensure that progress is reviewed at prescribed periods. Start with a pilot area and guarantee success. (Successful implementation of a project is a whole subject in itself. This has been briefly covered in Chapter 3, under system implementation.)

11. Set up a cross-functional team to implement and manage the realization of each service criteria selected. See that each team has a champion or sponsor. (Forming and successfully executing through teams is a whole subject in itself. This has been covered briefly in Chapter 1.)

12. Ensure that an effective material planning and product-tracking system is in place and maintained. Information must be easily available, timely, and accurate (see Chapter 3). Such information drives the whole process.

Dos and Don'ts

DOs

1. Do schedule monthly reviews of customer-service performance at a senior-management level.

2. Do understand that a critical element of success is top management actively championing the cause of customer satisfaction.

3. Do ensure that the entire supplier organization understands the key measurements and knows what drives them.

4. Do provide appropriate training to all personnel. Customer satisfaction is everyone's business, but planning personnel must be familiar with the manufacturing planning systems and shop supervision must understand *how to* improve lead times.

5. Do ensure that there is an integrated system that provides accurate timely information on order backlog and inventory availability and allows on-line customer order promise dates to be given.

6. Do set up electronic data interchange (EDI) with customers and suppliers, communicating order requirements, finished goods availability, and status of upcoming orders.

7. Do set up cross-functional teams to manage the key customer service drivers and measurements.

DON'Ts

8. Don't neglect to strengthen the customer contact level and ensure they are empowered to make most day-to-day decisions.

9. *Don't fail to review periodically the customer's expectations, the competitor's performance benchmarks, and update the company's targets.*

Questions for Improved Service

1. Who are my customers?
2. What are my customers' needs?
3. What are my customers' expectations?
4. Does my product or service meet my customers' needs?

5. How do I know that I am satisfying my customer?
6. How can I delight my customer?

Best Practices for Customer-Supplier Relationships

Overview

It was pointed out, in the section on customer service basics, that there is a tendency for market forces to push brand name products toward becoming generic commodities. How then does a company maintain its identity and retain or grow a core of regular customers? It has to form relationships with its customers—particularly its key customers. The relationship has to be built on knowing what the customer wants and providing it to them. The term being used is *mass customization*—providing individually customized goods and services, efficiently, to a large number of customers (Pine, 1933). Customers are not only buying products off the shelf, they are buying solutions—that is, a combination of products, capabilities, and services configured and ready to use.

There are several specific practices designed to develop a long-term customer relationship and provide better customer service. These are described below.

Supplier manufacturer partnerships are also common and have been covered in a separate section on purchasing (Chapter 10). In addition, all of the customer-related best practices can be and are applied to the supplier-manufacturer relationship, with the manufacturer being considered the customer.

Customer Best Practices

Customized Communication

Frequent and clear communication is the hallmark of good customer-supplier relationships. Several techniques are used today:

1. Electronic Data Interchange (EDI) is used to communicate customer demand, inventory levels at customer and supplier premises, replenishment requirements, and expected problems in supply or demand.

2. Periodic scheduled face-to-face reviews between customer and supplier.

3. Education of the customer through exchange visits, newsletters, and demonstrations.

4. Making a customer aware of manufacturing capacity and capability, including lead times and product availability.

5. Keeping detailed records on the needs, wants, likes, dislikes, and other individual information on key customers and updating these records periodically. This is the basis of mass customization.

Product Design

Where customers have products made to order or developed to meet customer specifications, it is becoming common to involve them with the design of the product. During this design process the customer can make trade-offs between price, reliability, and delivery. Involving the customer in the initial stages reduces the risk of making an inappropriate product that does not meet the customer's needs. Major customers are also routinely questioned for their views on future product development.

Demand Planning

Typically, customer demand is forecast. Based on the forecast, companies commit inventory dollars in raw material, work-in-process, and finished goods, and also invest in capacity. When the forecast does not materialize, or the product mix of the demand varies from that forecast, the company is left with excess, slow-moving, or obsolete inventory and with underutilized or overutilized capacity. Expensive products are exposed to the most risk as they usually take longer to make and have more inventory committed to them. Demand planning entails setting up an ongoing dialog with major customers and being informed of their changing product needs. In addition, efforts are being made to get a customer's commitment on what product will be required and when.

Reserving Capacity

This is a specific form of demand planning, where capacity is reserved for key customers. Planned orders are scheduled so that inventory and capacity are reserved in anticipation of receiving the hard order. The customer is expected to confirm the order with a sales order at a specified period of time (usually two or three weeks before the order is due). If this is not done, the planned order is canceled and the capacity is made available. The demand time fence and available to promise (ATP) calculation of the master schedule supports this arrangement.

Using the Master Schedule Effectively

There are several master scheduling best practices that enhance customer service. All of them have been described in detail in the section on master scheduling in Chapter 7. It is worth repeating some of the more commonly used techniques. These are:

1. Using a demand time fence (DTF). This is a point in time within which only customer orders (not forecasts) are considered for planning supply or product to be made. The DTF is usually set to equal the assembly lead time. Usually no change is allowed to orders within the DTF unless there is an exceptional reason. This helps to stabilize the master production schedule and allows the shop floor to have a firm immediate schedule to work to, thereby increasing the probability of on-time order completion.

2. Available to promise (ATP). This is a master schedule calculation that applies inventory, on an order-by-order basis, to booked customer orders, and provides a balance (ATP) that can be promised to new customers. It is a very useful tool to help a master scheduler accept and/or reschedule customer orders.

3. Planning bills of material. This is a listing of all the various features and options of a product or assembly, with an estimated consumption percent for each of the options. Instead of forecasting, making, and stocking all the variations of a product—a difficult and expensive proposition—options are initially made based on historical consumption and then replenished based on actual consumption. The planning bill can also be used to hedge against demand variability by making more of those options that are most likely to be consumed.

Special Supply and Delivery Practices

1. Direct supply. The supply chain is shortened by dealing directly with key customers. Product is stored for, and delivered to, customers based on their requirements. This arrangement eliminates dealers, distributors, wholesalers, and warehouses and ensures an ongoing dialog with the customer. For small products, supply can be made by having a daily "milk run" in which product level is replenished according to a prescribed method, such as a two-bin system or a min-

max level. Signals or kanbans are also used to determine replenishment intervals.

2. Ship to point of use. In addition to direct shipment, the need to receive and inspect product is being eliminated. Customers satisfy themselves that their supplier has adequate process and quality controls and then waive all in-house inspection. The effect of this practice is reduced supplier lead times.

3. Locating product. As an addition to a ship-to-stock arrangement, consignment inventory can be stored in a warehouse close to or even in a customer's factory, and the customer can consume the inventory as and when required by them. Replenishment is made by a simple signal, such as an empty bin or an electronic message. Accuracy of inventory balances is important.

4. Frequent delivery of small lot sizes. It has been recognized by most customers that frequent delivery of small lot sizes is the most effective means for them to reduce their manufacturing lead time, as less time is spent processing fewer parts. Such an arrangement also reduces the supplier's lead time and the total investment in the supply chain.

5. Packaging and delivery. Frequently the customer has to repackage product received from a supplier because of different multiples or packaging materials, or different labeling, and so on. Arrangements are now being made in which the supplier packages the product in the manner the customer would finally require it, thereby eliminating all unnecessary double handling and packaging.

Business Arrangements

1. Aggregate accounting. Numerous arrangements are established to ensure that the paperwork and accounting effort to track the product is minimized or eliminated. Product is kept on consignment at the customer's location and invoiced when used, usually by indirect means. For example, a sugar company supplies sugar in freight-car loads to the hoppers of a cereal company and is paid by the number of boxes of cereal manufactured, or a tire supplier is paid by the number of automobiles shipped. In both these cases there is no other paperwork. Replenishments are signalled by electronic messages, sending the suppliers a fax or linking them to the appropriate sections of the company's information system.

2. Joint agreements. An agreement is entered into annually between a company and its key customers in which details of the customer's requirements are spelled out and agreed to. Included in this agreement will be types of products, their required quantities, how changes are made, and the time limit within which change can be made only with potential financial repercussions. Price and escalation clauses are also agreed to for the short and long term.

Supplying a Complete Packet (Bundling)

Increasingly customers are buying a complete packet consisting of the product, its ancillaries, and all the connections necessary to make it effective. Customers are buying an end-to-end solution. This practice is referred to as *bundling*. The best examples of bundling may be found in the computer and telecommunications industries. Computers are sold with software installed or services such as a fax or modem configured. The customer merely turns on the computer and executes the program of his or her choice. Companies seek to satisfy customers by offering products that are branded and bundled.

Educating the Customer

Many perceived service failures arise because customer expectations differ from the service offered by the supplier. This is particularly apparent in make-to-order or engineer-to-order products. In order to minimize this, customer communications must be handled proactively and the customer must not be subjected to any surprises in the product or its performance. Educating the customer plays a major role in this process, and many innovative arrangements are being adopted to ensure that this occurs. These include:

1. Field visits to the customer's site and customer visits to the supplier's factory. Such exchanges have been found to be particularly informative and influential in developing an understanding of the product and the customer's expectations.

2. Providing a technical sales staff to complement the regular sales staff.

3. Developing a sales contract with the customer, so that special requirements and capabilities are understood and included.

4. Developing a customer service manual that defines the company's vision, objectives, and policies, and also includes details of products—

their options and performance. Information on who is to be contacted to deal with performance problems or troubleshooting a problem should also be included. A comprehensive manual is very valuable and can contribute greatly to the image of a service-oriented company.

Linking Customer Satisfaction with Company Business Success

1. Linking customer satisfaction to company performance. For the customer-service strategy to be strengthened, it is essential that the company be able to link the customer satisfaction approaches with marketplace successes. Employees will then understand that satisfying customers is not another corporate fad, but is linked to increased sales and profitability. Successful practices must be identified and reinforced.

   Example: Home Depot is renowned for its helpful floor staff. This in turn results in more people visiting the store thus generating higher sales.

2. Focusing on productive techniques. Resources and attention should be focused on successful practices. The results of such efforts must be measured. The process should be iterative.

   Example: Home Depot concentrates on recruiting and training floor staff to provide expert advice.

3. Reinforcing customer satisfaction successes. Successes must be communicated to all employees, thereby reinforcing their focus on customer satisfaction. The company can also use their success to reinforce their relationships with suppliers and customers.

Partnerships

All of the above arrangements involve some form of partnership. Today developing and forming partnerships with key customers is becoming increasingly more prevalent. Among the characteristics of effective partnerships are:

1. A win-win relationship. Both the customer and the supplier stand to gain from the arrangement.

2. Trust. The parties must believe in each other's intentions and this must be demonstrated. This is the most difficult and time-consuming aspect of the relationship to develop.

3. A long-term relationship. The partners are bound to each other for the long term, and are prepared to share each other's fortunes.

4. Preference. The customer will prefer a supplier partner and confer on them privileged treatment such as being a single source.

   Effective partnerships create a bond between customer and supplier, and the process is often referred to as *bonding*.

Creating a Customer-Service Climate

1. Supporting a customer's needs. In developing customer-service objectives, the customer's perceived service needs and their importance is first determined. It is essential that there is alignment between the customer's needs and the company's view of them. In other words, a company must develop policies and practices to support the customer's service priorities.

2. Internal alignment. Company personnel, particularly those who have direct contact with the customer, must be informed of what the customer considers important and what the customer wants. Too often surveys are conducted and an understanding of customer needs is developed, but this insight remains in the executive office or in the marketing department. It is essential that these insights be communicated to the frontline staff and that they be trained to develop the appropriate behavior and response to satisfy the identified customer's needs.

   Example: If customers of boutiques value empathy, then the boutique staff must be trained in and demonstrate empathy.

3. Practice and precept. A customer-service climate is not only created by upper management issuing a "vision" statement extolling the value of a customer. It is also created by demonstrated behavior, starting with top management, that reinforces the value of a customer. This is not easy. As late as February 19th, 1995, *The New York Times,* in an article on "The Bottom Line on 'People' Issues" quoted a study by Towers Perrin that pointed out that there is a gap between rhetoric and reality. Executives believed in the gospel of customer satisfaction, but did not understand how important employees were to its realization.

4. Employee satisfaction. For employees to delight customers, it is necessary that they be satisfied themselves. Customer-satisfaction goals

must be linked to performance evaluation, and required behaviors must be recognized and rewarded. At a management level, customer satisfaction should be linked to incentive compensation.

5. Company culture. The culture of customer service must pervade all levels of the organization. Everyone must be trained to view their jobs from a customer-satisfaction perspective. "If the key strategy of my company is to satisfy customers, how does my job fit in with this strategy?" is a question that should be asked of every position in the organization.

6. On-the-job reinforcement. There is need for continuous reinforcement of on-the-job performance. Trainers and observers must monitor and upgrade the performance of the staff dealing with customers. There should be frequent and visible recognition of good customer service. The reward system should complement the desired behavior. There must be an internal consistency in all the company processes supporting the service objectives.

Conclusion

The preface explains why I wrote this book. My motivation was twofold: first, my belief that one of the reasons for our lack of competitiveness is a lack of basic manufacturing knowledge; and second, because there is no easy means of obtaining this information, as there are few, if any, "how-to" books available in this area.

As a practitioner I addressed these concerns by writing *Manufacturing for Survival: The How-to Guide for Practitioners and Managers.* The book represents my experience and learning while working in U.S. factories for the last twenty years. It does not contain much referenced material. The book is not a comprehensive or definitive work. It is primarily a collection of the practices and techniques that I have used or heard of firsthand. Most of the practices are taken from the discrete product industry, dealing with the production of pieces.

The core of the book describes how manufacturing is planned and executed. Planning recounts the MRP II model of a computerized information system. This approach is fairly well known and extensively used. Describing manufacturing execution was more difficult. There are no clearly established preferred models on how to manufacture. This book used the premise that successful execution was synonymous with the reduction of lead time. Accordingly, a model was created that included major elements affecting lead time. The book then proceeded to describe how to effectively execute these elements. Overlying all the execution elements is the concept of variability and its effect on lead time. I have been singularly unsuccessful at trying to plan capacity at the shop floor. On the other hand, I have had success with controlling variability and managing constraints—both techniques featured throughout the execution sections.

In the competitive global economy, satisfying and retaining customers is critical. The final unit of the book provides some approaches, measurements, and best practices in customer service. It is an area that has been neglected in practice, and we can relate to the inadequacies in the service we experience. Companies that understand how to provide consistent customer satisfaction will always be leaders in their field. Here again, as with execution, there are no clearly established models of successful customer service.

Quality management, employee involvement, and manufacturing design are essential components of all effective manufacturing. Each has been extensively documented. I covered each area briefly and must state that my coverage only scratches the surface of these elements that I call the manufacturing infrastructure. I was fortunate to read and discuss the McKinsey Report on Manufacturing Productivity, and have used it as the basis of understanding present-day global competitiveness.

What of the future? What will determine manufacturing competitiveness? How will companies establish superior customer service and capture customers and markets globally? Here are the opinions of a practitioner:

The focus on customer satisfaction will intensify—gaining and retaining customers will become the critical business objective (it is already there in precept). I believe that quality is mandatory, but it is no longer a factor that will provide an advantage. Companies that do not have the highest quality will not exist. Product-design time-to-market cycles must and will continue to reduce. Again, companies that are late on significant technology will lose substantial market share. Currently we are not clear on how employee involvement is linked to productivity and what works. Our understanding is clouded by the bandwagon effect; everyone is writing and saying this is a must to gain competitive advantage. There is much work left to be done in this field.

We are left with successful planning and execution. Planning sets a factory up to build the right quantity of the right products at the right time. Execution enables a product to be made in the shortest possible lead time. A focus on minimizing lead time will drive superior performance of all manufacturing elements. This capability will strengthen a company's ability to meet customer expectations of product variety, price, and delivery. This is our circle of knowledge, or, to paraphrase Covey, our circle of influence. (Covey, 1989). I believe we manufacturing practitioners must know and apply superior techniques and practices so that we hold up our end of the competitive race.

I feel like a pioneer, as my approach to manufacturing, my focus on how to, and the organization of my book has covered new ground. I am sure there are many topics and elements that could have been covered differently. I hope there will be similar books that will fill these gaps. Till then, may this book serve the practitioner.

Blair R. Williams
Scotch Plains, NJ
May 1995

Bibliography

Abrikian. *New Product Innovation,* Stanford, CA: Stanford University Press, 1981.

Adler, Paul S. "Time and Motion Regained," Harvard Business Review Reprint no. 93101, 1992.

APICS Dictionary, 7th edition. Falls Church, VA: American Production and Inventory Control Society, 1992.

AT&T, "Statistical Quality Control," Charlotte, NC: Western Electric, 1956.

AT&T, "Quality Improvement Cycle," Berkeley Heights, NJ: AT&T, 1988.

AT&T, "The Memory Jogger," Indianapolis, IN: AT&T, 1988.

AT&T, "Root Cause Analysis," Indianapolis, IN: AT&T, 1990.

AT&T, "Total Quality Management," Palm Beach, FL: Qualtec, 1991.

Barnes, Ralph M. *Motion and Time Study,* New York: John Wiley & Sons, Inc., 1937.

Berry, Thomas H. *Managing the Total Quality Transformation,* New York: McGraw-Hill, 1990.

Blackstone, John. *Capacity Management,* Cincinnati, OH: South-Western, 1989.

Boothroyd and Dewhurst. *Design For Assembly,* 1987.

"The Bottom Line on 'People' Issues." *The New York Times,* Business Section, 19 February 1995.

Brandenburg, Johnson, and Johnson. Inventory Reduction Checklist. Internal memo.

Brocka, B. *Quality Management,* New York: Business One-Irwin, 1992.

Bruggeman and Haythornthwaite, "The Master Schedule," *Performance Advantage,* 1991.

Camp, Robert. *Benchmarking,* Milwaukee, WI: Quality Press, 1989.

Capron, Bill. "MTP II's Changing Face," APICS Int. Conf. Proceedings, 1994.

Chopra, Vinnie. "Final Assembly Scheduling," APICS Int. Conf. Proceedings, 1989.

Covey, Stephen. *The 7 Habits of Highly Effective People,* New York: Simon & Schuster, A Fireside Book, 1990.

Dertouzos, L. and Solow. *Made in America,* Cambridge, MA: MIT Press, 1989.

Deloitte and Touch. *Manufacturing,* New York: Irwin, 1990.

Dougherty, John R. "Getting Started with Sales and Operations Planning," Congress for Progress, 1992.

Dun & Bradstreet, annual publication.

Fogarty, Blackstone, and Hoffman. *Production and Inventory Management,* Cincinnati, OH: South-Western, 1991.

Fornell, C. "A National Customer Satisfaction Barometer," *Journal of Marketing,* 1992.

Frank. Don. "95% Inventory Accuracy Is Not Enough," APICS Int. Conf. Proceedings, 1985.

Frank, Don. "Frankly Speaking," Volume 2, No. 1, Fall 1994.

Frisby, Rick. Internal memo, Frisby Associates.

Gershwin, Stanley. *Manufacturing Systems Engineering*, New York: Prentice-Hall, 1994.

Goldratt, Eliyahu M. *The Goal and the Race*, Millford, CT: North River Press, 1984.

Goldratt, Eliyahu M. and Robert E. Fox. *The Race*, Millford, CT: North River Press, 1986.

Goldratt, Eliyahu M. *The Haystack Syndrome*, Millford, CT: North River Press, 1990.

Goldratt, Eliyahu M. *The Goal*, Millford, CT: North River Press, 1992.

Greene, James. *Production and Inventory Control Handbook*, New York: McGraw-Hill, 1987.

Griffin, Gleason, et al. "Best Practice for Customer Satisfaction in Manufacturing Firms," *Sloan Manufacturing Review*, Winter 1995.

Hall, Robert. *Zero Inventory*, New York: Dow Jones-Irwin, 1983.

Hayes, Wheelwright, and Clark. *Dynamic Manufacturing*, New York: Free Press, 1988.

Hayes and Wheelwright. *Restoring Our Competitive Edge*, New York: John Wiley & Sons, 1984.

Hendrix, Stan. "Manufacturing Concepts," *AT&T Technical Journal*, August 1990.

Hendrix, Stan. "Manufacturing Execution," *AT&T Technical Journal*, 1990, p. 36.

Hernandez, Arnaldo. *Just-in-Time Quality*, New York: Prentice-Hall, 1993.

Hunter, Mike. "How to Make a Realistic Production Plan," *Performance Advantage*, June 1982.

Imai, Masaaki. *Kaizen*, New York: McGraw-Hill, 1986.

Japanese Management Association, *Kanban*, Cambridge, MA: Productivity Press, 1986.

Leenders, M. R. and H. E. Fearon. *Purchasing and Materials Management*, 10th edition. Homewood, IL: Irwin, 1993.

Makridakis, S. and S. C. Wheelwright. *Forecasting Methods for Management*, 5th edition, New York: John Wiley & Sons, Inc., 1989.

Manufacturing USA. Gale Research, annual publication.

Mather, Hal. *Bills of Materials*, New York: Dow Jones-Irwin, 1987.

Mather, Hal. *Competitive Manufacturing*, New York: Prentice-Hall, 1988.

Mather, Hal. *How to Really Manage Inventories*, New York: McGraw-Hill, 1984.

McKinsey Report on Manufacturing Productivity. Washington, D.C.: McKinsey Global Institute, October 1993.

Monden, Yasuhiro. *Toyota Production System*, Industrial Engineering and Management Press, 1993.

Nakajima, Seiichi. *Total Productive Maintenance*, Cambridge, MA: Productivity Press, 1988.

National Academy of Engineering. *Manufacturing Systems*, Washington, DC: National Academy Press.

National Research Council. *Dispelling the Manufacturing Myth*, Washington, DC: National Academy Press.

Palmatier, George E. *Reducing Uncertainties in Demand*, APICS Int. Conf. Proceedings, 1988.

Pine, B. Joseph. *Mass Customization: The New Frontier in Business Competition*, Cambridge, MA: Harvard Business School Press.

Pine, B. Joseph, Bart Victor, and Andrew C. Boynton. "Making Mass Customization Work." Harvard Business Review Reprint no. 93509, 1993.

Plossi, George. *Production and Inventory Control Applications*, New York: Prentice-Hall, 1983.

Plossi, George. *Production and Inventory Control Principles and Techniques, Second Edition*, Englewood Cliffs, NJ: Prentice-Hall, 1985.

Plossi, George. *Orlicky's Materials Requirements Planning*, New York: McGraw-Hill, 1994.

Rosenthall and March, *Speed to Market*, Boston, MA: Boston Round Table, March 1991.

Schonberger, R. *Japanese Manufacturing Techniques*, New York: Free Press, 1982.

Schonberger, R. *World Class Manufacturing*, New York: Free Press, 1986.

Shingo, Shingeo. *Zero Quality Control.* Cambridge, MA: Productivity Press, 1986.

Shingo, Shingeo. *Non-Stock Production.* Cambridge, MA: Productivity Press, 1987.

Shingo, Shingeo. *Study of Toyota Production System*, Cambridge, MA: Productivity Press, 1989.

Skinner, W. *Manufacturing—The Formidable Competitive Weapon*, New York: John Wiley & Sons, 1987.

Taylor, Frederick W. *The Principles of Scientific Management*, New York: Harper & Row, 1911.

Tenner, A.R., and DeToro, I.J. *Total Quality Management*, Reading, MA: Addison-Wesley, 1992.

Tompkins, James. *Winning Manufacturing*, Norcross, GA: Industrial Engineering and Manufacturing Press, 1989.

U.S. Government Study commissioned by Don Ritter, U.S. Government Accounting Office. Washington, D.C.: U.S. Government Printing Office, 1991.

Vollman, Berry, and Whybark. *Manufacturing Planning and Control Systems*, New York: Irwin, 1992.

Walton, Mary. *The Deming Management Method*, New York: Putnam Publishing Group, 1986.

Watson, Theis, and Janek. "Design for Simplicity," *AT&T Technical Journal*, May/June 1990.

Welter, Theresa. "Designing for Manufacture and Assembly," *Industry Week*, September 4, 1989.

Wemmerlov, Urban. "Assemble-to-Order Manufacturing," *Journal of Operations Management*, 1984.

White, E.W. "Software Selection: Make the First Step a Good One," APICS Int. Conf. Proceedings, 1990.

Wiendhal, Peter. "Fundamentals and Experiences with Load Oriented Manufacturing Control," APICS Int. Conf. Proceedings, 1991.

Williams, Blair. "The Nemesis of Lead-Time Variability," APICS Int. Conf. Proceedings, 1991.

Williams, Blair. "The Realities of Empowering Teams," APICS Int. Conf. Proceedings, 1994.

Index

Total productive maintenance (TPM)
  dos and don'ts, 314–15
  implementation, 313–14
  objectives of, 311–13
  purpose and description of, 309–10
Total Quality Management (TQM), 13
  attitudes toward, 40–42
  important features of, 43
Towers Perrin, 433
Toyota, 216, 237, 274, 298, 304–5, 316
Training
  ISO standards and, 21
  for manufacturing information system
      implementation, 77–78
  quality and, 16
  team, 30
Transfer lots, 269–73
Transportation inventory, 348

-U-
U layout, 231–33
Uniform scheduling, 166
  cycle times, 247
  dos and don'ts, 250
  process, 247–50
  purpose and description of, 246–47
Unions, employee involvement and, 26
U.S. General Accounting Office (GAO), 13,
      28, 42–43
Unit of measure (UOM), 83
Utilization ratio, 221, 272

-V-
Value-added time, 319
Variability
  calculation of, 342–43
  defined, 321

  description of, 221–23, 246
  effects of, 321
Variable control charts (X bar and R charts),
      288, 338–39
Vision
  employee involvement and, 26
  employee teams and, 28–29
  slogans, 26

-W-
Wal-Mart, 413
Warehouse, difference between storeroom
      and, 362
Waste, defined, 216–17
Weighted moving average method, 107
Work centers
  dos and don'ts, 94
  purpose and description of, 81, 93
  setting up, 93–94
Worker-machine activity charts, 238
Work flow, 226
Work-in-process (WIP), 347
Work-order packet, 242–43
Workplace organization
  dos and don'ts, 236
  process, 234–35
  purpose and description of, 233–34
Work teams, 29

-X-
X bar charts, 288, 293, 338–39
Xerox, 34

-Y-
Yield, 290
Yield coverage stock, 348
Yield loss, 348

TMAC
7300 JACK NEWELL BLVD. S.
FORT WORTH, TEXAS 76118
(817) 272-5922   Fax: (817) 272-5977